AF556675

Testing of Communicating Systems

IFIP – The International Federation for Information Processing

IFIP was founded in 1960 under the auspices of UNESCO, following the First World Computer Congress held in Paris the previous year. An umbrella organization for societies working in information processing, IFIP's aim is two-fold: to support information processing within its member countries and to encourage technology transfer to developing nations. As its mission statement clearly states,

> IFIP's mission is to be the leading, truly international, apolitical organization which encourages and assists in the development, exploitation and application of information technology for the benefit of all people.

IFIP is a non-profitmaking organization, run almost solely by 2500 volunteers. It operates through a number of technical committees, which organize events and publications. IFIP's events range from an international congress to local seminars, but the most important are:

- the IFIP World Computer Congress, held every second year;
- open conferences;
- working conferences.

The flagship event is the IFIP World Computer Congress, at which both invited and contributed papers are presented. Contributed papers are rigorously refereed and the rejection rate is high.

As with the Congress, participation in the open conferences is open to all and papers may be invited or submitted. Again, submitted papers are stringently refereed.

The working conferences are structured differently. They are usually run by a working group and attendance is small and by invitation only. Their purpose is to create an atmosphere conducive to innovation and development. Refereeing is less rigorous and papers are subjected to extensive group discussion.

Publications arising from IFIP events vary. The papers presented at the IFIP World Computer Congress and at open conferences are published as conference proceedings, while the results of the working conferences are often published as collections of selected and edited papers.

Any national society whose primary activity is in information may apply to become a full member of IFIP, although full membership is restricted to one society per country. Full members are entitled to vote at the annual General Assembly, National societies preferring a less committed involvement may apply for associate or corresponding membership. Associate members enjoy the same benefits as full members, but without voting rights. Corresponding members are not represented in IFIP bodies. Affiliated membership is open to non-national societies, and individual and honorary membership schemes are also offered.

Testing of Communicating Systems

IFIP TC6 9th International Workshop on Testing of Communicating Systems
Darmstadt, Germany 9–11 September 1996

Edited by

Bernd Baumgarten

Heinz-Jürgen Burkhardt

and

Alfred Giessler

GMD – German National Research Centre for Information Technology
Darmstadt
Germany

SPRINGER-SCIENCE+BUSINESS MEDIA, B.V.

First edition 1996

Originally published by Chapman & Hall in 1996
MyCopy version of the original edition 1996

DOI 10.1007/978-0-387-35062-2

A catalogue record for this book is available from the British Library

Printed on permanent acid-free text paper, manufactured in accordance with ANSI/NISO Z39.48-1992 and ANSI/NISO Z39.48-1984 (Permanence of Paper).

www.springer.com/mycopy

CONTENTS

Part Eight Contributions to Theory

Part Nine Test Practice

PREFACE

This volume contains the proceedings of the IFIP WG 6.1 International Workshop on Testing of Communicating Systems (IWTCS'96), held in Darmstadt, Germany, on September 9 through 11, 1996, continuing the IFIP WG 6.1 series of International Workshops on Protocol Test Systems (IWPTS). In this series of working conferences, held annually since 1988, many valuable contributions have been presented with an emphasis both on the OSI conformance testing methodology and framework as well as the standardization effort on formal methods in conformance testing. While there are still open issues and divergencies between theory and practice in these fields, the scope of the series started expanding to related fields of growing practical relevance, for example to testing with regard to the B-ISDN protocol reference model, in particular ATM, the ODP reference model, and the Internet protocol suite, as well as to interoperability testing and performance testing. To reflect the extended scope, the program committee agreed on a new name for this series of working conferences, to be applied the first time to the 1996 conference. However, in order to emphasize the tradition of IWPTS, the numbering of this series was continued, such that IWTCS'96 is counted as the 9th International Workshop on Testing of Communicating Systems.

As in the years before, the workshop aims at bringing together researchers and practitioners, furthering the exchange of views, and correlating the work of both sides. It consists of 18 regular and 4 short reviewed papers, 2 invited papers, a standardization report, all of which are reproduced in this volume, as well as of a panel discussion and tool demonstrations.

IWTCS'96 was organized under the auspices of IFIP WG 6.1 by GMD, the German National Research Center for Information Technology. It was financially supported by the Commission of the European Communities, Deutsche Telekom, DeTeSystem, Mannesmann Mobilfunk, Siemens, Stiftungsfonds IBM Deutschland, and Vebacom.

We would like to express our thanks to everyone who has contributed to the success of the conference. In particular, we are grateful to the authors for writing and presenting their papers, the reviewers for assessing and commenting these papers, the members of the program committee, some of whom travelled long distances to help in the selection process, A.R. Cavalli and S. Budkowski, who shared their experience in organizing IWPTS'95 with us, and all the cooperative people participating in the local organization.

Bernd Baumgarten
Heinz-Jürgen Burkhardt
Alfred Giessler

Darmstadt, Germany, September 1996

CONFERENCE CO-CHAIRS

Bernd Baumgarten, Heinz-Jürgen Burkhardt, and Alfred Giessler
GMD – German National Research Center for Information Technology
Rheinstr. 75
D-64295 Darmstadt, Germany
{baumgart, burkhard, giessler}@darmstadt.gmd.de

PROGRAM COMMITTEE

G.v. Bochmann	University of Montreal, Canada
E. Brinksma	University of Twente, The Netherlands
R. Castanet	University of Bordeaux, France
A.R. Cavalli	INT, France
S.T. Chanson	Hong Kong University of Science & Technology, Hong Kong
A.T. Dahbura	Digital Equipment Corp., USA
J. de Meer	GMD, Germany
R. Dssouli	University of Montreal, Canada
J.-P. Favreau	NIST, USA
T. Higashino	Osaka University, Japan
D. Hogrefe	University of Lübeck, Germany
M.C. Kim	Korea Telecom Research Labs., South Korea
H. König	Technical University of Cottbus, Germany
J. Kroon	KPN Research, The Netherlands
J.A. Mañas	Technical University of Madrid, Spain
P. Marchese	CSELT, Italy
T. Mizuno	Shizuoka University, Japan
K. Parker	Telstra Research Labs., Australia
A. Petrenko	CRIM, Canada
M. Phalippou	France Télécom CNET, France
O. Rafiq	University of Pau, France
B. Sarikaya	University of Aizu, Japan
M.L. Shi	Tsinghua University, P.R. China
W. Stoll	IBM, Germany
K. Stöttinger	Deutsche Telekom, Germany
K. Tarnay	KFKI, Hungary
J. Tretmans	University of Twente, The Netherlands
H. Ural	University of Ottawa, Canada
S.T. Vuong	University of British Columbia, Canada
N. Yevtushenko	Tomsk State University, Russia

FURTHER REVIEWERS

Y. Benkhellat
L. Boullier
E. Cantu
O. Charles
D. Cypher
P.R. D'Argenio
A. Fazio
R. Gerin-Lajoie
J. Grabowski
S. Grande
R. Groz
A. Guerrouat
L. Heerink
O. Henniger
S. Heymer
M. Higuchi
G. Huecas
M. Hunter
S.W. Kang
D. Larrabeiti
T. Macavei
P. Maigron
D. Ouimet
R. Pignatta
T.C. Ruys
M. Sanders
F. Sato
S. Some
M. Tabourier
Q.M. Tan
R.E. Toense
M. Törö
A. Ulrich
H. van der Schoot
M. Vecchiato
A. Williams
S.J. Yoo
G. Zörntlein

ORGANIZATION CO-CHAIRS

Gunther Gattung and Olaf Henniger
GMD – German National Research Center for Information Technology
Rheinstr. 75
D-64295 Darmstadt, Germany
{gattung, henniger}@darmstadt.gmd.de

PART ONE

Testing Today

1

How could research on testing of communicating systems become more industrially relevant ?

R. Lai
School of Computer Science and Computer Engineering,
La Trobe University, Victoria 3083, Australia.
E-mail : lai@latcs1.oz.au

Abstract

Academic research has made significant advances in the generation of test sequences from formal specifications and in the development of computer-aided test tools with the aims of improving the effectiveness of testing communicating systems. However, this state-of-the-art research is not necessarily the state-of-the-practice; these methods and tools are seldom used in the communications industry. As academia tends to work on problems that are not too industrially related and does not quite address the problems facing testers in the industry, testers generally regard academic testing techniques impractical and irrelevant to solving real problems. There is a big gap between testing practice and research results published in journals and reported at conferences. This paper argues that academic research on testing of communicating systems needs to become more industrially relevant, describes the means by which this change can be facilitated, and suggests some research topics that are relevant to the industry. It aims to help effect a change in the direction of academic research on testing of communicating systems.

Keywords

Formal Method, Formal Description Techniques, Conformance Testing, Test Sequence Generation

1. INTRODUCTION

In the last decade, some of the research work on testing of communicating systems include conformance testing, interoperability testing, test case generation, test execution, analysis of results, theoretical framework, and theories of testing. In particular, there have been much research work conducted on testing using formal methods. Three Formal Description Techniques (FDTs) - SDL (CCITT, 1987), Estelle (ISO, 1989a) and LOTOS (ISO, 1989b) - have been standardised. One of the main aims in developing FDTs is to have a formal specification acting as a sound basis for such a testing. Successes have been achieved by academia on automatic test case generation from formal specifications to optimize the number of test cases in a test sequence so that minimal effort is required to obtain an acceptable level of confidence in the communication software (Sidhu and Leung, 1989).

The communications industry, with its own method of testing, is shipping to the customer communication software that still has bugs, even after it has been tested in the factory. With the merits of the academic testing methods, one would expect that they would be accepted widely in the industry. However, this state-of-the-art research is not necessarily the state-of-the-practice; they seem to have had very little impact on testing conducted in the industry (Lai and Leung, 1995). There is not much progress in the use of test sequence generation techniques for practical testing of communication networks. Test design is still largely performed by testers by interpreting the specifications written in a natural language.

Academia tends to work on problems that are not too industrially related and does not quite address the problems facing testers in the industry. The academic community likes to work on theories, with very little attention paid to real application and problems. There are very few researchers who are working on real projects and trying out new methods. Testers have to face real problems like deadlines, resource and economic constraints. For a researcher, he deals with a communicating system usually described in a finite state machine and tries to come up with a technique for testing the system on paper; for a practitioner, he is presented with an implementation of a communicating systems and needs to figure out ways to uncover most of the bugs within some economic, resource and time constraints. This is the fundamental difference between the work of a researcher and a practitioner. Testers therefore generally regard academic methods being impractical and irrelevant to solving their real problems. On the other hand, the industrial community only wants applications and have not much interests in the underlying theories. Academia and practitioners basically operate in two different cultures. There is a big gap between testing practice and research results published in journals and reported at conferences. This big gap between academic and industrial testing practices and that academia has not been addressing the real-life testing issues and problems account for the fact that academic testing methods are seldom used in the industry (Lai and Leung, 1995). There is a tremendous need to find a bridge between the theories and practices of testing.

This paper argues that academic research on testing of communicating systems needs to become more industrially relevant, describes the means by which this change can be facilitated, and suggests some research topics that are relevant to the industry. It aims to help effect a change in the direction of academic research on testing of communicating systems.

2. WHY CHANGE ?

How much longer can academia afford doing research that are not very industrially related ? The answer is "Not very long". The circumstances under which universities function in societies have changed. Gone were the days that academia could hide themselves in ivory towers to do research and afford not interacting with outsiders. Due to these changes, conducting researchs that are more industrially relevant is absolutely important for academia. The reasons are listed below.

Students : our direct customers In Australia and many countries, the level of funding that a university receives is proportional to the number of students that it has. The reputation of a university is built over a long period through its graduates, courses, and research output. More and more, universities are required to be accountable to the communities for their qualities of teaching. The contents of courses academics teach are usually related to their research interests. Slowly, communities get to know which universities' courses are more practical and which ones produce graduates that are more employable. Students tend to shy away from universities which offer courses that are less practical. As a result, such universities will receive a lower level of funding from the government; the job securities of some academics will be jeopardised. Students are the customers of academics; providing a good service to students is their duty and a means for their survivals in today's society.

Industry : our indirect customer Graduates of university courses which are too theoretical tend to have a harder time in getting their preferred jobs. Managers and practitioners want to hire graduates who are grounded in theories and also able to tackle real problems in the industry. Students having a good exposure to industrial issues during their university courses tend to perform better at interviews and hence be employed. If a company is happy with the graduates of a particular university, it usually comes back to hire some more. Producing graduates that are more employable and effective at their work is a key to building a good reputation for universities. If a university has a good perception from the industry about its graduates, more students will be attracted to its courses and subsequently a higher level funding will be received from the government.

A barometer for academic research quality What are the means to judge the quality of academic research ? Some barometers that have been suggested are : how many times a research paper is cited by other researchers and the reputation of the conference proceedings/journals in which a paper is published. In this fast moving communications field, a paper is usually cited by others once or twice. After a couple of years, the work would probably become obsolete. Is there a way to judge which journal/conference is better ? The answer is "Yes" because by and large, papers published in journal and presented at conferences are judged by academic standards; in terms of how useful the papers are in practice, the answer is probably "No". Would it be better that a different criterion be used for assessing the quality of research ? It needs to be judged by how suitable the technique is for use in the industry and how extensive it is used. This would be a more practical and objective way of measuring the usefulness and effectiveness of research work.

Expertised knowledge not just located at universities Gone were the days that universities were the only places where intellectural knowledge are located. A large percentage of the community's intellectural strength is not currently located in universities. Academics are not the only people that have the expertised knowledge. In fact, much of the leading research and development are conducted in the defence and aerospace industries; unfortunately, very seldom these reports are available to public use. The notion that the links with the wider community end when students graduate is really inappropriate. Academia can benefit by staying in touch with professionals in the industry. By working closely together, academia and practitioners can generate new intellectual products for the next century.

Time for a change For more than a decade, academic researchers have been doing work that are largely academic and have very little relevance to the real problems encountered in the industry. Is academia finding the work very interesting ? Maybe, the real problems are far more interesting than the problems discussed in the literature. Solving real problems could be more interesting and challenging for academic. There is already a huge backlog of research results not being put into practice. Therefore, it is doubtful if there is much value in continuing to stack more research results on top of existing ones which are not being used. It is high time that researchers do more work on how existing techniques can be applied to solving real problems and on deriving new methods to address industrial issues.

Government research funding In industry, the focus is on short-term results. In the academic community, conventional funding structures are rapidly disappearing, with a similar focus on short-term relevance, as opposed to basic research. In this modern down-sizing and leaner world, more and more government funds for research are required to be accountable to taxpayers. These changes make it harder for "pure researchers" to obtain government research grants. Millions of dollars have been spent on research on protocol engineering and alike in the last decade around the world. Can any one say that we have got the value for money and we have generated a worthwhile return ? Unless we address the real issues, we will not be able to get more research grants and will not deserve further funding.

Demonstrate or die The old saying for the survival of academia is "Publish or Perish". However, the system has changed to some extent for some academics. A new saying, "Demonstrate or Die", has sprung up. The ability to attract research grant from commercial firms is another way of judging the standing of a academic researcher. For instance, in the Boston area where the headquarters of many high technology computer and communications companies are located, researchers at institutions, like Massachusetts Institute of Technology, need to demonstrate to companies that their ideas are going to work on real-life systems before companies would give them research grants. Having a good idea is not good enough for attracting research funding from commercial firms. To gain promotion, academics are now very much pressurised to obtain research grants which are getting more and more competitive. Without the demonstration of the applicability of theories, there is very little hope that researchers are able to attract research funding from commerical companies.

Contribution to a nation's economy Academia in the western countries might be publishing more papers than their counterparts in countries like Taiwan. But what are really the contributions of publication to a nation's economy if the theories are not put into practice ? Sadly, there is not really a lot. This is evidenced by the trade deficits that some countries like Australia have. This is partly due to the fact that theories are seldom put into practice. On the other hand, Taiwan has one of the biggest trade surplus in the world. Taiwanese computer firms capture a substantial portion of the personal computer market of the world. One of the factors that contributes to the technological and economic successes of Taiwan is the government-initiated research infrastructure for helping the computer, electronics and communications industries. In Taiwan, research results tend to get more practised in the industry, hence making a contribution to the nation's economy.

3. CATALYSTS FOR THE CHANGE

With the huge gap between research and practice and the facts that academia are more interested in publishing papers and that practitioners are too busy to meet deadlines, making academic research on testing of communicating systems to become more industrially relevant would not happen naturally. Some catalysts are required to help bring about this change. These catalysts are discussed below.

3.1. Paper reviews

It is largely academics who review papers and control conferences, workshops, and publications. Academic reviewers do not like simple approaches and tend to reject papers which use simple techniques, even if they are useful. They usually do not like papers that only deal with real problems and large systems as they generally consider such papers not being original enough, even though practitioners have to solve problems in a real world environment. Vast majority of reviewers are in academic positions. So the standards by which papers are judged are by and large academic. Reviewing a paper from an academic point of view is not too helpful towards judging the true value of a paper. Their attitudes towards industrial papers and reviewing them somehow need to be changed.

Recently, it is encouraging to see that some journals emphasis the applications of theories to real problems. An example is the Journal of Systems and Software published by Elsevier Science; the editor-in-chief spells out that papers should contain an evaluation of the theories used and those that do not meet this criterion are returned to the authors without being sent off for review. We need more editors and reviewers who take a different view of reviewing and judging the values and the qualities of those papers that discuss real problems and systems.

3.2. Bridging agents

Academia are too busy writing papers and industrial people are under much pressure to finish their projects by the deadlines. Neither academia nor practitioners would have the time for each other. Agents are required for practitioners and researchers to interact, assuming that they have the genuine desire to interact. Who should act as these agents ? Professional bodies are the most promising agents to accomplish this task as many conferences are sponsored by organisations like IEEE, ACM and IFIP. Professionals in the industry need to take advantages of the availability of such conferences and journal publications to examine the work of academia more closely. An important mission of these societies should be to narrow the ever-growing gap between research and practice and to reconcile these two worlds. These professional societies need to take a better lead to improve the dialogue between academia and developers in the industry.

In general, the majority of participants at conferences are university researchers, and professionals are the very minorities. Part of the problem is due to the conference/symposiu program committee in the way they select papers and organise the program. IWTCS must serve as an agent for practititioners and researcher to interact. Its purpose should be to influence research directions, stimulate new approaches, promote collaboration, and generate interactive exchanges among researchers and practitioners. IWTCS should aim at being a meeting place for researchers who want to work with real problems and developers will have a hope that researchers will have something relevant to say.

3.3. Fitting new technologies into companies

The main goal of companies is to make profits. Productivity is an important factor that helps achieve this goal. Productivity is largely determined by the methods, tools and skills used in software development. A company, with the goal of making profit, always searches for new and better methods and tools to improve its productivity and competitiveness This is also a survival strategy. Only if substantial improvements to productivity can be guaranteed, a company will seriously consider new methods and will invest money and time to introduce them into practice. It will invest in technology, provided that it pays. Researchers need to demonstrate to companies how their new techniques can hence productivity. Academics need to show the benefits and demonstrate its success to the industry Successful uses of new techniques on large systems needs to be reported at conferences and to be published in journals.

To ensure success of business, the choice of technology is an important decision. From a manager's point of view, trying a new idea poses risk. Managers are only interested in new ideas if they do not pose significant risks, do not cost too much to develop, and can be introduced gradually. The old approach may not be good, but it is predictable. This is good from a business point of view. A new technique might be good but not necessary predictable. New techniques need not be perfect, but in order to be used industrially they must fit in with the organisation and the way it has been working.

3.4. Consumable mathematics

Unfortunately, the methods offered by academia so far have been mathematical, and there has been excessive amount of effort focussing on proofs of theories. There is nothing wrong with mathematics which is simply a tool that supports analysis. Engineering personnels like civil and electrical engineers use mathematics for modelling and solving problems; likewise, mathematical methods are a key to improving effectiveness and productivity in testing of communicating systems. However, it s worth noting that engineering is a more matured discipline than computer science. Moreover, testers are not mathematicians, and they do not have the same type of knowledge as academics have. It may take a tester much longer to learn the method than to do the tests, so that in practice it may be just as easy for a tester to make mistakes in the use of mathematics as in testing. This slows down the process of testing without increasing its quality. Researchers need to make mathematical parts more consumable for testers in order for the new techniques to be in wider use. There are different levels of mathematics required by different personnels. Mathematical details need to be hidden from testers when they use the new techniques.

3.5. Education

The gap between theory and practice is partly due to the fact that testers have not been educated in formal methods for testing. Ultimately, future design and testing of communicating systems using formal methods will be initiated by engineers and testers who have been educated in formal methods. Education will be a key factor influencing the adoption of formal methods by the industry. This educational gap could be filled by university education.

Tertiary courses that focussed on formal methods of development of communicating systems became available in the late 80's (Lai, 1992). Apart from teaching theories, such courses need to aim at helping students apply formal methods in real life complex situations. Ideally, students are required to pass a practical component on development of communicating systems. Academics needs to actively solicit involvement from companies in the set-up of a curriculum. Universities should set up industrial advisory task forces from which input to courses can be sought. At times, practitioners and developers could be invited to deliver a couple of guest lectures. Unfortunately, course on formal method and with input from the industry on the development and testing of communicating systems is not common enough and needs to be more widespread.

3.6. Engaging in real projects

One of the practical ways forward is to have researchers working closely with practitioners on real projects. They can be housed in the same building. Then researchers and practitioners can interact regularly, and researchers will then be more aware of the industrial issues. A typical research project can be carried out to experiment the new idea in parallel to the actual project. Researchers can then present results back to the practitioners who are working on real projects. This will take the pressure off the practitioners as most of the risks involved in using new technique are removed.

Both sides will benefit from this type of collaboration. University researchers are not usually rewarded for their ideas that are put into practice on real projects. They are rewarded for publications of their research results and for getting research grants. If an idea works well, the technique can be adopted. If it does not, researchers still can write a paper explaining what went wrong. However, if the idea works, researchers can write several papers. In the end, university researchers still have papers published and do not lose any thing. In fact, they will have more to gain : they will have the credentials to attract more research funding as their ideas have been demonstrated to work on real projects. As for the practitioners, they will appreciate how new techniques can be of practical use to their problems and how their productivities could be increased. The gap between theory and practice will then become narrower.

3.7. Better tools

Tools are not very frequently used in the industry, but they play an important role in the industry as they will be an effective means of reducing human effort and will cover more test cases. A number of test tools have been developed by academia (Grassl et al., 1990; Shiratori et al, 1989) but such tools are seldom used commercially. They are usually prototypes and tend to be small and unreliable as normally there is not enough man power involved in developing them. Often such a type of work is performed by postgraduate students as part or all of their theses. Could we trust small and cheap tools ? The general perception is that it can not be good if it is small and cheap. Low cost tools are generally useful for a small or one-person project. Most academic tools are not of industry strength and therefore can not be commercialised. They tend not to measure up to expectation, and cannot be extensively used in the industry. For a large scale project, an industry strength tool that can support a large problem is required. Industrial developers tend to favour large and expensive tools. For instance, a jumbo 747 aeroplane cannot be designed with a small tool. The size and feature of a tool need to be matched to the complexity of the project. Academia should seek the human resources and financial sponsorship of companies in developing industry strength tools.

4. SOME INDUSTRIALLY RELEVANT TOPICS

Some industrially relevant research topics are suggested in this section. These topics are by no means less challenging and of lower standard than those that have already been conducted by academia. As not much research results for these topics can be found in the present literature, university researchers can easilly published some papers on each of these areas if publication is still a concern for them. Should academia work on these topics, the gap between the theories and practices will become even narrower.

Resource constraint As the industry has to operate in a profitable manner, constraints on resources often hamper the effectiveness of testing (Genuchten, 1991). One can use coverage percentage as an indicator of when to move the testing resources onto another area.

Conflict resolution Internal politics are often involved : designers see testing as a threat, while test managers want to test the product thoroughly. Developers can get upset if lots of discrepancies are found. They perceive that the the test team is slowing them down. However, they should be delighted at each discrepancy found, as this allow them to improve the quality of the product. Conflict thus arises. Conflict resolution and negotiation techniques are critical to software success. There is not too much research that recognises these realities.

Completion criteria Test completion criteria are hard to define. Communication software is often tested until the probability of failure is believed to be small, or until the deadline for the product release is reached (Sherer, 1991). There is nothing wrong with financial and economic criteria as the industry has time and financial constraints. Perfection can not be a completion criterion as software can never be perfect and will always have bugs. The usual practice in the industry is that after releasing a product to the market, suppliers will fix the bugs reported during the warranty period. Knowing how much of the source codes has been covered by the test suite can help estimate the risk of releasing a piece of software to users and thus could form a basis for completion criterion.

Point of diminishing return Inexperienced testers tend to execute down the same path of the software, and no new information is thus yielded. Points of diminishing return need to be identified. There is not much research work on determining where best to concentrate the testing efforts and pointing out where additional testing would be fruitful.

Prioritising Since it is impossible to test every possible case and there is time constraint, priority has to be given to testing the most important and vulnerable parts of the communicating system. Prioritising is necessary, as it is not too fruitful to set identical test coverage goals for all components of an implementation and some components are more critical than others. There is a need to identify the high risk areas so that more testing effort is devoted to these areas and a higher test coverage goals is set for them.

Test case value Since the number of test cases is enormous, there needs to be an optimal way of designing a test suite. By assigning a value to a test case, the relative importance of test cases can be ranked and an optimal test suite can then be designed. The value of a test case needs to be measured in economic terms which could be based on the probability that a particular case will occur and the its consequence of disaster if it fails. Some failures could bring a much larger catastrophic effect than others. Again, there is not much research work on this area.

Code review and measurement Bad codes depend heavily on testing to uncover errors and good codes can only come from good methods. There are not much research on code review and measurement for testing purpose. Statistics need to be collected about the codes in order to get some hints as to what codes need to be improved and tested more rigorously. The gathering of statistics is the start for getting measurement in practical use (McCabe, 1976); the aim is to predict error-proneness by measurements. Conducting a measurement of the codes gives an objective method of locating the error prone codes. Designers and testers usually have a fairly good idea of where the problems might lie, but measurement results affirm their judgements and support their arguments.

Functional aspect Most research focus on the control part of a communicating system. More efforts need to be spent on its functional and robust aspects such as multiplexing, boundary cases, timing, invalid parameter, inopportune events and capacity tests. These types of tests are even more important, without which communication software will not be too reliable.

Empirical comparative Studies There are so many techniques published by academia. Which one should testers adopt ? Empirical comparison of different techniques are seldom conducted and will be useful. A forum must be provided for researchers and practitioners to report both original and replicated studies, varying from controlled experiments to field studies. This experience can be used to evaluate and reveal the relationship between testing theories and practice. Replicated studies, both successful and unsuccessful, highlighting what can be learned from them for improving future work, need to be encouraged for publication.

Test coverage tool The design of a test suite is a slow, labour intensive and costly exercise. The higher the coverage is, the higher is the cost; coverage and cost need to be compromised. The use of tools for test coverage measurement is essential for advancing test coverage methods. Test selection is something that cannot be completely automated. Test sequences generated are not usually related to test purposes and are therefore not cost effective. Human insight continues to be necessary for cost-effective testing. Human guidance and intuition are important to its success. Good test tools should provide an interface that enables a tester to guide the test sequence generation, display test coverage choices and compute the test coverage attained by a particular test suite. Such tools should include input like PICS (Protocol Implementation Conformance Statement), protocol specification, and test suite.

Cost Estimation From experience, cost of testing is very high. There are many contributors to this high cost. Some of them are number of lines of codes, number of modules, number of interfaces, number of requirements and number of function points. There have been some models established for estimating the cost for software development (Boehm, 1981; Putnam, 1978), but none has been established for testing a communicating system yet. It would be nice if there is a simple formula into which we could plug the values for the cost contributors and which would, as a result, provide an estimate of the testing cost.

Others The failures of the relevant application of much of the theoretical work for testing on communicating systems are due to the failure to acknowledge other factors that are critical to projects. Some examples are risk control, functionality, user confidence, predictability, and early deliverable of visible. These areas have been hardly researchered into yet.

5. CONCLUSIONS

This paper has presented the argument that academic research on testing of communicating systems needs to become more industrially relevant and describes the means by which this change can be facilitated. The primary reason for this need is that academia can no longer afford not to work on real problems due to the changes in societies and the ways research grants are awarded to academia. This paper has also suggested some research topics that are useful to the testing community. To help industrialisation of academic techniques, empirical study and a new orientation towards real issues need to be given higher priorities by researchers, and real-life test coverage and field experience need to be more frequently reported . An effective academic testing method applicable to the industry must deal with the practical problems encountered in industrial testing.

It is the wish of the author that there will be a change in direction of academic research on testing of communicating systems after this IWTCS. To help effect such a change, academia must interact with practitioners in the industry. There are massive benefits as well as the costs of having such an interaction. It requires an enormous amount of efforts by both sides, as well as the resources to support the efforts. On the other hand, developers must recognise that they have to change their work habits if they wish to exploit research results. This process is difficult and expensive and it takes up time. Unless both sides are willing to work together despite the cost, the chance of succeeding is low; and we cannot be blind to the difficulties and costs. In conclusion, we still have to do it as the costs of not doing it is even higher than the cost of doing it.

6. REFERENCES

Boehm, B., (1981) *Software Engineering Economics*, Prentice Hall, Englewood Cliffs, N.J.

CCITT (1987) Recommendation Z.100 : Specification and Description Language SDL, Com X-R15-E.

Genuchten, M. van, (1991) Why is software Late? An Empirical Study of Reasons For Delay in software Development, *IEEE Transaction on Software Engineering*, vol.17, no.6.

Grassl, W. and Kossmann, H. (1990) Automated Implementation and Testing of SDL Protocol Specification, *Proceedings of the 2nd International Workshop on Protocol Test Systems*, North-Holland.

ISO (1989a) Information Processing Systems - Open Systems Interconnection - ESTELLE - A Formal Description Technique Based on an Extended State Transition Model, ISO 9074.

ISO (1989b) Information Processing Systems - Open Systems Interconnection - LOTOS - A Formal Description Technique Based on the Temporal Ordering of Observational Behaviour, ISO 8807.

Lai, R. (1992) Teaching Protocol Engineering in Honours Year, *Lecture Notes in Computer Science*, Vol 640, Springer-Verlag.

Lai, R. and Leung W. (1995) Academic and Industrial Protocol Testing - The Gap and the Means of Convergence, *Computer Network and ISDN Systems*, 27(1995) 537-547.

McCabe, T.J. (1976) A Complexity Measure, *IEEE Transaction on Software Engineering*, 2(4):308-320.

Putnam, L., (1978) A general empirical solution to the Macro software sizing and Estimating Problem, *IEEE Transactions on Software Engineering*, Vol. SE-4, no. 4, 345-61.

Sherer, S.A. (1991) A Cost-Effective Approach to Testing, *IEEE Software*, March, 1991.

Shiratori, N., Takahashi K. and Noguchi, S. (1989) An Intelligent Use-Friendly Support System for Protocol and Communication Software Development, *Protocol Specification, Testing and Verification VIII*, North Holland.

Sidhu, D.P. and Leung, T.K. (1989) Formal Methods for Protocol Testing: A Detailed Study, *IEEE Transactions on Software Engineering*, vol.15, no.4.

PART TWO

Means of Testing

2

Testing the broadband network : a quickly shifting demand for test equipment solutions

Ing. L. Vandevenne
Project Manager
Siemens Atea, Atealaan 34, B-2200 Herentals
e-mail : p82453@vnet.atea.be

Ir. B. De Sutter
Broadband, Networks and Protocols
Siemens Atea, Atealaan 34, B-2200 Herentals
e-mail : p82581@vnet.atea.be

Ing. E. Goossens
Verification Testing and Validation
Siemens Atea, Atealaan 34, B-2200 Herentals
e-mail : p82835@vnet.atea.be

Abstract

In the area of test needs there is a quickly shifting demand for test-equipment. This is initiated, not only by the new transmission techniques (such as ATM), but also by network sizes. Nowadays national specific variants loose interest, and cross-border standardizations result in more transparant and cost-effective telecommunication networks.

This paper describes the research results of a project of Siemens Atea in the area of test needs. It has mainly two segments : the first describes the test world where signaling protocols are involved, and its segmentation. The second part describes a model of a solution. It contains also some of the software design techniques used for designing a prototype.

Keywords

testing, ATM, broadband, B-ISDN, telecommunication, protocols, interfaces

1 INTRODUCTION

The booming demand for communication bandwidth has created the need for broadban networks. The established base of narrowband technology is unable to cope with th exponentially increasing data throughput, which is to be delivered by networks intended fo multimedia applications, requiring high- capacity video-, audio- and data-streams.

Within this rapidly evolving domain of broadband networking, the phrase "Asynchronou Transfer Mode" has truly become a cliché. The entire communications industry seems to b aware of the major advantages offered by the ATM technology. The main benefits can basicall be subdivided into 3 categories:

- ATM can support a large variety of integrated application requiring large and flexible bandwidths.
- ATM is scaleable and can be placed on virtually any type of physica carrier, ranging from low-end 2 Mbit/s copper links (E1/DS1) up t multi-Gbit/s optical fiber.
- ATM offers highly efficient and easily manageable networking capabilities.

However, these advantages are haunted by the inevitable ghost of the "testing netherworld". Clearly, testing and troubleshooting this kind of high performance ATM-based networks is not a task to be taken lightly. Human abilities to control and observe the immense amount of exchanged data involved are totally insufficient.

Therefore, the purpose here is not to elaborate on the "why" and "how" of broadband network design and evolution, but to present a point of view on how these networks can be tested without having to deal with a mental test hangover.

2 DEFINING THE KEY TEST ELEMENTS

Power and flexibility are not the only players in the game to ensure that the ATM-based network technology will be successful. In order to check and guarantee their stability and reliability in network environments, ATM systems have to be subjected to extensive testing at all levels, starting from the design and implementation phases up to daily under-operation checking and trouble-shooting.

ATM is destined at making its appearance in network solutions of varying scale, from local corporate utilization up to pan-European or even global interconnected high-performance networks. This implies automatically a multi-vendor, multi-operator environment. As a result, ATM network testing activities have to take these abilities into account.

Globally summarizing the common factors within any test process, results in following elements:

- A testable entity.
- A selection of properties to be tested.
- suitable test tools

The scope and purpose of the test processes can differ significantly. What we want to do here, is to define the above listed key elements for ATM network-related testing.

2.1 A testable entity

Depending on the desired scope of testing, this can either be:

- a single ATM access or network node (= switch or concentrator)
- multiple, inter-operable ATM network nodes (= the network itself)

The tests carried out on the single ATM switch are essential for the manufacturer or vendor involved. It allows them to verify their system during the design and implementation phase, in an environment where failures or bugs cannot influence a critical operational network. The network operator can then decide, afterwards, to perform a series of acceptance tests on this single switch, also without endangering the already established network.

Once a stable system is integrated into the ATM network, additional day-to-day operational tests remain necessary for maintenance or troubleshooting purposes. These tests can only be carried out within the (potentially global) operational network, thereby possibly triggering the need for efficient remote testing.

2.2 A selection of properties to be tested

The core of each test process consists of the system or network properties that need to be tested. The earlier mentioned power and flexibility, which ATM technology brings to broadband switching systems and networks, certainly do not reduce the number of test paths and interworking functionalities to be covered.

The most important properties that have to be subjected to testing in ATM networks are the following:

Signaling protocols:
Certainly the primary aim in any switched ATM network test. The presence of signaling is an absolute prerequisite in any network. Many protocols can, conform to the OSI concept of layers, ride on the ATM protocol. Although bodies like ITU and ATM Forum define standards, such as the ITU-Q.2931 in case of UNI-signaling, the appearance of variants cannot be avoided. These variants bring incompatibility problems along, which are impossible to trace without decent testing and monitoring.

Variety of physical line interfaces:
As stated before, ATM can be considered independent of the physical carrier it's placed upon. E3/DS3, E1/DS1, TAXI, STM-1, STS-3c, STM-16, ... the list of supported carriers grows almost every day. Again great for flexibility, but it adds additional complexity to the testing process.

High performance traffic capabilities:
The roots of broadband are to be found in the quest for flexible admission of virtually unlimited bandwidth. Performing tests to verify if the network, or its separate nodes, deliver the switching capacity and throughput as expected, is imperative. Not only performance is the key issue here but the reaction of the network to conditions of overload and congestion needs to be verifiable as well.

Reliability:
Broadband networks are not only famous for their performance, but also for their reliability and stability. Even this claim has to prove itself under stringent test conditions. Tests on behalf of long duration stability and error-free data transmission over the U-plane connections should not be absent in any self-respecting test-suite. Signaling alone (an aspect most covered in tests) does not make up an entire network, since the single most important purpose is to transport data.

2.3 Suitable test tools

Where it all boils down to is the availability of suitable test tools. Each of the previous aspects could be covered by a dedicated test tool. Regarding the earlier mentioned and inseparable "complexity-flexibility" twins, this would lead to a rather large pool of different test tools, each catering for a small, specialized domain of ATM testing. Make no mistake, these tools are necessary for those experts who are continuously working in these domains, in the form of research or development.

However, there is also a need for more universal test tools. These tools have to cover a larger area of testing, albeit not with the same in-depth attention to all possible domains as their dedicated counterparts provide. These tools would allow operators and manufacturers alike, to perform a significant part of their tests by means of one, easily usable device.

Ideally, such a device would adhere to the following requirements:

- Multi-interface support, to cover the wide variety of physical interfaces.
- User-alterable multi-protocol support, to counter problems with the inevitable protocol variants.
- Multi-role capabilities (conformance TTCN testing, inter-operability with narrowband, performance/BHCA/throughput testing).
- Provide for combined C/U-plane testing, not only to test signaling but error-free data transport as well.
- Non-intrusive line monitoring abilities.
- Suitable for stand-alone, LAN and long-distance testing.
- Modular expandability of all above mentioned requirements.

This versatility precludes practically all potential deficiencies as encountered in diverse problematics of broadband network testing. Pursuing this "broadband test utopia" is a goal that may never be achieved. However, the research results indicate that very well performing solutions can be found.

3 SEGMENTATION OF THE TARGET

Clearly, due to the complexity of broadband network configurations and to the variety of ATM equipment, actual test tools have to provide the necessary versatility, power and flexibility to ensure that proper inter-operability between new ATM network components can be tested in a uniform and user-friendly way.

Research activities at Siemens Atea resulted in the development of the ATS ('ATM Traffic Simulator'), which is intended to fill in these requirements as much as possible.

4 ATM TRAFFIC SIMULATOR

4.1 History

End '93, when the research activities started, no appropriate ATM test equipment was found on the market. Therefore the research was started as described above. Of course when the needs are defined, an answer to these needs could prove the correctness. And if this answer turns out to be a good solution, it could also turn into a commercialisable product.

At first a solution was looked for in the adaptation of existing narrowband test equipment. However, to avoid being limited by the non-ATM based technology of the existing product, the effort was undertaken to start a new product from scratch, which is called ATS.

Commercial hardware for the rack and the processor module is used as a basis. The PC is taken as software development platform. A commercially available operating system is used on the test unit, complemented with an own operating system shell to implement the functions of a real time operating system.

In order to meet the demand of conceptual flexibility for allowing the operation with a user specified test scenario for the signaling layer under test, a Protocol Description Language (similar to textual SDL) is designed. The test scenario's can be modified in run time, without having to re-compile and re-link the complete load. From this test scenario, other layers/functions (f.i. data path testing) can be activated. To meet the demands for a test environment for conformance and interoperability tests, a complete TTCN workbench will be developed in the near future.

4.2 System description

Introduction

The ATS consists of a GUI (Graphical User Interface) and one or several SU's (System Units). The GUI software packet, running on a commercial PC, presents to the ATS user the interface to control the application software of the SU's. It is connected to the SU's via RS232 or Ethernet LAN interface, based on TCP/IP. The SU consists of an industrial PC cabinet with a CPU module, harddisk, network interface and currently up to 8 DIU's (Digital Interface Units). Each DIU contains the necessary hardware for testing one or more broadband interfaces.

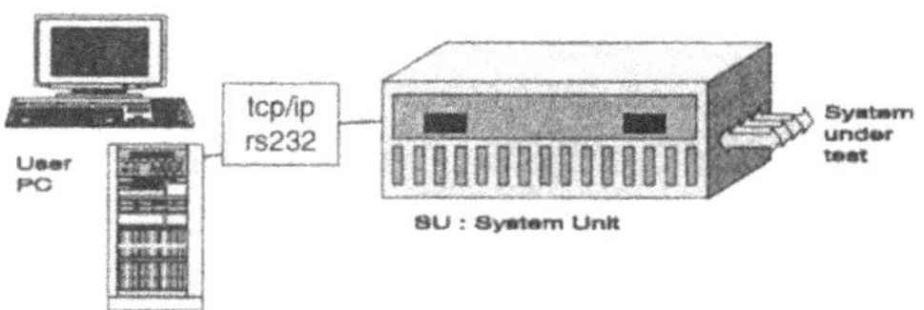

Figure 1 Local system configuration

Hardware of the SU

In today's maximum configuration, an SU consists of up to 8 interface boards (DIU's) to connect the system(s) under test, and one processing unit for the necessary test activities to be executed. The SU is VME compatible, the CPU board is PC compatible and uses an Intel processor. The system is transportable, the interfaces simple to exchange.

The DIU's interface to the system(s) under test, they are completely realized in hardware. The DIU blocks are available in different types depending on the interface requirements. Each DIU block contains a number of ATS test ports depending upon the DIU type (STM1 electrical, STM1 optical, E1/DS1, E3/DS3, ...). Each DIU consists of 2 cards : a BLM (Broadband Load Module) and a BIM (Broadband Interface Module). In B-ISDN, negotiation of bandwidth for the datapath may lead to different bandwidths per call. Therefore, each BLM contains programmable datapath testers, which are able to receive/send the required bandwidth. The BIM encapsulates/decapsulates the ATM cells. The ATM cells are exchanged between BLM/BIM via the standardized UTOPIA interface.

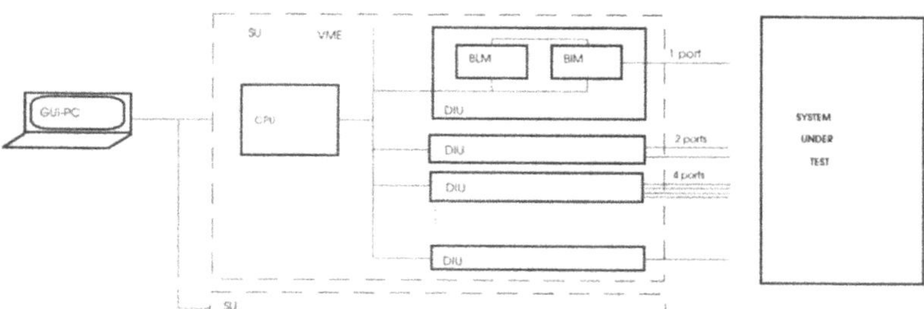

Figure 2 System architecture

Software

The software of the test equipment is mainly structured into the following blocks : on one hand the GUI and off-line tools for preparation of the test scenario's, on the other hand the test execution software on the SU.

- **GUI and Test Scenario Preparation SW**:

The GUI provides the access to the test system. It is responsible for the configuration, the administration and the supervision of the test units. It provides the capabilities to log events and diagnostic information on screen, printer or disk. Windows with pull-down menus, pop-up menus, window panes, toolbar, icon buttons and context sensitive help functions will guide the user. Off-line tools (i.e. the Sdtool) are offered for test scenario preparation.

- **Test execution SW**:

This software is responsible for the execution of the signaling testing, line monitoring, load testing and data connection testing.

The software part which is responsible for signaling, can be subdivided into 2 parts:

- A fixed part, representing the underlying protocol stack, is only defined for B-ISDN testing. It is based on AAL5.
- A programmable part, representing the signaling layer under test.

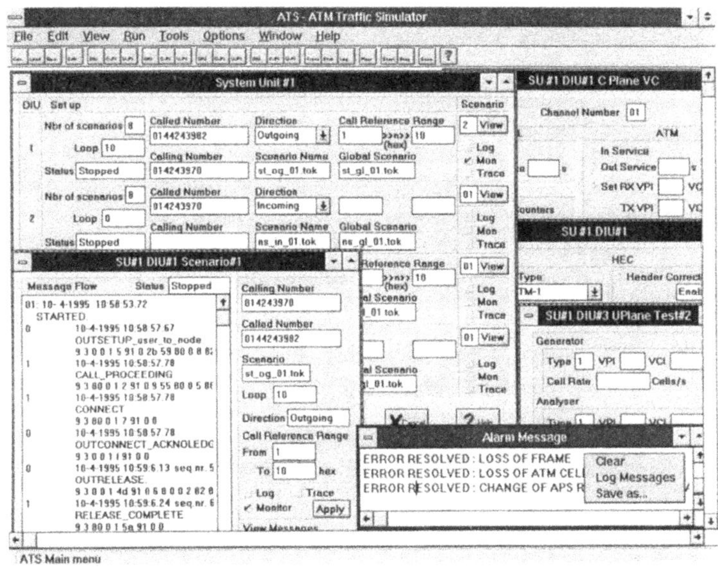

Figure 3 SU edit window on the GUI

The chosen operating system shell provides scheduling, timers, inter-process and inter-processor communication using the ACM (Asynchronous Communication Machine), MAE (Manageable Entity) and Timers concepts. It ensures the portability of the software on a different platform. The OMT (Object Modeling Technique) is applied as software engineering methodology, with C++ as programming language.

4.3 Test configurations

Protocol and Load Test

Figure 4 Protocol and Load testing configuration

The ATS is simulating a node and/or a terminal. The signaling protocol under test is simulate by replacing the implementation of a defined protocol layer by a test scenario. This scenari allows to test the protocol implementation of the SUT by enabling the tester to send ou sequences of user defined protocol messages in order to verify the protocol behaviour in both normal and error situations. In this way, the ATS can be used to perform manually conformanc and interoperability tests.

Line Monitoring

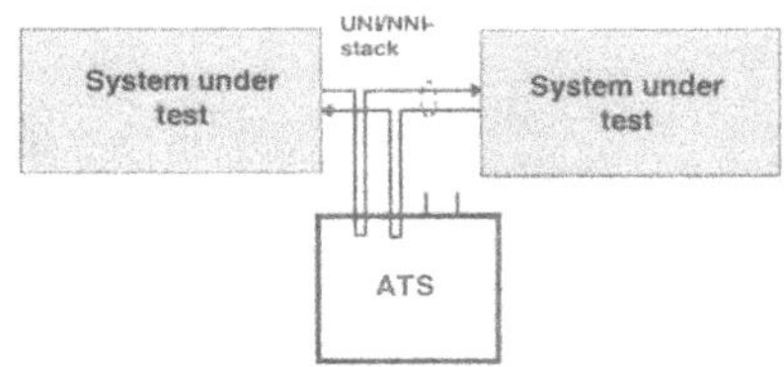

Figure 5 Line monitoring testing configuration

In line monitoring configuration, the ATS is looking at the signaling data passing on the line, without changing the data. The monitored data can be analysed on the fly or stored on the hard disk for a later off-line analysis.

Data Path Testing

The same test configuration that is used for protocol and load test applies. The testing of the data connections implies sending cells over the connections and checking the received information. Datapath testing can be triggered from within the scenarios (call related). It can also be used for testing PVC's (without any signaling relation).

4.4 Description of the scenario concept

Using a test scenario involves two separate phases, as shown in the next figure.

The first phase is the creation of a scenario which is done in the GUI. All means to create or change test programs with a behaviour fully specified by the user are provided. Elements of a protocol (incoming messages, outgoing messages, timers and state/event processing) may be specified with a large degree of freedom.

The second phase is the set-up of the on-line form of one or more instantiations of a scenario created during the first phase.

Phase-1: Creation of a test scenario.

Test case generation starts with off-line *test scenario preparation.* Composing a scenario is making a behavior description in terms of scenario building components (SBC) i.e. messages, states, timers and actions, including the possibility of following certain branches in scenario dependent on conditions.

This phase is completely handled off-line. A test scenario is created/changed in three steps:

- **Step one**

A scenario script file (filename.scn), which is a text file, is created or changed using a set of rules described in the Protocol Description Language. The user is offered the option to define/change the name and/or contents of the SBC's.

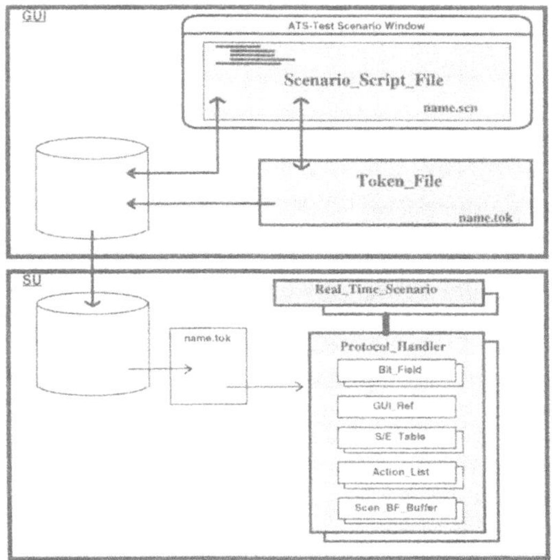

Figure 6 Scenario preparation

In the declaration of outgoing messages, there are no limits on the introduction of errors in the layout or content of messages. The actual content of a message sent in a running scenario can be dependent upon the contents of a received message. In the scenarios one can define several types of error injection. The ATS has a number of predefined SBC's giving the possibility of composing scenarios rather quickly. The SBC pool is structured according to the different protocols under test supported by the ATS (ATM Forum 3.1/4.0 (UNI), ITU Q.2931 (UNI) Q.2761-64 (NNI)).

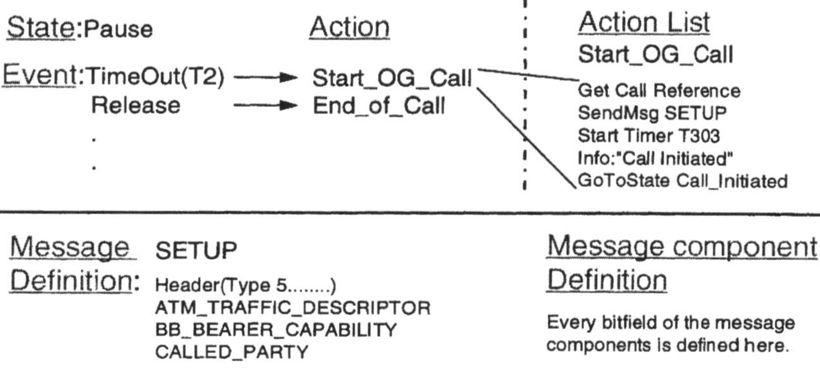

Figure 7 The scenario definition can be subdivided into message definition, description of action lists, and a state/event description.

- **Step two**

A token file (filename.tok) is created with the Sdtool, ready for processing by the real-time software. During the conversion, several tests are performed to detect possible errors in the protocol specification. The conversion, done off-line, speeds up the processing on the real-time processor. The tools used for the processing of the textual representation are YACC and LEX. The first stage verifies the syntactical and semantic correctness of the input specification. During the second stage the textual description is disassembled into a set of inter-dependent data structures (bitfields, components, messages, states, timers and GUI conversion table). In the last stage the inter-dependent data structures are translated into a stream of tokens representing the necessary real time objects to simulate the specified protocol. In here you can find all necessary information how to identify a message, which bitfields must be checked and which actions must be taken. The first and second step are processed simultaneously, while the token generation starts at the end of file of the textual description.

A scenario token file has several parts in a specific order. The order of appearance is essential for the setup of the real time scenario.

- **Step three**

The created/changed token file (filename.tok) is copied from the DIU disk to the SU disk.

Phase-2: Set-up of an on-line scenario.

The SU receives the instructions for the set-up of one or more on-line scenario's in one or more DIU's from the GUI. An on-line scenario is the executable form of a scenario token file. A simplified form of on-line scenario's is presented by the figure below. As one can see there are two parts involved in on-line scenario's : the RTScenario (Real Time Scenario) and a ProtocolHandler. A RTScenario is created for each simultaneous instantiation of a scenario and contains all information which is specific for each running copy of a scenario.

A ProtocolHandler is created with all the information of a token file and contains three types of building elements : buffers, action trees and state/events:

- All data of a running scenario, is located in buffers.
- An action tree is installed and executed as a queue of different actions where each action in a tree returns a reference to the next action or an end indication.
- A state/event is created for each state of a scenario. Each state/event is programmed with the information to recognise the user specified incoming messages and events, as well as with the references to the action tree specified for each case.

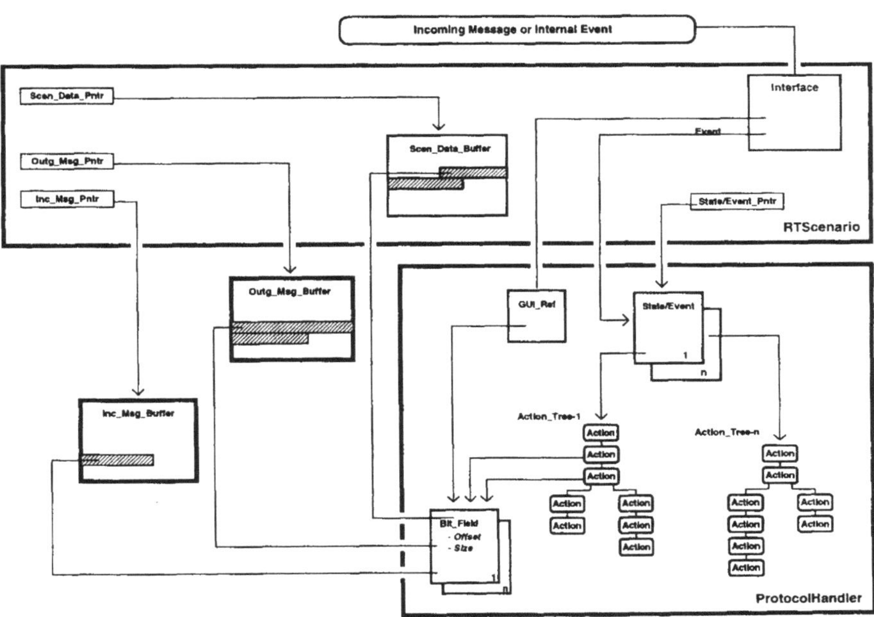

Figure 8 Setup of a real time scenario

5 CONCLUSION

The lack of versatile test equipment on the ATM market, triggered research activities at Siemens Atea. The subsequent "R&D"-effort spawned the first prototype of the ATS, which in its turn has matured and proven itself as a viable test tool for ATM networks. From the earliest stages on, the most important key points in the development of the ATS have been *flexibility*, *expandability* and *user-friendliness.* A test tool that is too rigid to cater for the evolving needs of the systems it is destined to verify, is not worth the resources it dissipated. Neither should the ease of use be dwarfed by the power the device actually harbours.

The ATS does not aspire to be everything to everyone in the world of ATM testing, but it does offer a workable solution for significant areas of ATM network testing.

6 ABBREVIATIONS

ATM	Asynchronous Transfer Mode
ATS	ATM Traffic Simulator
DIU	Digital Interface Unit
DS1	Digital Signal 1, method for digital transmission at 1.5 Mbit/s
DS3	Digital Signal 3, method for digital transmission at 44 Mbit/s
E1	ITU standard for digital transmission at 2 Mbit/s
E3	ITU standard for digital transmission at 34 Mbit/s
GUI	Graphical User Interface
LAN	Local Area Network
NNI	Network Node Interface
PDL	Protocol Description Language
PVC	Permanent Virtual Channel
SBC	Scenario Building Component
SDL	Specification and Description Language
STM-1	Synchronous Transport Module level 1, a 155Mbit/s interface
SU	System Unit
SUT	System Under Test
UNI	User to Network Interface
UTOPIA	Universal Test and Operations Physical Interface for ATM

7 REFERENCES

- Martin de Prycker (1991), *Asynchronous Transfer Mode, solution for broadband ISDN*, ISBN 0-13-053513-3
- Daniel Minoli, George Dobrowski (1995), *Principles of signaling for cell relay and frame relay*, ISBN 0-89006-708-2
- The ATM Forum (1993), *ATM User-Network Interface Specification version 3.0*, Prentice Hall PTR, ISBN 0-13-225863-3

8 BIOGRAPHY

- Ing. Luc Vandevenne
  Luc Vandevenne graduated as industrial engineer electronics in 1989 at the K.I.H.L. in Diepenbeek. His professional carreer started at Siemens Atea as protocol test engineer. From 1992 he coordinated the system test process for several projects (ISDN and GSM) and steered the introduction in the field of several releases in this area. Since 1995 he is project manager for broadband test equipment.
- Ir. Beatrix De Sutter
  Bea De Sutter graduated as civil engineer physics at University of Ghent in 1989. In 1989 she did research in nuclear physics at KFA in Julich (Germany). In 1990 she joined Siemens Atea as protocol designer in the area of SCCP and TCAP. From 1995 on she moved to broadband protocol design and now coordinates the activities in the ACTS INSIGNIA project for Siemens Atea.
- Ing. Erik Goossens
  Erik Goossens graduated as industrial engineer electronics (chip design) at K.I.H.L Diepenbeek in 1994. Immediately after his studies he joined Siemens Atea as sytem verification engineer. Now he is involved in the testing of the Belgian releases for narrowband telecommunication systems.

3

Conformance testing trials on ERMES receivers

P. Kostarakis, A. A. Alexandridis, K. Dangakis, A. Vlahakis, P. Katrivanos, F. Lazarakis, A. Paschalis and N. Gaitanis
Inst. of Informatics and Telecommunications, NCSR "Demokritos", 153 10 Aghia Paraskevi, Athens, GREECE,
Tel. No.: + 30 1 65 20 847, Fax No.: + 30 1 65 32 175,
E-mail: aalex@iit.nrcps.ariadne-t.gr

Abstract

A test system for checking the receivers of the digital pan-European paging system ERMES for conformity to the standard specification for air interface and receiver operation has been developed according to the ISO 9646 standardized methodology for conformance testing services. The test procedures, various test setups and test execution software, implemented and integrated in a fully operational test system, are described in this paper. Complete test suites have been carried out on ERMES receivers. A part of the results of test trials is listed here, demonstrating the test system functionality.

Keywords

Paging system, ERMES, conformance testing

1 INTRODUCTION

Conformance testing is the verification that an implementation meets the formal requirements of the referenced standard. Its primary objective is to ensure interoperability among different product implementations. The need for a variety of interconnection and interoperability schemes among different implementations has led to the establishment of a standardized methodology and modeling approach for the description of the operation of the introduced systems and services, namely the Open System Interconnection (OSI) model.

The operational capabilities of any system or commercial product, derived through the realization of a new concept, are fully described in the relevant standard specification which may be considered as national, international, licensed, etc. Such implementations, which usually come from various manufacturers and, sometimes, even technologies, can be tested uniformly to determine whether they conform to these standards. Consequently, authorized organizations should ensure the existence of accredited laboratories to provide Conformance Testing Services (CTS) for the industry.

In this paper, we describe the development of a Conformance Test System* for the air interface and receiver operation of the European paging system ERMES (European Radio MEssage System), as specified by the European Telecommunications Standards Institute (ETSI) (prETS 300 133-1, prETS 300 133-4, prETS 300 133-5, 1991). In particular the test procedures, with the various test setups and the test execution software, are described. Finally, a detailed list of results, obtained from trials covering key test cases, is included.

2 ERMES OVERVIEW

ERMES is allocated on a common frequency band throughout Europe comprising sixteen 25 kHz spaced channels between 169.425 MHz and 169.800 MHz. The ERMES network operates at a data rate equal to 6.25 kbits/s, which is reduced at a symbol rate equal to 3.125 kbaud by the use of 4-PAM/FM modulation. Normally, an ERMES receiver will receive calls on a single channel used by its home network. However, it will scan all the 16 channels in the case when that single channel is lost (e.g., during roaming).

The structure of the ERMES transmission protocol is based on a sequence of 60 minutes total duration comprising 60 paging cycles. Sequences are coordinated with the UTC (Universal Time Coordinated) so that on the hour a new sequence commences. Each cycle (one minute long) is divided into five subsequences commencing at 12 second intervals. Each subsequence is further divided into 16 batches and each batch has four partitions (synchronization, system information, address and message partitions). The first fifteen batches in every subsequence have a length of 154 codewords and the final batch has a length of 190 codewords (standard information unit of 30 bits length). The receiver population is divided into 16 groups and each receiver is allocated to one of the 16 batch types according to the 4 least significant bits of its basic Radio Identity Code (RIC).

The ERMES system supports the following four paging categories:

- Tone only: the receiver shall respond to at least one of eight alert types.
- Numeric: the receiver shall provide for the reception of a 20 digit numeric message.
- Alphanumeric: the receiver shall provide for the reception of a 400 character text message.
- Transparent data: the receiver shall provide for the reception of an arbitrary data message.

In addition, a wide range of supplementary services is also supported, e.g., acknowledgement, protection against loss of messages, three levels of priority, long messages.

*This work has been co-funded by the European Union and carried out under the CTS-3/ ERMES project, contract No. 90/45067.

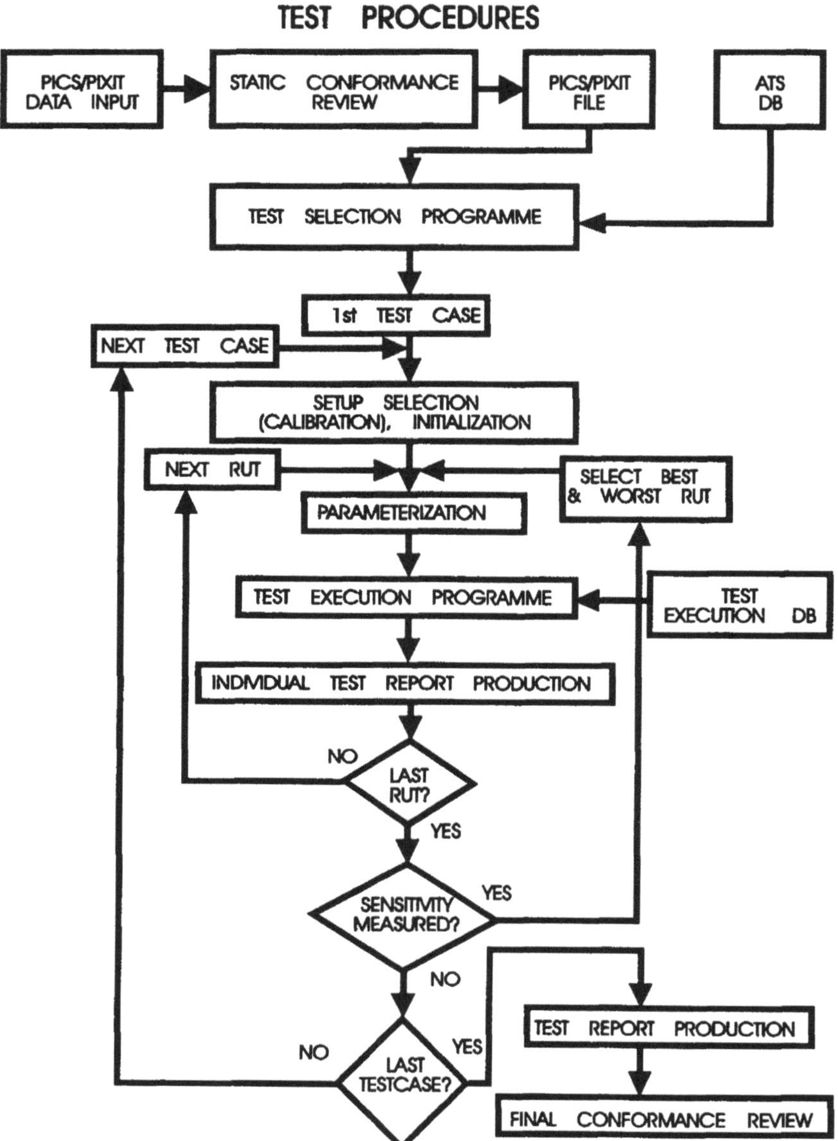

Figure 1 Complete Test Procedure Flow Chart.

3 TEST PROCEDURES

The procedures carried out by the test system in order to run the tests are shown in the flow chart of Figure 1. A more detailed description is given in (CTS-3 ERMES Project 45067, Deliverable 3, 1992) and (Dangakis, 1992). PICS (Protocol Implementation Conformance Statement) and PIXIT (Protocol Implementation eXtra Information for Testing) contain the required information for test selection and parameterisation (CTS-3 ERMES Project 45067, Deliverable 5, 1992). The data included in the PICS and PIXIT proformas will be stored in a file in order to allow an automated selection and parameterisation procedure to be performed. The programme that enables the data input performs also some consistency checks ("static conformance review"). The whole ERMES Abstract Test Suite (ATS) is organized in an ERMES database (CTS-3 ERMES Project 45067, Deliverable 1, 1991). The "test selection programme", based on the PICS and PIXIT information, handles the ATS database and produces the selected ATS containing all the test cases relevant to the particular Receiver Under Test (RUT). For each selected test case, the file produced by the "test selection programme" contains all the necessary information for the test execution. First the setup, that is appropriate for the particular test case under consideration, is selected and the necessary installation and preparations are done (ETSI PT8, Final Report, 1990).

Several parameters must be specified before running a test. Some of them are derived from the RIC that is specified, for each one of the six receivers to be tested, by the ERMES standard (prETS 300 133-5, 1991). Most of the test cases require that only the best and the worst (with respect to the measured average usable sensitivity) receivers of the six supplied shall be used. This means that the sensitivity measurement tests shall be carried out first, and then their results shall be taken into account during the test case selection and parameterisation. Some parameters are specified according to the input information from PICS / PIXIT. Others are set according to the requirements of the particular test case and their values are retrieved by the "test execution programme" from the "test execution database". Finally, there are some parameters that need to be set and updated in real time, while running the tests.

The different setups that are used and the "test execution programme" (software tool actually running the tests) are described in the next two sections.

4 SETUPS OF THE TEST SYSTEM

According to the standards (prETS 300 133-5, 1991), some tests are carried out on an Open Area Test Site (OATS) while the rest are carried out using a test fixture (TEM-cell). For extreme test conditions a test fixture placed in a climate chamber is also required. So, the following four setups were used in the test system:

- setup A for OATS measurements;
- setups B, C and D for measurements using a TEM-cell.

The block diagrams of these setups are given in Figure 2, where a RF signal generator simulates the ERMES base station. The calibration of the test site and test fixture (ETSI PT8, Final Report, 1990) establishes a relationship between the output level of the RF signal generator and the field strength applied to the RUT. This relationship for each one of the setups is given below.

Test setup A

The following formula was applied, giving the field strength *e* at the RUT position:

$$e = u + \mathrm{AF} + \mathrm{AL} + \mathrm{CL} \quad (\mathrm{dB\mu V/m})$$

where

u	: substitution antenna output measured with spectrum analyzer (dBμV)
AF=12.9 dB	: substitution antenna factor
AL=0.2 dB	: substitution antenna loss
CL=0.9 dB	: cable loss, from substitution antenna to spectrum analyzer

Test setup B

According to the calibration of the TEM cell, the average usable sensitivity (dBμV/m) equals to the average of recorded values minus 9 dB.

Test setups C & D

RF generator A (wanted signal) output level, *r*, adjustment:
The results from the measurements of the average usable sensitivity, in the OATS and the test fixture, are used to calculate the *r* (dBμV) corresponding to a field strength of 25 dBμV/m, in the test fixture. The following formula was applied:

$$r = l + d + \mathrm{A}$$

where

- l: RF Generator A output level (dBμV) obtained from the measurement of the average usable sensitivity in the test fixture.
- d: Difference (in dB) between the measured average usable sensitivity in OATS and the reference figure (25 dBμV/m).
- A: Additional attenuation (in dB) due to combiner and extra cables (equal to 7.7 dB for setup C and to 10 dB for setup D).

5 TEST EXECUTION PROGRAMME

This software runs on the main controller (PC) and communicates with the other devices through digital I/O, serial and IEEE interface cards. Before running a specific test case, it initializes the setup by:

- Reading the ambient temperature and relative humidity values and verifying that they are within the limits specified by the standard for normal test conditions.
- Setting the climate chamber to the extreme temperature and relative humidity values specified by the standard for extreme test conditions (if needed).

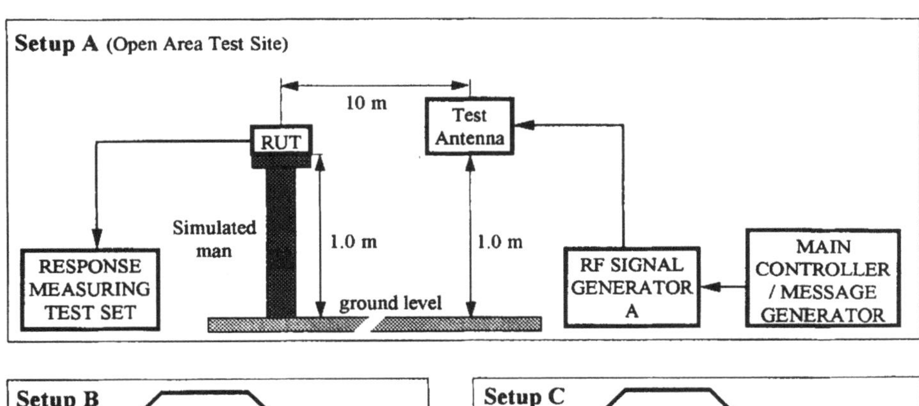

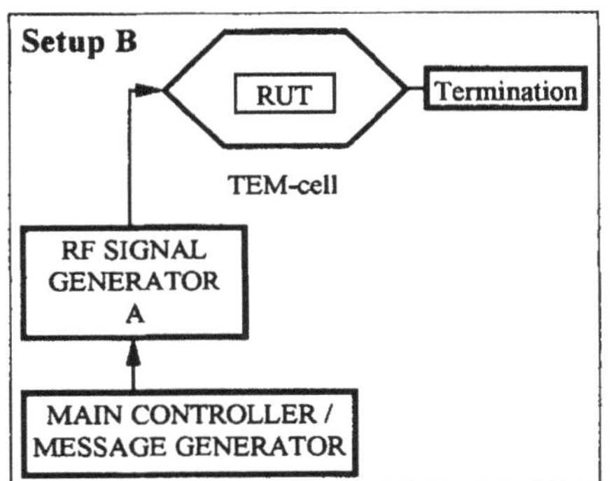

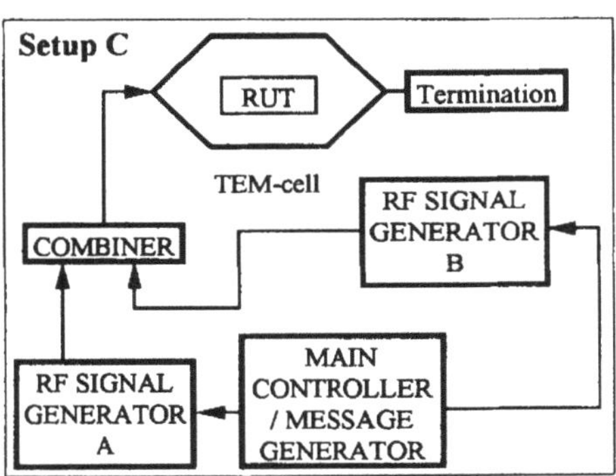

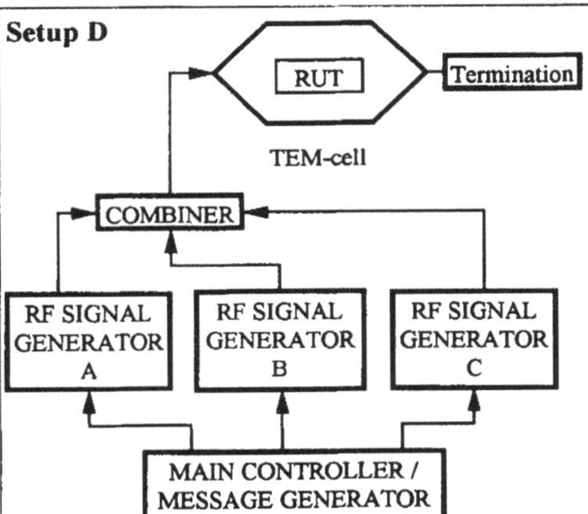

Equipment used in setups	Manufacturer Type
Open Area Test Site	DEMOKRITOS
Simulated man	DEMOKRITOS
Antenna mast OATS	Anritsu MB19A
Message Generator	DEMOKRITOS
RF Signal Generator A	R / S SME-03
RF Signal Gen. B & C	R / S SMG
Antenna (ERMES signals)	EMCO 3121C
Antenna for substit. meas.	EMCO 3121C
RF combiner	R / S DVU4
TEM cell	DEMOKRITOS
Response Measuring Test Set	HP 8595A spectrum analyzer

Figure 2 Block diagram of Test System Setups and list of equipment.

- Setting the power supply voltage to the nominal or extreme (maximum or minimum) value.
- Setting the center frequency of the RF signal generator to the specified RF channel.
- Setting the RF signal generator output power level to the proper value.

During most of the considered test cases, the common and basic task next carried out by the test execution software is the transmission of a burst containing a message at a specified time slot (batch). In order to do this, the software must step through the following:

- It forms the information content that is to be transmitted, by combining information retrieved from PICS / PIXIT and the test execution database or entered by operator.
- Time related information (date, hour, cycle, subsequence and batch numbers) is also inserted and updated in real time.
- Error correction coding, where the information bits are grouped in codewords, is done followed by codeword interleaving.
- The resulting bit-stream buffered in the PC RAM, is then passed, at the nominal bit-rate and through the digital I/O card, to the modulation data input of the RF signal generator performing ERMES modulation. The software uses for synchronization a time reference obtained from a UTC source (Odetics GPStar).

The above steps, resulting in the transmission of an ERMES message (single batch transmission) at a specified RF channel frequency, may be repeated several times while running a single test case, depending on the requirements. This procedure is automatically repeated but, after each transmission, the software waits for the operator's input denoting if the response was successful or not.

The ERMES conformance testing services may be divided in two categories: the RF parameter tests and the protocol conformance tests (CTS-3 ERMES Project 45067, Deliverable 5, 1992).

The first category includes all test cases directly related to Radio Frequency operational characteristics of ERMES receivers. The value of the RF parameter under test is derived statistically from the RF signal levels recorded during repeated single-batch transmissions. This value is then compared with the corresponding verdict criterion, in order to assign the verdict (Pass or Fail) to the specific test.

To illustrate such a procedure, we consider, as an example, the test case concerning the measurement of average usable sensitivity that corresponds to a "message acceptance ratio of 80%", as defined in Annex A, subclause A.1.2. of (prETS 300 133-5, 1991). The software uses the algorithm shown in the flow chart of Figure 3. This procedure is repeated for the 8 positions, 45° apart, of the receiver. The corresponding average values of the generator output are determined and noted. Using the calibration of the test site, the 8 field strengths corresponding to the 8 average values are calculated and the average measured usable sensitivity is derived as a mean field strength.

The second category (protocol conformance tests) includes all test cases referred to the essential or optional features of ERMES receiver operation, with respect to its particular paging type. The test execution software manages more complex message transmissions covering multiple batches, subsequences (16 batches) and RF channels. In this category, generally, one message transmission occurs and the verdict is derived by comparing the receiver response to the expected one.

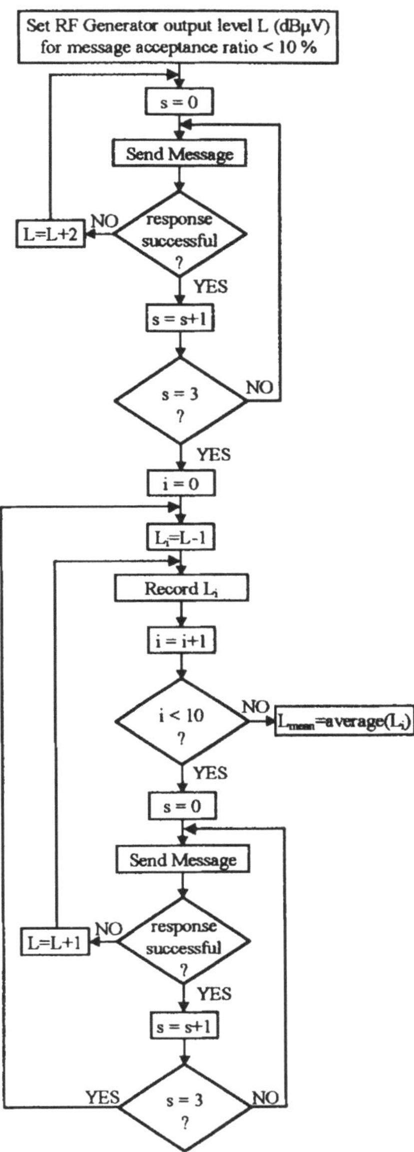

Figure 3 Flow chart of the algorithm for the implementation of a message acceptance ratio of approximately 80%.

Table 1 List of covered test cases with the corresponding ATS No. and setup

ATS No.	*SETUP*	*DESCRIPTION*
001-003	A	Average Usable Sensitivity channel 8
004-006	B	Average Usable Sensitivity channels 0, 8, 15
007-009	B	Average Usable Sensitivity with tx freq. offset
010-012	B	Average Usable Sensitivity (extr. conditions)
013-015	B	Average Us. Sens. (extr. cond.+freq. offset)
016-018	B	Average Usable Sensitivity channel switching
020-022	A	Usable Input Level Range
023-025	C	Co-channel Rejection
026-028	C	Adjacent channel Selectivity (norm. cond.)
029-031	C	Adjacent channel Selectivity (extr. cond.)
032-034	C	Spurious Response Immunity
035-037	D	Intermodulation Immunity
038-040	C	Blocking Immunity
044-046	B	First Message in Batch
047-049	B	Last Message in Batch
050-052	B	Message in Next Batch
053-055	B	Message in Last Available Batch
056-058	B	Tenure of Message
059-061	B	Two Messages in the Same Batch
062-064	B	Message in the next Subsequence
065-067	B	Message reception on all ERMES channels
068-070	B	Recognition of Zone Code
071-073	B	Two Messages in the Same Batch
080-081	B	Maximum length message
082-084	B	Alert (Normal operation)
085-087	B	Alert (Silent mode)
088	B	Termination of numeric messages
089	B	Termination of alphanumeric messages
090-095	B	Repeated Call Indication
096-098	B	Group Call
099-101	B	Tenure of Group Message
102-103	B	Low Battery Indication
111-113	B	RSVD bits
116-119	B	Indication of lost message
120-122	B	Urgent message indicator

Table 2 List of results for test cases concerning RF parameters measurements (alphanumeric paging category).

ATS No.	*Measured Value*	*Verdict Criteria*	*Verdict PASS*	*Verdict FAIL*
022	4 correct messages out of 4	4 correct messages out of 4	✓	
025	7.6 dB	≤ 10 dB	✓	
034	79.2 / 83.2 dBμV/m	≥ 76 dBμV/m	✓/ ✓	
037	73.9 / 73.7 dBμV/m	≥ 70 dBμV/m	✓/ ✓	
040	103.4 /109.0 dBμV/m	≥ 84 dBμV/m	✓/ ✓	
003	21.8 dBμV/m	≤ 25 dBμV/m	✓	
006	22.7 / 24.3 dBμV/m	≤ 25 dBμV/m	✓/ ✓	
009	24.1 / 20.5 dBμV/m	≤ 25 dBμV/m	✓/ ✓	
012	No RUT response	≤ 31 dBμV/m		✓
015	No RUT response	≤ 31 dBμV/m		✓
018	No RUT response	≤ 25 dBμV/m		✓
028	54.8 dB	≥ 60 dB		✓
031	No RUT response	≥ 50 dB		✓

Table 3 List of results for test cases concerning Protocol Conformance tests (alphanumeric paging category).

ATS No.	*Verdict Criteria*	*Verdict PASS*	*Verdict FAIL*
046	Correct reception without error	✓	
049	Correct reception without error	✓	
052	Correct reception without error	✓	
055	Correct reception without error	✓	
058	No reception	✓	
061	Correct reception of both messages without error	✓	
064	Correct reception without error	✓	
067	Correct reception on all ERMES channels without error	✓	
070	No reception	✓	
073	Correct reception of message 2 without error	✓	
081	Correct reception without error	✓	
084	Correct reception without error giving the appropriate alert	✓	
087	Correct reception without error (no audible alert)	✓	
088	Presentation of only the first 4 characters	✓	
089	Correct reception without error	✓	
095	Correct reception without error, giving a repeated call indication		✓
098	Correct reception without error	✓	
101	No reception of group message	✓	
103	≥ 1.08 V	✓	
113	Correct reception without error	✓	
119	The receiver shall indicate that messages are missing	✓	
122	Correct reception without error, with an urgent message indication	✓	

6 TEST RESULTS

When the execution of a test case has finished and a verdict (Pass, Fail) is assigned to it, the test results are stored. This information is updated at the end of execution of each test case and is used to produce the test report at the end of the test campaign (Figure 1).

The verification and the calibration of the ERMES test system is performed using the ERMES Reference Implementation (RI), (CTS-3 ERMES Project 45067, Deliverable 7, 1993). The ERMES RI is a device that consists a reference system implemented in accordance to the ERMES standard. It proves the equivalence of the testing tools as well as the functionality of the CTS centers, prior to approvals and commercial testing.

The ERMES test system presented here is currently operational in our laboratory. Test suites covering the test cases listed in Table 1 have been carried out on ERMES receivers coming from various manufacturers. In order to demonstrate the operation of the implemented ERMES test system, we present (in Tables 2, 3) the results of some test cases applied on a specific prototype ERMES receiver for alphanumeric paging category. The information that this receiver failed in some tests, could be useful in modification of the design of the particular product, in order to meet the ERMES standard requirements completely.

7 CONCLUSION

In this paper, the development of a Conformance Test System for the ERMES air interface and receiver operation was presented and the procedures for running the tests were described. Complete test suites have been carried out on ERMES receivers coming from various manufacturers and a list of indicative results was given, demonstrating the operation of the system in accordance to the ISO 9646 standardized methodology for conformance testing services (ISO/IEC 9646, 1993) and (EWOS / ETSI PT5, 1990).

8 REFERENCES

CTS-3 ERMES Project 45067, Deliverable 1, Version 2, (1991) "ERMES Test Specification, Abstract Test Suite".

CTS-3 ERMES Project 45067, Deliverable 3, Final Version, (1992), "ERMES Test Specification, Test Methodology and Procedures".

CTS-3 ERMES Project 45067, Deliverable 5, Final Version, (1992), "Test Executive Software".

CTS-3 ERMES Project 45067, Deliverable 7, Final Version, (1993), "ERMES Reference Implementation".

Dangakis, K., Tombras, G., Paschalis, A., Papavramidis, A. and Kostarakis, P., (1992) "Development of a Conformance Test System for ERMES Receivers", Proceedings of The Third IEEE International Symposium on Personal, Indoor and Mobile Radio Communications, Boston, Massachusetts.

ETSI PT8, Final Report, (1990), "Methods of Measurement for Mobile Radio Equipment".

EWOS / ETSI PT5, (1990), "Technical Report on OSI Conformance Testing Methodology and Procedures in Europe".

ISO/IEC 9646, (1993), "IT - OSI - Conformance Testing Methodology and Framework".

prETS 300 133-1, (1991), "Paging Systems; European Radio Message System (ERMES), Part 1 - General Aspects".

prETS 300 133-4, (1991), "Paging Systems; European Radio Message System (ERMES), Part 4 - Air Interface Specification".

prETS 300 133-5, (1991), "Paging Systems; European Radio Message System (ERMES), Part 5 - Receiver Conformance Specification".

4

TSE-P - a highly flexible tool for testing network management applications using simulation

Dr. Winfried Geyer, Stephan Hesse, Dr. Norbert Newrly
Telenet GmbH
Rhine-Main Office, Marburger Straße 14
D-64289 Darmstadt
phone +49 6151 733-361, -334, -360
fax +49 6151 733-325
email geyer@telenet.de , hesse@telenet.de, newrly@telenet.de

Abstract

In this article we describe TSE-P, a tool for testing network management applications developed by Telenet by order of the Deutsche Telekom AG. TSE-P is specially designed for supporting tests of management applications based on CMIP, but also other interfaces as FTAM and RPC have been integrated. Series of operations to be sent to or expected from the IUT can be formulated in a special test script language. TSE-P holds and maintains the same MIB as the IUT. This makes it possible to generate realistic operations automatically without specifying every detail.

Several information models have been implemented, e.g. customer administration information models, information models for managing ATM, etc. A special feature of the TSE-P are gateway tests, where the reaction of the IUT at an interface is tested when an operation is sent to another interface. Different information models at those interfaces are supported simultaneously.

New information models for CMIP applications can be implemented very quickly by using specially developed ASN.1 and GDMO compilers. For the implementation of behaviour a new prolog-like language called RDL (rules description language) is used, which supports nondeterminism in its decisions.

1 TESTING OF NETWORK MANAGEMENT APPLICATIONS

Within the area of OSI network management the functionality of the communicating systems is partially described by the OSI or national communication standards, i.e. protocol specifications etc. The communication standards describe methods how a managing system (*manager*) queries and modifies a data base (*Management Information Base — MIB*) in the managed system (*agent*) and how the agent notifies the manager of spontaneous changes in its information structure.

The behaviour of the agent when performing operations on the data base is described together with the structure of the data base in a document called the '*Information Model*' or '*Object Model*'. Since this behaviour must describe real world implementations it can be very complex. For example, operations on an object are allowed to have side effects on other objects.

This potentially complex application behaviour requires testing facilities that concentrate on the application and use the communication protocols only as a vehicle to access the implementation under test. (The term 'implementation under test' — IUT — is used to reference that part of the 'system under test' — SUT —, that is the target of the test. For example, the communication protocol stack is part of the SUT, but not of the IUT. See also [X.290] for a definition of these terms.)

1.1 Testing by Simulation

The testing strategy of TSE–P is based on simulation. That means that the behaviour of the test system is mainly controlled by a simulator of the real application. This is in contrast to other testing approaches where the behaviour of the test system is only controlled by a sequence of (automatically or manually generated) test scripts. While this seems appropriate for testing low level protocols, its weakness becomes evident when testing complex behaviour of applications:

- In some cases, the described behaviour can be achieved in different ways. For example changing two attributes in an object can either be done with a single SET-operation giving a new value to both attributes or with two SET-operations, each changing the value of one attribute. Traditional testing based on [X.290] (ISO 9646) requires all possible combinations to be specified in the test scripts.
- The effect of an operation depends on the state of the MIB. When performing tests using a fixed sequence of test scripts, any possible state of the MIB has to be taken into account. Since the MIB is a 'potentially infinite' finite state machine, much effort has to be spent to describe the PDUs (protocol data units) to configure the MIB correctly. In contrast, testing by simulation keeps track of the state of the MIB by holding and

maintaining a copy of the data base. This also enables the TSE-P to automatically generate test operations based on the current state of the MIB.

- The international standards describe methods for event reporting and log control. TSE–P has a module for generating appropriate event reports and log records in accordance to the current MIB. The expected behaviour of a IUT regarding event reporting and log control doesn't need not to be specified in appropriate test scripts.

The role of test scripts in TSE–P is reduced to influencing the behaviour of the simulator. This leads to a significantly reduced number of test scripts and to less complex test scripts. Nevertheless, if necessary TSE–P test scripts may be detailed up to a level which is used by other test approaches.

2 TEST CONFIGURATIONS

The test system can be used in three different configurations: test of a manager, test of an agent and gateway test.

2.1 Testing a Manager

The principal test configuration for testing a manager is shown in figure 1. In this configuration the test system contains a component which simulates the agent upon which the IUT works.

The generating capabilities of the TSE–P can be used to specify (under test script control) valid operations to be executed by the test operator on the IUT's user interface.

This test configuration may also be used to simply execute operations sent by the manager. This application is particularly useful for implementation support (or basic interoperability testing) where the manager only needs a communications partner ('sparring partner') without the need to control the whole test flow by the test system.

All operations are checked against the specifications for protocol and application and — after passing all the checks — are used to modify the state of the agent.

In general, testing comprises the following steps:

- The test script specifies an action that the manager is expected to perform on the agent. This action is potentially complex so that the manager needs more than a single operation to perform the action. In most cases the exact sequence of operations is not well-defined and may be chosen by the manager.

  The action specification in the test script may be incomplete in that some or all of the

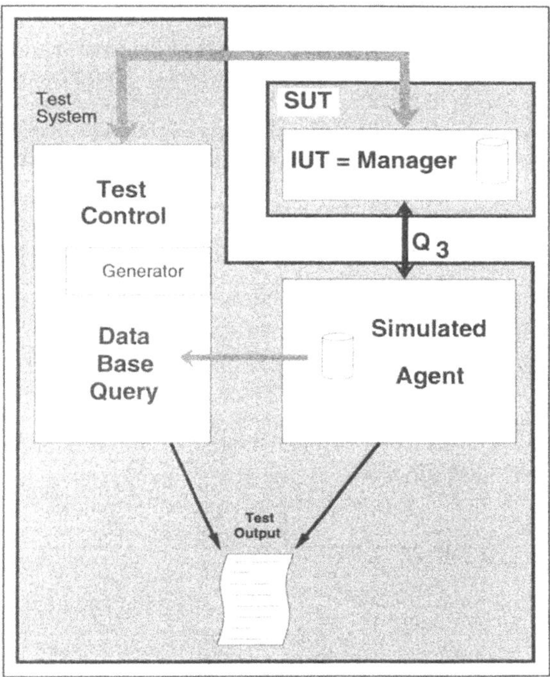

Figure 1 Test configuration for testing a manager.

parameters can be left unspecified. A lookup in the data base of the simulated agent is used to complete such action specifications.

- The request to perform the action is communicated to the manager. This may be done — if technically possible — by data communications (online) or through the help of an operator.
- The request is transformed by the IUT into a sequence of operations which is transferred via the Q_3-Interface to the simulated agent (Q_3 is a reference point between manager and agent in OSI network management).

- The simulator checks and executes the operations, which causes a change on its state represented in its data base.
- The simulator generates appropriate results and sends them back to the IUT. Test scripts may be used to verify the receipt of special operations or to predefine special results.
- On receipt of some trigger event like the success message from the manager the test control component initiates a query on the data base of the simulator to verify that the changes in the data base correspond to the expected ones.
- Differences between the expected and the actual state of the data base as well as errors detected during the check of the operations are flagged in the test output as potential errors.
- At the end of a test a summary of the encountered errors and warnings is generated.

2.2 Testing an Agent

In the test configuration for testing an agent, TSE-P acts in the role of a manager (see figure 2). Although it looks quite different from the test configuration given in figure 1, the main components are the same.

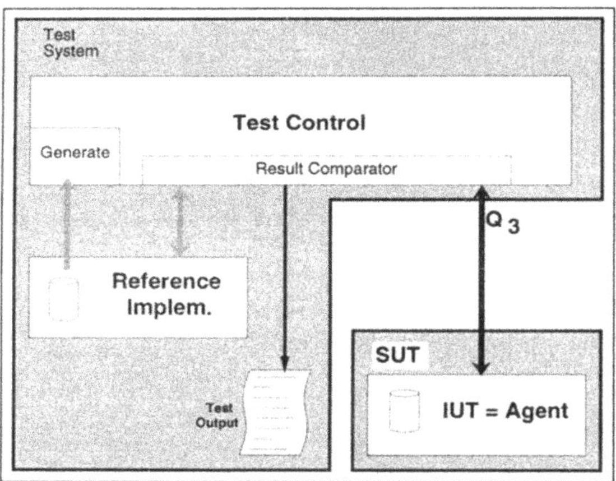

Figure 2 Test configuration for testing an agent.

The simulated agent is used in this test configuration as a reference implementation. It keeps track of the real agent's state information. This is accomplished by executing each operation on the real agent and on the reference implementation.

Testing an agent is always controlled by a test script. The elements in the test scripts are essentially the same, but they are interpreted in a slightly different way:

- The test script specifies an action which has to be performed on the agent. This action may be realized as sequence of operations on the agent's data base.
  The action specification in the test script may be incomplete in that some or all of the parameters can be left unspecified by the test script author. A lookup in the data base of the simulator is used to complete such incomplete action specifications.
- The action is converted into a sequence of operations. These operations are executed on the IUT via the Q_3-Interface and on the reference implementation using an internal interface.
- Both agents generate results to the operations which are compared. Differing results indicate deviations of the behaviour of the IUT from the specification. Such differences are flagged as potential errors in the test output.
- At the end of a test a summary of the encountered errors and warnings is generated as test result.

2.3 Gateway Testing

In the gateway test configuration, both TSE–P and the IUT have to play simultaneously the role of a manager on one interface and the role of an agent on the other interface. In general, a test comprises the following steps:

- First, TSE-P plays the role of a manager, sending operations to the IUT at interface 1. The IUT receives these operations in the role of an agent.
- The IUT transforms the received operations into other operations, which are sent in the role of a manager to the TSE-P at interface 2.
- The TSE-P receives these operations in the role of an agent, processes them, and sends results to the IUT.
- The IUT receives the results, and, taking them into account, decides what to do with the original operations which it received as an agent. It processes these operations and sends the results to TSE-P on interface 1.

When deciding if the IUT acts correctly, the TSE-P could simply generate the operations it expects from the IUT and compare these with those actually received. However,

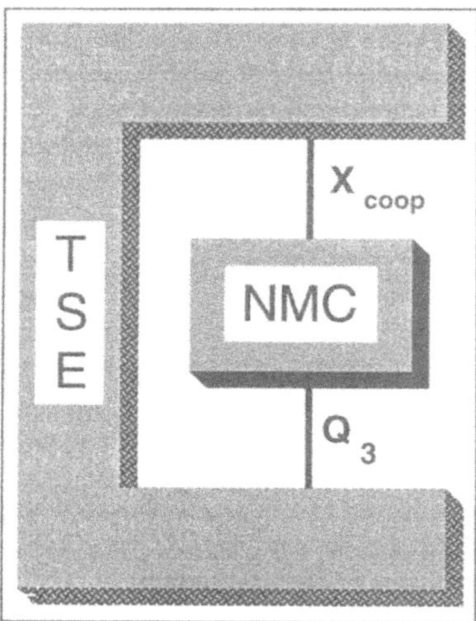

Figure 3 Example of a gateway test configuration (see chapter 4).

these operations as well as their order are not explicitly specified. It depends on the implementation of the manager and not on the interface specification. Hence, it was decided not to compare the operations, but the effects they have on the MIB.

The gateway test is therefore performed like this:

- When the TSE-P sends the operations to the IUT in the role of a manager, it generates operations which the IUT is meant to generate and send to the TSE-P at interface 2. These operations are processed on a so called "reference MIB", which is kind of a copy of the current MIB reflecting the MIB of the (interface 2) agent.
- The operations sent by the IUT on interface 2 are processed on the current MIB.
- After all results have been sent and received, the reference MIB and the current MIB are compared. Relevant differences are displayed.

3 RUNNING A TEST

Telenet TSE–P runs on a UNIX machine with a colour graphics terminal. It uses X window and OSF/Motif for presentation of the user interface.

After startup the TSE–P main window is presented (see figure 4). It allows various

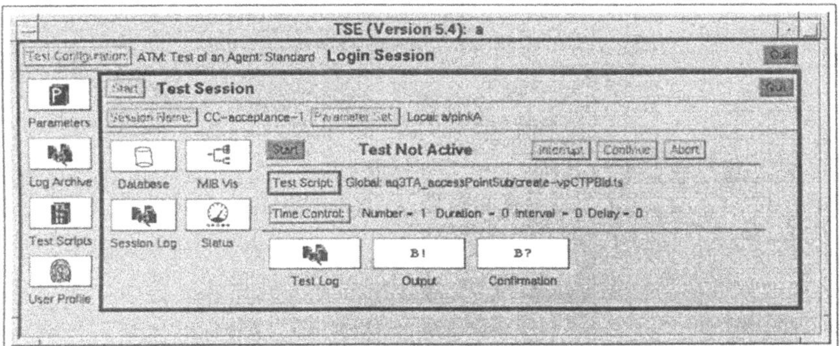

Figure 4 Main window of TSE–P application.

manipulations on different objects in the test system like test scripts, data bases, test documentation archives etc.

3.1 Test preparation

Before starting a test it has to be verified that the MIB of the TSE-P and the MIB of the SUT are equivalent. If necessary, an appropriate data base has to be loaded or generated. The address parametrisation has to be adjusted to the address of the SUT by using either a text editor or the TSE–P parameter set generator.

Furthermore a test script has to be selected from the set of predefined test scripts or has to be written in the user's work area. TSE–P test scripts are normally quite simple since the user may (and normally will) choose to specify only very few parameters. All unspecified parameters will be filled in during the execution.

Figure 5 shows a typical TSE–P test script. The header serves to document the test script. The body simply says 'create an object of class vpCTPbidirectional with any name

```
HEADER {
  PURPOSE =
        "Create a vpCTP-Bidirectional object "
        "          MANAGED OBJECT CLASS = vpCTPbidirectional "
        "          MANAGED OBJECT INSTANCE will be generated "
        "          Attribute list will be generated ";
  KONFIG  = AGENT;
};


STEP "create vpCTP-bidirectional" (REPEAT=0) {
  BLOCK () {
    CMISE () CREATE {
      MANAGED OBJECT CLASS = "vpCTPbidirectional";
      MANAGED OBJECT INSTANCE = *;
      ATTRIBUTES = *;
    };
  };
};
```

Figure 5 Typical TSE–P test script.

fitting to the current MIB and with any attributes that are valid in the current state of the IUT'.

Test scripts may either be edited using a normal text editor or using a special interactive tool, the TSE test script generator (see figure 6).

3.2 Test execution

During test execution the test system processes the test script and generates valid values for all incompletely specified parameters (e.g. an asterisk means 'any or omit'). The generated operation is transmitted to the IUT and also processed by the test system itself. The result received from the IUT is compared with the own result.

Figure 7 presents an example of the produced test output. It consists of a basic output window with a one-line informative message per event. More detailed information is available on every test output line. It can be viewed by clicking on it. A special navigation tool aids in quickly locating the relevant test output lines.

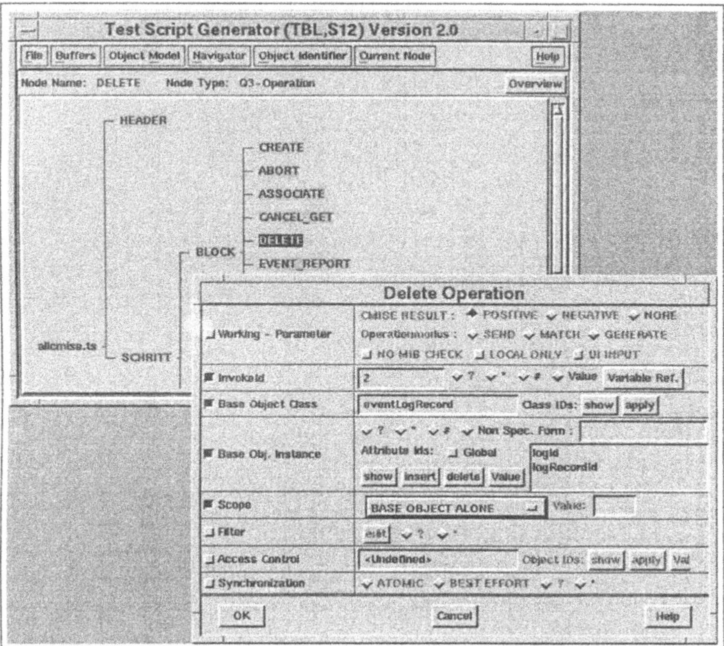

Figure 6 The TSE test script generator.

3.3 Interactive MIB browsing

During or after the execution of a test it is possible to browse interactively through the MIB. The MIB visualisation (see figure 8) presents the tree structure of the MIB and allows the user to focus on specific objects and to display their attributes.

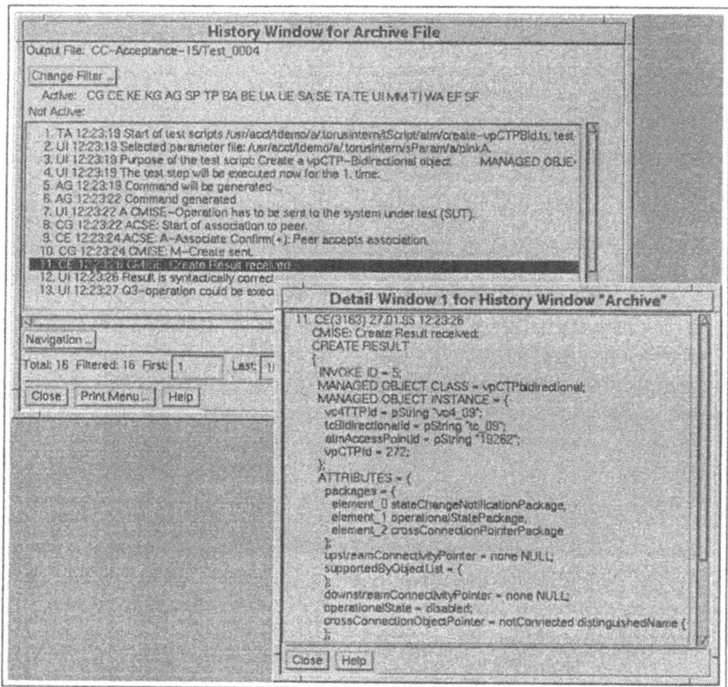

Figure 7 Sample test output of TSE–P.

4 TECHNICAL DEVELOPMENT, USAGE, PRACTICAL EXPERIENCES

Development of the TSE-P started in November 91, when Telenet was asked to support the Deutsche Telekom AG (Deutsche Bundespost at that time) to specify an information model for customer administration, embedded in a project called SuB ("Systemunabhängige Betriebsführung"). An appropriate test system, a first version of the TSE–P, was implemented until June 92. In addition to the facility of testing agent or manager, it

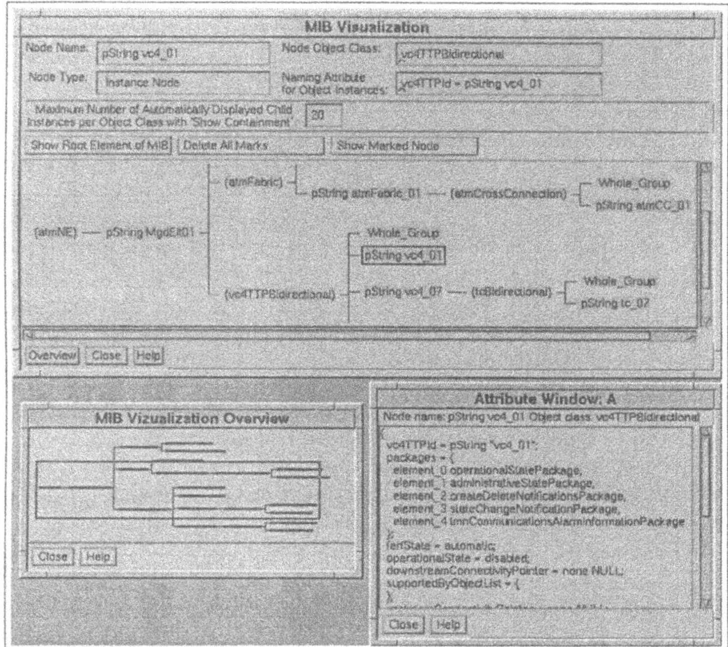

Figure 8 TSE–P MIB visualisation window.

already had the capability to perform gateway tests. Customer administration requests were sent in a customer-specific format to the IUT and the transformation into correct Q_3-operations was checked.

The test system was not only used by the Deutsche Bundespost for acceptance testing, but also by the providers of the manager and the agent systems.

Both the information model and the behaviour of the information model were implemented hard coded in C++.

For the next information models to be implemented, the new prolog-like programming language RDL (rules description language) was designed to implement the behaviour. In contrast to prolog, RDL supports nondeterminism for generating test cases randomly. The information model itself (supported classes and their attributes, etc.) still was coded in C++. The information models delivered with this new version in November 93 were TBL ("Technik für Betriebslenkung"), MML ("man machine language") and VMZ ("Verkehrsmeßzentrum"), dealing with configuration and performance management.

In April 94, Telenet was asked to develop a test system for network management within a bilateral ATM pilot project of the Deutsche Bundespost and France Telecom. For flexibility reasons, it was decided to provide the knowledge about the information model not as compiled C++ code, but as precompiled input to be interpreted during runtime. These interpretable files are generated by specially developed ASN.1 and GDMO compilers.

NMCs (network management centers) and ATM-NEs (network elements) managed by them were to be tested. The NMCs communicate via an CMIP interface called Xcoop, and with the ATM-NEs managed by them via a Q_3 interface. The information models mainly deal with the reservation of bandwidth. Besides testing an ATM-NE and each interface of an NMC separately, also a gateway test of an NMC was implemented.

TSE-P for ATM was delivered in July 95. Using ASN.1 and GDMO compiler and parts of the already coded behaviour, different versions of these information models could be implemented very quickly.

The TSE-P for ATM was ordered in a very early stage of development of the information model. While implementing th information model for the test system, many errors, questions and obscurities have been detected within the specifications, which led to changes in the involved interface specifications. So the TSE-P not only served as a test tool or as reference system for the real system, but also as a pilot implementation of a new information model.

New versions for customer administration information models were implemented, now also using the new mechanisms described above. Further developments also deal with other interfaces, as FTAM and RPC.

While implementing the information models mentioned above, several standards had to be supported, such as ISO 9596/ITU X.711 CMIP, ISO 10040/ITU X.701 Systems Management Overview, ISO 10164-1/ITU X.730 Object Management Function, ISO 10164-2/ITU X.731 State Management Function, ISO 10164-4/ITU X.733 Alarm Reporting Function, ISO 10164-5/ITU X.734 Event Management Function, ISO 10164-6/ITU X.735 Log Control Function, ISO 10164-13/ITU X.738 Summarization Function (partially), ISO 10164-11/ITU X.739 Work Load Monitoring Function (partially), ISO 10165-2/ITU X.721 Definition of Management Information, ISO 8650/ITU X.227 ACSE, ISO 9072-2/ITU X.229 ROSE, ISO 8823/ITU X.226 Presentation, ISO 8826/ITU X.225 Session, ISO 8073/ITU X.224 OSI Transport Class 0, ISDN S_0, ISDN S_{2m}, X.25, etc.

As described above, ASN.1, GDMO and RDL compilers have been developed. There is a large set of RDL rules which covers behaviour which is contained in every information model following [X.711]. Moreover, there is a mechanism to generate RDL rules automatically, which cover the part of information model knowledge formulated in GDMO. New rules only have to be written for information model specific informally formulated behaviour (see figure 9).

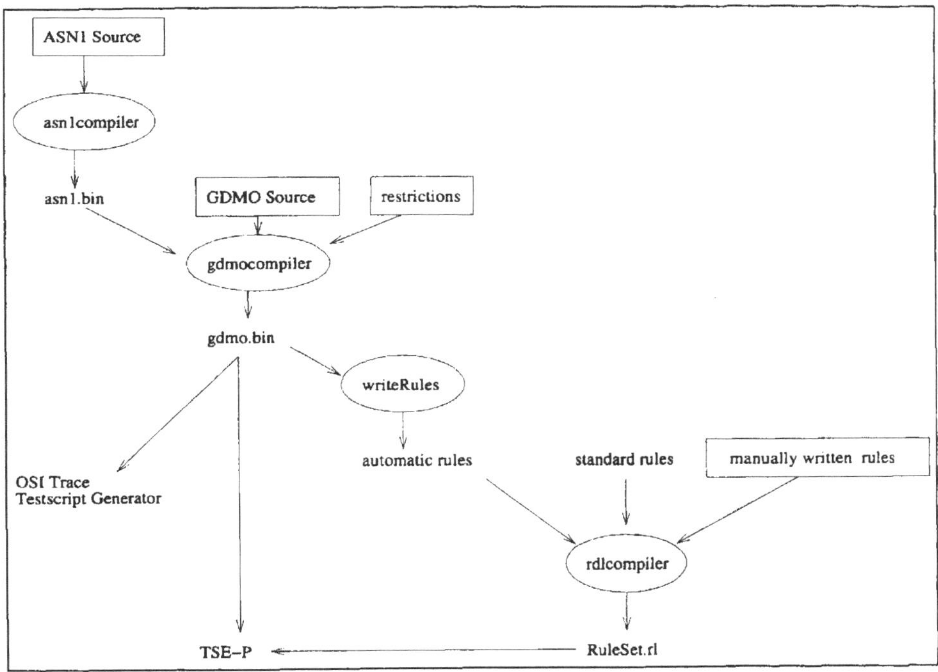

Figure 9 Implementing new information models.

For the implementation of a new information model (see chapter 5) these tools and mechanisms permit an easy and quick implementation of the formal part of the GDMO description. After this implementation step formal tests are already possible. In a following implementation step RDL rules have to be written representing the informally described behaviour. Some work has to be done here, but once this behaviour is implemented, it does not need to be formulated in eventually many test scripts (which also can lead to mistakes), but is tested and simulated automatically by the test system.

5 IMPLEMENTING NEW INFORMATION MODELS

Implementing an new information model means to adjust the TSE–P to all application specific informations. This information may be divided into three different parts:

- abstract syntax information (usually described in ASN.1),
- MIB data base schema layout (usually described in GDMO) and
- MIB behaviour specifics (usually described in the non-formal parts of GDMO).

TSE–P represents the behaviour description as a set of rules expressed in RDL. These rules are used for generating valid behaviour as well as validating behaviour of the IUT.

TSE–P encompasses tools that read ASN.1 and GDMO information and generate the relevant internal descriptions. The rule set for an information model consists of three parts:

- General rules for every information model (e.g. a CMISE-operation should have an invoke id),
- automatically generated rules from the formal GDMO description (e.g. there is no SET operation allowed on a specific attribute),
- and manually written rules in which the behaviour specifics are encoded (e.g. the creation of an object instance assumes the existence of another instance with specific properties).

A compiled rule set can be generated without having to write rules manually. This rule set only covers the behaviour as described in the standard rules and in the rules generated automatically. It can be seen as a prototype of the desired implementation. New rules can be added describing parts of the informally described behaviour, until the implementation is complete. In this way, rapid prototyping is possible.

REFERENCES

[X.25] CCITT (1988), Interface between DTE and DCE for terminals operating in the packet mode and connected to public data networks by dedicated circuit (ISO 8208).

[X.208] CCITT (1988), Specification of Abstract Syntax Notation One (ASN.1) (ISO 8824).

[X.209] CCITT (1988), Specification of Basic Encoding Rules for Abstract Syntax Notation One (ASN.1) (ISO 8825).

[X.224] CCITT (1988), Transport protocol specification for open systems interconnection for CCITT applications (ISO 8073).

[X.225] CCITT (1988), Session protocol specification for open systems interconnection for CCITT applications (ISO 8327).

[X.226] CCITT (1988), Presentation protocol specification for open systems interconnection for CCITT applications (ISO 8823).

[X.227] CCITT (1988), Association control protocol specification for open systems interconnection for CCITT applications (ISO 8650).

[X.229] CCITT (1988), Remote operations: protocol specification (ISO 9072-2).

[X.290] CCITT (1988), Conformance testing (ISO 9646-1).

[X.291] CCITT (1992), Abstract test suite specification (ISO 9646-2).

[X.292] CCITT (1992), TTCN (ISO 9646-3).

[X.701] CCITT (1992), Systems management overview (ISO 10040).

[X.711] CCITT (1991), Common management information protocol specification for CCITT applications (ISO 9596-1).

[X.721] CCITT (1992), Definition of Management Information (ISO 10165-2).

[X.722] CCITT (1992), Guidelines for the definition of managed objects (GDMO) (ISO 10165-4).

[X.730] CCITT (1990), Object Management Function (ISO 10164-1).

[X.731] CCITT (1990), State Management Function (ISO 10164-2).

[X.733] CCITT (1992), Systems Management: Alarm Reporting Functions (ISO 10164-4).

[X.734] CCITT (1992), Event Report Management Function (ISO 10164-5).

[X.735] CCITT (1992), Systems Management: Log Control Function (ISO 10164-6).

[X.738] CCITT (1993), Summarization Function (ISO 10164-13).

[X.739] CCITT (1993), Workload Monitoring Function (ISO 10164-11).

[Prolog] W.F.Clocksin and C.S.Mellish, Programming in Prolog, Springer Verlag, 1984

PART THREE

Special Types of Testing

5

A workstation's communication performance benchmark

Kurt Maly[a], *Ajay Gupta*[b], *and Satish Mynam*[c]

[a]*Old Domininion University, Department of Computer Science, Norfolk, Virginia 23529-0162. Fax: (804) 683-4900. email:* maly@cs.odu.edu
[b]*Old Domininion University, Department of Computer Science, Norfolk, Virginia 23529-0162. Fax: (804) 683-4900. email:* ajay@cs.odu.edu
[c]*Old Domininion University, Department of Computer Science , Norfolk, Virginia 23529-0162. Fax: (804) 683-4900. email:* mynam@cs.odu.edu

Abstract

In this paper we propose a communication benchmark suite: BTU(Bits To the User). In the BTU benchmarking process, the vendor supplies a workstation running a UNIX* operating system, we install our benchmark program and connect the workstation to a testbed which emulates a LAN/WAN environment. The benchmark run will submit the workstation to a carefully designed combination of tests. The result is a predictor of what the user, at the application level, can expect in terms of bits sent to or received from a remote host. Our benchmark takes into account concurrent activities such as CPU and I/O activities which compete for resources on the test machine. The combination of these activities and concurrent activities on the network will interfere with the test machine's communication performance. The automated BTU benchmark test suite produces results at various levels of abstractions ranging from a single number, characterizing average performance in the style of SPEC92, to a TCP time sequence chart for abnormal behaviors. We intend to serve the user of a workstation by providing information which should enable the users to make a reasonable judgment when acquiring a workstation with a specific configuration within certain cost constraints. Our results when used along with existing benchmarks will make a powerful combination for predicting overall performance. Neither price nor SPECmarks are as good a predictor of communication performance as that measured by BTUs. We conclude this paper with a discussion of extensions to faster

---

*TM of USL, Inc.

networks and our plans for having the community accept this proposal as a communication benchmark.

Keywords
Communication Benchmark, TCP/IP, Ethernet, ATM , FDDI, SPECmarks.

1 INTRODUCTION

In this paper we report on our efforts to develop a benchmark which will test the performance of a workstation communicating with other workstations across a network. That is, we will measure how a particular vendor's choices in hardware design and operating system implementation influence that workstation's performance with regard to communication. We are interested in providing a predictor to actual performance in a heterogeneous environment when concurrent activities occur on a workstation and on the network and, in particular, provide a predictor at the user level. Hence we call our benchmark the BTU benchmark which will give an indication as to what the typical user can expect a specific workstation to deliver to an application in terms of B(its) T(o the) U(ser), measured in Mb/s. We restricted the domain of investigation to UNIX in an Ethernet LAN connected to an arbitrary WAN.

At this time there exist a number of benchmarks, e.g., SPEC, which predict how fast workstations will compute. These benchmarks are widely referenced by administrators and users to determine which workstations will best suite their needs. A number of studies have been performed on the maximum performance of various protocol stacks. Typically these studies use ttcp and experiment with various system parameters such as buffer management and window sizes. Invariably, these studies are designed to elicit under which parameter configuration the workstation can push the most data to the network. Although quite useful in themselves these studies do not tell us what we can expect in terms of data in and out of the machine at the user level when competing processes are present on the workstation and the network is loaded with contending traffic.

The client/server computing paradigm is becoming pervasive, distributed computing's importance is increasing, and information accessed over the Internet is becoming a dominant application for workstations and PCs(for the rest of the paper we shall use the term workstation to mean both). In this environment compute power is only one factor, and not necessarily the most important, in determining the real time an application needs to complete. Very possibly, the dominant factor may be communication. For example, how many people use low end Sun workstations as routers in an Ethernet LAN without realizing that they limit themselves to fairly low level throughput?

Thus, we believe having a communication benchmark which will rank various vendors' workstations will be of great use in todays world of network computing. We need to be aware though that communication time depends on two factors: bit network flow rate on the physical channel and bit host processing rate inside the host from the time bits arrive at the interface and are handed over to the application. The first factor depends on whether the physical network is a 9.6K baud modem line, an Ethernet, an 155 Mb/s ATM network, how many routers are between source and destination and similar characteristics; all are totally independent of the the workstation. The second factor depends on such things as the protocol stack used, the bus architecture, and the operating system. If the network flow rate is small compared to the bit host processing rate the user will probably see not much difference in total communication time for different workstations even if they are ranked differently by the benchmark. If, however, the network flow rate is sufficiently high (say > T1) than the host processing time becomes the dominant aspect of the communication time and the ranking will be relevant. Consider workstation A rated at 2.5 BTUs (2.5 Mb/s to the user - to be explained in the rest of the paper) and workstation B rated at 1.6 BTUs, both executing Netscape (installed on a server in an Ethernet LAN) to access a page in England (with a 56Kb link somewhere in between). Bringing up a Netscape window involves intensive network communication between the server and the client and might take 20s by workstation A but .will take about 30s on workstation B. However there will be not much difference in retrieving the page from England because the dominant factor will be the slow network flow rate. Another example would be FTP over a T1, one would expect workstation A to do significantly better because host processing is the dominant factor.

In this paper we address the lack of such a communication benchmark by proposing an automated benchmark suite with results for a specific workstation structured by levels of details:

it will provide a **single BTU** number reflecting the performance of a number of applications in a typical network environments with typical concurrent activities on a host machine together with a penalty factor measuring how much the performance of a workstation is degraded when compute and communication processes exist concurrently; it provides a **specification for the testbed** together with source code for replication of the test results; it provides BTU information on the **performance for individual, common application classes** together with their penalty factors; it provides detailed data on the **performance of individual components of the benchmark test** for the knowledgeable user and software and hardware architects; for every **abnormal test result** it provides a TCP time sequence number chart for detailed analysis.

In Section 2 of this paper we give some of the background on benchmarking in general and performance testing; in Section 3 we outline our approach to the design of our proposed benchmark test suite. Section 4 describes the implementation and the automation process we developed. Section 5 provides a discussion of the statistical methods involved and the results of applying our benchmark to a set of machines of various vendors with varying operating systems and configurations(SUN, SGI, PC) over a fixed Ethernet. We conclude with a discussion and outline of future work and possible models for operating communication benchmarking in Section 6.

2 BACKGROUND

The PARKBENCH, a comprehensive set of parallel benchmarks (Hockney, 1994), concentrates mainly on CPU performance and includes only one component on network performance. It is a test of how well a machine can send data to another host while varying the message length. A number of programs and suites for MACH OS emphasize either low level system calls or study CPU intensive applications like SPECINT, CINT 2.0 and CFP 2.0 and as such are only relevant as far as general benchmark principles are concerned. A great deal of work has been done to determine the performance of networks. Netperf (Rick, 1996) is a public domain database containing network performance data implemented on various workstations. Also, there have been results reported for transport protocols (e.g. TCP) (Nettest, 1992) over various media (e.g. Ethernet, ATM (Allyn, 1993)) and tests on Network File Servers (NFS). A study on WAN congestion control and performance evaluation of TCP (Ahn, 1994) provides methods to benchmark TCP performance and to visualize the differences between the Tahoe, Reno and Vegas versions of TCP. It uses sequence number versus time plots of the wide area connections. The study reveals the performance of different versions of TCP and the improvement in congestion control using adaptive algorithms in TCP. All of these benchmarks used and defined a large number of inconsistent terms. RFC 1242 attempts to define a set of terminology that vendors can use to measure and report performance of network systems. This provides users with comparable data from different vendors. The emphasis in all these studies, however, is on maximum performance under ideal conditions and no correlation is made with background network, CPU, and IO activities.

3 APPROACH

The difficulty with designing a benchmark suite for communication is the many factors which influence overall performance of communication. For example, communication is influenced by the media access protocol, the bus architecture of the workstation, the implementation of the higher level protocols, the operating system and most importantly by activities occurring concurrently on the host machine and the network. No benchmark can test all the combinations of these factors and we were forced to make choices in our design. First and foremost we decided on a most common situation in the networking environment: UNIX workstations running TCP/IP in a LAN with connections to other hosts in a MAN or WAN. Secondly, we decided early on to concentrate our first effort on a widely distributed network, albeit slow speed: Ethernet. Our intention is to extend the benchmark suite to work for highspeed networks(ATM, FDDI, Fast Ethernet) as well. Thirdly, since this is a proposal for a communication benchmark we limited our study to a small number of well known vendors in the workstation arena: SGI, DEC, SUN, and Intel processor based personal computers(in this paper we give only results for various Sun models, SGI, and one PC). For applications competing for resources we considered ftp, telnet, rlogin, X windows Clients, CPU intensive and disk I/O intensive applications. Applications which are CPU intensive we considered are software compressed/decompressed video and realtime audio applications. Bulk Data transfer applications such as client/server distributed databases, storage and retrieval of multimedia streams from storage devices come under I/O intensive and network intensive applications.

In this paper we will compare and rank a number of workstations. We intend to serve the user of a workstation and provide data on how a particularly configured workstation can be expected to perform in a realistic network environment. The information should enable the users to make a reasonable judgment when acquiring a workstation with a specific configuration within certain cost constraints. Thus, the results of the benchmark test are given together with a detailed specification of the test machine and its list price.

The second audience group we are addressing is the hardware and software designers of workstations which can use our results to fine tune their operating system and identify hardware and software bottlenecks in the architecture of their workstations. To satisfy both audiences we provide for each machine a report structured by levels of abstractions. One is a greatly abstracted set: the bit rate a user(BTU) can expect, on the average, to communicate in a realistic setting and the penalty incurred by combining net with host activities. Part of this set is also the BTUs and penalties for specific application classes which can be useful if a user needs to have certain applications run efficiently. The second set is a detailed tabulation of the performance of each component of the benchmark test suite and TCP time sequence charts for abnormal cases.

Briefly, our approach is to have workstations from various vendors, preloaded with their implementation of UNIX operating systems and TCP/IP, subjected to our benchmark. The benchmark will create a controlled, fair and replicable test environment. It consists of two major components: the Benchmark Testbed and the Benchmark Suite.

The benchmark testbed: it is used to emulate a communications environment. To emulate a communication environment with a slow link somewhere in its route, the testbed will introduce pre-set packet delays. Similarly, to emulate a lossy environment, the testbed will drop packets(or a sequence of packets) with a pre-set probability at set intervals. Details of the testbed are given in (Maly, 1996). The major components of the testbed are as follows. **Test Workstation (W_{test}):** this is the workstation which is being tested and monitored to assess its performance. **Active Black Box Workstation (W_{active}):** this workstation has its operating system kernel modified such that it emulates various communication environments including packet loss and delay. Network traffic is generated between W_{active} and W_{test} in each of the cases specified in the Benchmark Suite section. This Black Box is called active because of its inherent functionality namely, generating load between itself and W_{test}. Almost all the network activity of the benchmark is caused by W_{active} except for the two Black Boxes participating in the tests together(as described in cases 5 and 6 of the Benchmark Suite). At the end of the test W_{active} also post-processes the raw results to produce the final report. **Passive Black Box Workstation ($W_{passive}$):** this workstation is used to generate background network traffic between W_{test} and itself and is used to emulate simultaneous sending and receiving from two competing workstations to and from W_{test}. $W_{passive}$ also has its kernel modified to allow for WAN emulation for the multiple connections to W_{test}. Each of W_{active} and $W_{passive}$ is called a "Black Box" because these workstations present a uniform and replicable communications environment. The only component that is varied in this testbed is W_{test}. As a result, all the workstations under test will undergo the same level of test. All modifications to emulate a new communications environment is done in the Black Boxes ($W_{passive}$ and W_{active}) while W_{test} is not changed at all to reflect the real communication efficiency of the test machine. The network between W_{active}, $W_{passive}$ and W_{test} for this study is Ethernet. In the implementation section we shall describe the details on the communication between W_{active}, $W_{passive}$, and W_{test} necessary to obtain a fair and fully automated benchmark test.

The benchmark suite is the set of programs to be executed on the testbed to test a workstation's communication performance. We have categorized the programs as "Network Activities" and those which compete for resources on W_{test} as "Host Activities". *Network Activities* are the core activities of the entire benchmark. Measurements made during these activities are used to compute the BTU number for W_{test}. These activities are classified according to: type of data transfer, multiplicity of connections, direction of traffic, LAN and WAN topology. *Host Activities* include CPU, I/O and communication intensive applications such as FTP which compete with network activities for shared resources such as bus, memory, media and disk.

In summary, the benchmarking process consists of the following steps: i)initiate network to be emulated; ii) execute Network, CPU, I/O, and FTP activities separately to obtain baselines; iii) execute Network and CPU activities concurrently; iv) execute Network and I/O activities concurrently; v) execute Network and FTP activities concurrently; vi)compute BTU as the normalized average of all Network activities(while competing for resources); vii)compute compound penalty on all activities.

4 THE IMPLEMENTATION

In this section, we describe our implementation of the benchmark we discussed in the previous section. First, we shall describe the technical details of the testbed followed by a description of the components which make up the benchmark suite. Finally, we provide details on the automation of the benchmark process. This latter part proved to be important because of the large number of runs and the time they consume in obtaining the BTUs for a particular machine.

4.1 TestBed Description

Our current testbed, consists of the three machines W_{active}, $W_{passive}$ and W_{test}. W_{active} is a SUN SPARC 10 with 1.2GB disk and 32MB RAM, $W_{passive}$ is a SUN SPARC 2 with 1.2GB disk and 32 MB RAM. Both of the workstations run SunOS 4.1.3 UNIX operating system. The specific selection of workstations for the blackboxes obviously will influence the performance of the test machine but the test bed will remain the same for a period of at least six months. All machines to be tested will have the same, replicable testbed and the machines selected as blackboxes represent machines commonly found in heterogeneous LANs.

We modified the IP layer in W_{active} and $W_{passive}$ to emulate various communications conditions. To emulate a WAN environment we have modified the kernel IP layer to allow for packet delays and drops. To maintain efficiency in the black boxes we did not add a timing functionality to the kernel but used simple counters(of IP packets) to implement the drop and delay of k sequential packets at certain intervals. A major factor in communication performance is how a particular implementation of TCP/IP handles time-outs and acknowledgments, particularly, when losses occur for an entire sequence of packets(such as in buffer overflow at a switch or router). In our implementation we allow for a recurring pattern of dropping k_1 packets at packet number n_1 mod 100, k_2 sequential packets at packet number n_2 mod 100,..., and k_m sequential packets at packet number n_m mod 100, all the drops adding up to an overall drop probability of x%.

4.2 Benchmark suite

The causes for bandwidth limitation and latency at the user level are twofold: network latencies and host latencies. Network latencies are host independent but dependent on the network load situation and the speed of the physical medium of the network. Host latencies are independent of the network but dependent on the host's CPU speed, hardware architecture, and the operating system. Various factors influence the overhead incurred by a message moving upstream/downstream in the system. They include: copying of data, buffer management, protocol processing, interrupt handling, system call handling, unpredictable overheads obtained due to different message sizes, internal TCP path taken(avoid redundant buffer copying). All these cause reduction in CPU cycle available for other applications. The concurrent network and the host activities contend for network bandwidth and CPU cycles. Host activities include a CPU activity, FTP activity and an I/O activity:

$$\text{Host\_activities} = \{\text{CPU, FTP, I/O}\} \tag{1}$$

Host activities: Network activities to be measured are run concurrently along with other host activities such as CPU, I/O and other network activities. On W_{test}, a CPU activity is a process which keeps the CPU busy 100 percent of the time. This job has two parts; the first part performs complex arithmetic calculations which acts as a basic CPU intensive processing step and the second part allocates and frees chunks of memory, copying the results of the complex calculations in the first part to the additionally allocated memory thus loading the MMU (memory management unit) and occupies a significant bandwidth on the bus interface. This two steps are repeated over a number of times so that the overall baseline CPU time is comparable with the baseline network activity's time. The I/O activity constantly reads a large file and copies it into another file which is repeated a number of times, again to make its time comparable to the baseline network activity's time. Disk I/O contends with network data processing for bus, memory and context switches. Network activities are performed between W_{test}, $W_{passive}$ and W_{active} workstations. At the core of all our components of the benchmark suite is a modified TTCP, we call COMM, which acts as the sending and receiving agent for the user on W_{test}.

In order to emulate the heterogeneous nature of networks and the likelihood for a test machine to encounter communication hosts with small window sizes we kept the window sizes on the blackboxes at 4K. To achieve our objective of giving not only average overall performance but also results specific to certain application classes, we used the buffer size to characterize the classes and representative applications to name these classes:

$$\begin{aligned}\text{Application\_set} = \{&(\text{Telnet, 100 bytes}), (\text{FTP, 1024 bytes}), (\text{Audio, 1250 bytes}),\\ &(\text{Netscape, 4K bytes}), (\text{Video, 64K bytes})\}\end{aligned} \tag{2}$$

These five applications represent an interactive telnet application, ftp or Mosaic or any typical application which uses a standard 1K buffer size, a typical 16KHZ audio sample data captured for every 1/8th of a second (1250Bytes) or a compressed 160 x 120 sized video frame, a typical intelligent application which uses 4K to optimize the buffer management, and some bulk data application which needs to transfer large amount of

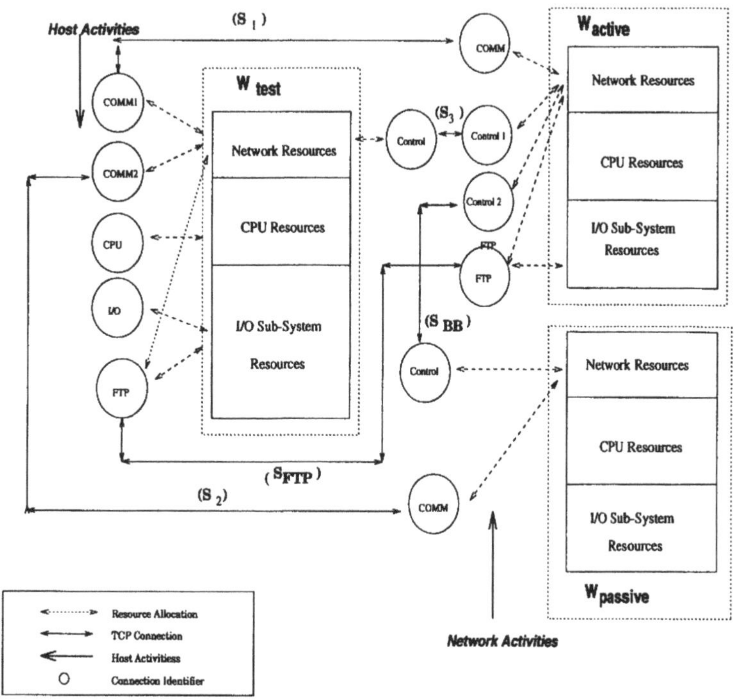

Figure 1 Connectivity of the different workloads and activities in Automated Benchmark

data. The seemingly odd size of 1.22KB (1250Bytes) was added because, when analyzing software resident in our system we found applications with odd message sizes executed quite frequently.

Benchmark suite components: The benchmark suite consists of six components (described below) which taken together represent a large percentage of network activities in a typical LAN environment and which will exercise most factors influencing communication performance.

$$\text{Test\_suite} = \{\text{SEND}, \text{RECV}, \text{SEND+DL}, \text{SEND\_RECV}, \text{2RECV+DL}, \text{2SEND+DR}\} \tag{3}$$

Each component will send 32Mb of data according to the description below. In the cases of the benchmark suite, we make reference to the delay introduced, DL, and the overall drop probability of packets, DR. The range of values for DL is from 0.1 ms to 1s and for DR from 0.0001 to 0.01 according to patterns defined earlier(for example: for DR = 0.01 we drop 1 packet at 200, 3 consecutive packets at 400, and repeat this for every 400 packets. The acronym at the end of each case will be used in later discussions to refer to that case: i) execute COMM with W_{test} sending the data to W_{active}($SEND$); ii) execute COMM with W_{test} receiving the data from W_{active}($RECV$); iii) execute the COMM with delay added to the IP of W_{active} and W_{test} sending the data ($SEND$+DL); iv) execute COMM with two connections to W_{test} from W_{active} one receiving and one sending ($SEND\_RECV$); v) execute COMM with two connections from W_{test}, one to W_{active} and another to $W_{passive}$ along with delay added to the IP of the blackboxes; the test machine is receiving half the amount of total data from each of the blackboxes (2 $\times RECV$+DL); vi) execute COMM with two connections from W_{test}, one to W_{active} and another to $W_{passive}$ along with packet dropping added to the IP of the blackboxes; the test machine is sending half the amount of total data to each of the blackboxes(2$\times$ $SEND$+DR).

4.3 Automation of the Benchmark Process

We have automated the entire benchmark processing software including data collection and post-processing. The user has to start the server on the two blackboxes and a client program on the test machine. For the typical machine it takes from 1 to 2 hours to complete the benchmark; upon completion, W_{active} collates the results by collecting the data from the test machine. In order to do so, five connections are made between the three machines namely W_{active}, $W_{passive}$ and W_{test}. The automation connections which are shown in Figure 1, are: i) connection between W_{active} and W_{test} (S_1); ii) connection between $W_{passive}$ and W_{test} (S_2); iii) a connection between W_{active} and W_{test} to get control information from W_{test} (S_3); iv) a connection between W_{active} and $W_{passive}$ to get control information (S_{BB}); v) a connection between W_{active} and W_{test} to obtain FTP control (S_{FTP}).

Each process interacts with the processes on the other machines. Processes are concurrently started on the test machine whenever load (CPU, I/O, FTP activities) is introduced concurrent with the net activity. The network and host activities have been chosen so that the baseline execution times of these activities are comparable. When a host load activity such as a CPU job is run concurrently with a network activity, both of them will obviously finish later than they did in the baseline case. We included a restart capability which became necessary when some machines could not complete a particular component.

The W_{active} benchmark process is started with the test machine name as an argument, which then makes connection S_3 with the test machine and S_{BB} with $W_{passive}$. These two connections synchronize the benchmark process till the end of the test. Whenever FTP load is applied on W_{test}, S_{FTP} is used and remains idle in the remaining span of time; idle TCP probe packets are neglected in the calculation of the results. At each step of the test (SEND, RECV, ...) the results of the activities are collected from W_{test} using S_3, S_{FTP} and from $W_{passive}$ using S_{BB}. A simple token mechanism is used for the proper synchronization of the testbed which occupies a minor amount of network bandwidth. This amount is negligible since the synchronization packets are sent only after each test is performed.

5 EXPERIMENTS AND RESULTS

Calculation of BTU: To define the calculations necessary to arrive at the final BTU number for a specific machine we shall use the tables in the appendices in (Maly, 1996). Assuming definitions (1), (2), and (3), we define the time for COMM to complete a case x, for an application y, when a competing host activity is run:

$$COMM_{CPU}(x,y) \quad \text{where } x \in \text{Test\_suite}, y \in \text{Application\_set}$$
$$COMM_{IO}(x,y) \tag{4}$$
$$COMM_{FTP}(x,y)$$

Finally, let M be the amount of data transferred(measured in Mb/s) during one execution of one component of the test suite and taking (1) through (4), let us define the BTU number for a particular host activity z and a particular application y as:

$$BTU_{z,y} = \text{M} \,/\, GM_{x \in Test\_suite}\, COMM_z(\text{x,y}) \tag{5}$$

where GM refers to the geometric mean. Given (5), we can define the BTU number for a particular application y as:

$$BTU_y = GM_{z \in Host\_activities}\, BTU_{(z,y)} \tag{6}$$

Using (1) through (6), we can define the composite BTU number for a particular machine as:

$$\text{BTU} = GM_{y \in Application\_set}\, BTU_y \tag{7}$$

measured in Mb/s.

Calculation of Penalty Factor: We have introduced the term 'penalty' to provide a predictor for how much a machine will be slowed down when concurrent net activities occur. We arrive at a dimensionless factor by comparing to a base case of runs done in isolation. To define this factor we introduce terms analogous to (4) which measure how long a host activity takes when COMM is run concurrently:

$$CPU_{COMM}(x,y) \quad \text{where } x \in \text{Test\_suite}, y \in \text{Application\_set},$$
$$IO_{COMM}(x,y) \tag{8}$$
$$FTP_{COMM}(x,y)$$

Akin to the BTU definition, we introduce the term $Penalty_{z,y}$ to denote the percentage a host activity z takes longer for an application y averaged over the test suite componenents.

$$\text{Penalty}_{z,y} = (((\text{GM}_{x \in Test\_suite}\ z_{COMM}(x,y)\ )/\ z_{base})\ \text{-}\ 1)\ *\ 100 \tag{9}$$

Given (9), we can define the penalty factor for a particular application y as:

Table 1: Final BTU Ranking

BTU Rank	*Machine Architecture*	*BTU(Mbps)*	*Penality*	*SPEC'92*	($)
1	Sparc20	2.87	20	76.9	15,195
2	PC (Linux)	2.72	21	32.2	2,900
3	Sparc5(Solaris)	2.45	33	65.3	7,745
4	Sparc5(SunOs4.1.4)	2.19	28	65.3	7,745

$$Penalty_y = GM_{z\epsilon Host\_activities}\ Penalty_{z,y} \quad (10)$$

Finally, taking (8) through (10), we can define the penalty factor for a particular machine as:

$$\text{Penalty} = GM_{y\epsilon Application\_set}\ Penalty_y \quad (11)$$

measured as a percentage increase of host activities.

Statistics: In the final report (Maly, 1995) we provide for each machine the arithmetic(AM), geometric(GM), and harmonic(HM) means for each of the tables. For statistical analysis, we chose the unweighted geometric mean(GM) as the measure for BTUs and the penalty factor: (5) through (11). Though the harmonic mean preserves the measure by giving better consistency, the comparison of two systems is independent of the reference system only for the weighted HM unweighted GM (Ran, 1995 and Jain, 1995). A small mix of efficient components instead of a large mix of components can rate the systems provided the weights and cases are properly selected, independent of the reference system. Since we do not rate the weights for each application and the cases in the test suite, we chose the unweighted GM as our standard measure of network performance.

5.1 Key results:

In (Maly, 1995) we reported on the results for BTU95 where we tested four different machines(Sparc10, SGI, DEC Alpha, PC) with our benchmark suite. The SPECmarks(65, 36, 158, 32 respectively) gave quite a different ranking among the machines than the BTU ranking(2.4, 0.99, 2.37, 2.07) which clearly showed that CPU performance is not the only factor affecting the network throughput. Neither was price an effective predictor for communication performance. For BTU96, the one reported on in this paper, we have made significant changes based on feedback from the community.

For this paper we include the results of testing four different workstations. The specific configuration of each machine can be found in (Maly, 1996). In Table 1 we provide the BTU numbers, the BTU ranking of these machines along with the SPECint '92 numbers (PDS, 1996), the penalty factors, and the cost of each machine tested. Finally, we provide in Figures 2 and 3 the summary of the detailed test reports for each machine; these reports are given in (Maly, 1996). The bits to the user chart shown in Figure 2 thus represents the bits, (7), users of four different machines can expect to communicate in a typical LAN for different types of applications. The x-axis of Figure 2 represents various applications; on the left size of the diagram we measure BTUs in Mb/s(this should be compared to the 10 Mb/s capacity of Ethernet). On the right side we provide the absolute time an average application takes measured in seconds. The penalty chart, Figure 3 provides a bar chart of the penalties, (10), each bar representing the penalty factor for a particular machine for a particular application.

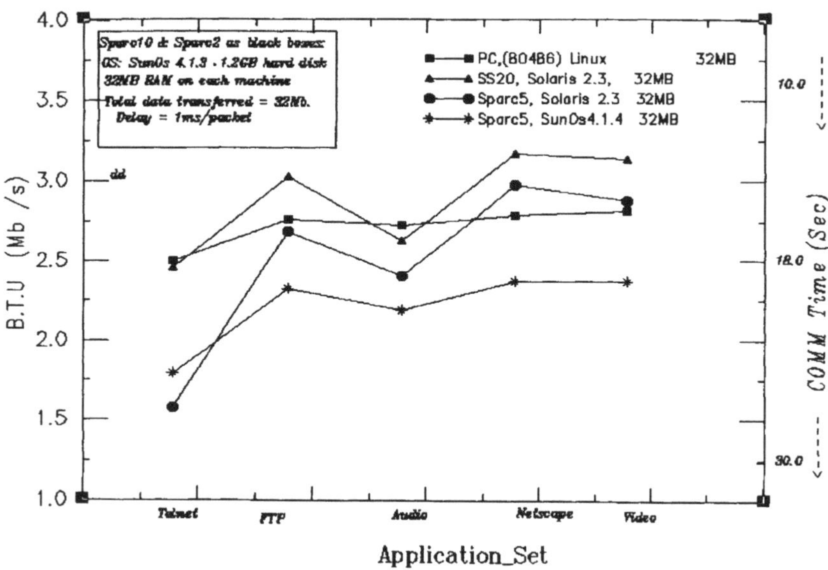

Figure 2 BTU Benchmark: Ethernet

In Figure 2 we have plotted the resulting BTUs for different applications for the tested machines. We observe a low values for Telnet-like applications which is due to more overhead and context switches involved in sending small sized buffers. The Audio-like applications also show a decrease in BTUs for all but the PC due to the odd application buffer size of 1250 Bytes. We have observed this phenomenon in other System V akin UNIX implementations(SGI, DEC) but not for BSD akin PCs(Linux and FreeBSD). Thus, applications which generate messages of this size suffer some throughput problems.

Consider some of the odd cases in (Maly, 1996); the interaction between the send and receive buffers of the two-ends of the TCP connected system could be easily traced using the TCP time sequence plots. We perform a special test suite case for the abnormal behaviour to study the affects of TCP end-point interactions. The tests sometimes produce

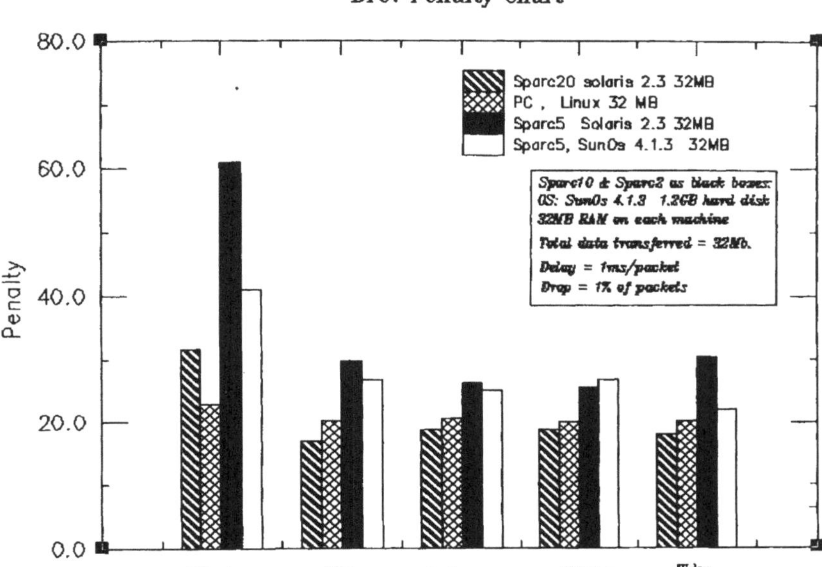

Figure 3 BTU Penalty Chart

long pauses in the sequence numbers because of the acknowledgement-retransmission interaction - this has been solved in the Vegas implementation of TCP but not in SunOs 4.1.2 which is running the Tahoe version of TCP (which lacks the adaptive congestion control). The penalty graphs in Figure 2 show clearly that even with the same architecture the operating system has a significant impact.For example, compare the numbers for the SPARC 5s running different operating systems, particularly, as it concerns Telnet-like applications. The most striking result is that the PC has a communication performance close to a SPARC 20 although its cost is only a fraction of that of a SPARC 20 and similarly the SPEC perfomance is much smaller than that of the SPARC 20. Comparing the send case with the receive case of the test machine, the send case takes more time; this can be ascribed to the different window sizes of the test machine and the blackbox and the implementation differences (eg. mbufs and streams) along with the well known transport level protocol effects such as SWS, delayed acknowledgements, Nagle's algorithm and slow start.

6 CONCLUSION AND FUTURE PLANS

In this paper we introduced a concept for a communication benchmark: BTU(Bits To the User). Our benchmark does not test: What is the maximum performance of a workstation in a homogeneous network under ideal circumstances? Rather, the benchmark will submit the workstation to a carefully designed combination of activities on the host to be tested(e.g., CPU, I/O) and the blackbox network and produce an indicator for what the user can expect in terms of bits sent/received to/from a remote host. Secondary results are the penalty individual activities experience under the influence of net activities. On the one hand we intend these results for the workstation buyer who can use the data together with the configuration specification and the list price to make a well reasoned decision whether or not the workstation meets the user's needs when compared to the results of other workstations, on the other hand we believe these results can provide insights to the architects as to the location and nature of communication bottlenecks of a particular workstation.

We have implemented the concept and described in this paper an automated process which will produce for each machine tested a structured report with results of varying detail. The results from testing four different workstations indicate this benchmark produces data sets which are not necessarily related to CPU speeds but are truly an indication on how a particular hardware architecture and operating systems handles network communication. Neither the speed of the CPU(measured in either Mhz or SPECmarks) nor the price was a predictor of communication performance as measured in BTUs. Thus, we believe we have made the case that this proposed benchmark satisfies the stated objective.

Several issues relating to the testbed remain unresolved to a certain degree. It is not obvious how easy it is to adapt the design of the benchmark such that it will enable us to test a machine's performance over several networks of different speed(ATM, FDDI, and T based 100 Ethernet). By the time these highspeed networks become prevalent though, we feel confident that we will have such a benchmark suite. We performed a number of experiments to see the dependence of the results on the actual configuration of the testbed. We used various workstations as testbed machines and compared the results. In general, the actual numbers were indeed different but they did not affect the ranking of the tested workstations. Neither did the problem cases for the tested workstations change: the same components for the same message size produced results outside the normal range. This supports our contention that the type of the testbed machines is not important as long as we keep them representative of current installed bases. It leads though to the problem we are facing: How should such a benchmark suite process be run? We see two clearly different models: one is similar to the testing procedure used by (Scott, 1996) to evaluate network routers and switched internetworking products and the second would emulate SPECmarks. In the first model, an independent organization such as NIST, MITRE or OSF would maintain an up-to-date testbed and periodically publish rankings of machines supplied by vendors for testing. In the second model, we would explicitly specify in great detail the current testbed and provide the software throughout the Web for anybody who can duplicate the testbed. These individuals or organizations could then run the benchmark test themselves. This latter process implies that the testbed would be certified by some central organization.

Acknowledgment: we are indebted to the help Rajeshkumar Vangala provided in the automation process of the BTU benchmark development.

REFERENCES

Allyn, Romanow. (1993). Preliminary Report of Performance Results for TCP over ATM with Congestion. *Technical Report*, SunMicrosystems Inc.

Hockney, R. and Berry, M. (1994). Public International Benchmarks for Parallel Computers: PARKBENCH Committee Report-1. *Scientific Programming*, vol3 (2), pp. 101-146.

Nettest. (1992). Nettest a network Benchmarking tool. *Technical Report, Cray Research Inc.*

Clark, D. David. (1982). RFC 813, Window And Acknowledge Strategy in TCP. *MIT Laboratory for Computer Science*, CSCG.

Stevens, W. Richard. (1994). TCP/IP Illustrated Volume 1. *ISBN 0-201-63354-X*, pp. 325 - 330; pp. 263 - 285, Addison-Wesley Professional Computer Series.

Jones, Rick. (1996). Network Performance Home Page. http://www.cup.hp.com/ netperf/NetperfPage.html.

PDS. (1996). *PDS: The Performance Database Server.* http://performance.netlib.org/ performance/html/spec.html.

Bradner, Scott. (1996). Networks Consulting Services. http:// www.snci.com/netrpt.htm.

Maly, K., Gupta, A., Mynam, S. and Khanna, S. (1995). B(its) T(o the) U(ser): a Communication Benchmark Proposal. Technical Report. Department of Computer Science, ODU, http://lite.ncstrl.org:3803/Dienst/U1/2.0/Describe/ncstrl.odu_cs

Maly, K., Gupta, A., Mynam, S. and Khanna, S. (1996). BTU: A Host Communication Benchmark, Technical Report, http://www. cs.odu. edu/~btu/papers/ TechReport96.ps, Department of Computer Science, ODU, Norfolk, April 96.

Ran, Giladi and Niv, Ahituv. (1995). SPEC as a Performance Evaluation Measure. *IEEE Computer*, pp. 33-42.

Jain, R. (1991). The Art of Computer systems Performance Analysis: Techniques for Experimental Design, Measurement, simulation as a Performance Evaluation Measure. *Wiley and Sons*, New York.

Ahn, J. and Danzig, P.B, et.al. (1994). TCP Vegas: New Techniques for Congestion Detection and Avoidance. SIGCOMM 94.

BIOGRAPHY

Kurt J. Maly received the Dipl. Ing. degree from the Technical University of Vienna, Austria, and the M.S. and Ph.D. Degrees from the Courant Institute of Mathematical Sciences, New York University, New York, NY. He is Kaufman Professor and Chair of Computer Science at Old Dominion University, Norfolk, VA. Before that, he was at the University of Minnesota, both as faculty member and Chair. His research interests include modeling and simulation, very high-performance networks protocols, reliability, interactive multimedia remote instruction, Internet resource access, and software maintenance. His research has been supported by DARPA, NSF, NASA, CIT, ARPA and the U.S. Navy among others.

Ajay Gupta graduated with a B.E. in Electronics Engineering from Bangalore University. He received the M.S. degree in Computer Science from Old Dominion University.

Currently he is Director of Computer Resources, Department of Computer Science, Old Dominion University as well as an adjunct faculty in the same department. His research interests include high speed networking and enterprise security.

Satish K.Mynam is an M.S. student in Computer Science at Old Dominion University. His research interests include Computer Networks, Operating Systems, Distributed Systems and Systems Performance Evaluation. He obtained his B.Tech in Computer Science and Engineering from Jawherlal Nehru Technological University, India.

6

Automatic checking of the compatibility of the ISDN user-network interface (layer 3) using the tool VACCIN

Pierre-Yves DANET
France Telecom CNET
Dept. LAA/TCS/TRA
Technopôle Anticipa, 2 avenue Pierre Marzin, F-22307 LANNION Cedex FRANCE
Tel : +33 96 05 11 55, Fax : +33 96 05 02 42
E-Mail : pierre-yves.danet@lannion.cnet.fr

Abstract

This paper presents the results of a study which aim was to develop a tool able to detect automatically incompatibilities between ISDN terminals already used and a futur version of the network.

Keywords

ISDN, SDL, user-network interface

1. INTRODUCTION

The introduction of new versions of software for the ISDN switching system presents a compatibility problem with the terminals already in operation. The aim of the telecom operator is the non-disruption of the operating terminals during the introduction of new services on the network. The testing of terminals (even with certification) is not sufficient to provide a total guarantee.

Instead of considering very expensive corrective measures - the development of terminal software or the introduction of network corrections - this study attempts to provide preventive measures using simulation techniques.

Previous experience has shown the limits of a "manual" analysis between two versions of a specification. CNET has therefore examined the possibility of automatically verifying this compatibility by using a tool capable of "exhaustively" simulating all the terminal-network exchanges (layer 3 of ISDN : CSE P 22-30A [1] for the terminal side & ETS 300 403-1 [3] for the network side). But the high number of simulations required make it impossible to carry out an exhaustive study.

This new approach consists in verifying the compatibility of the network-terminal interface specifications by using formal specifications (SDL 1992 [5] - Specification and Description Language). The first objective of this study involves the development of a simulator on the basis of the GEODE tool (v.2) from Verilog. This will make it possible to achieve simulations with reference to the formal descriptions included in Terminal and Network specifications.

The various problems which are likely to arise are:

- message not known,
- information element not known,
- field value of information element not known,
- different interpretation of a value,
- automaton modification.

On this basis the tool should be capable of achieving the simulation of the establishment and release of a call by systematically scanning all the coding combinations allowed by the standard (Ref.4.2, page 6). The possible combinations are so numerous that we have introduced an option which allows only the scanning of those coding combinations corresponding to the existing terminal types (telephone, adapter, fax, videophone, ...).

2. TECHNICAL DESCRIPTION

2.1 Development of a simulator

This model is based on the GEODE simulator and uses in particular the SDL descriptions for the network and the terminal specifications, as follows :

- SDL specification of version Vn of the network side protocol,
- SDL specification of version Vn-1 of the terminal side protocol,
- Description of the message coding and of the information elements of version Vn,
- Description of the message coding and of the information elements of version Vn-1.

To obtain significant simulation results the initial data listed above must be of the highest quality.

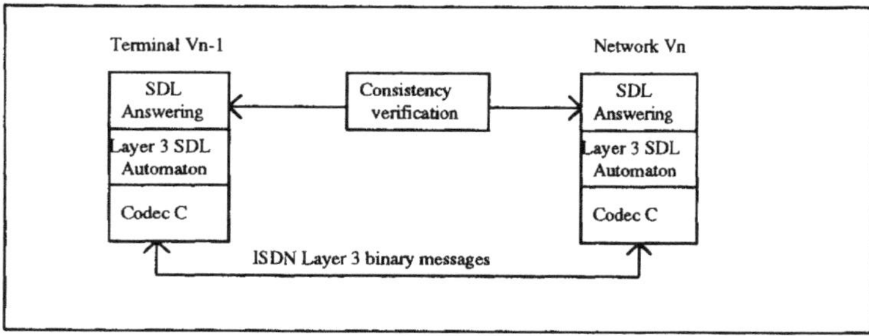

Figure 1 Functional description of the system.

Role of each object:

The SDL answering systems activate the SDL automata using predefined scenarios. Each scenario consists of a series of layer 3 service primitives. These primitives indicate the ISDN messages to be transmitted or received by the automaton. It is possible to fix or vary the value of the information elements for each service primitive. The answering systems, the service primitives and the information elements are implemented in SDL. At present the scenarios involve call establishment and release. At a later stage they could be designed to simulate the implementation of supplementary services.

The layer 3 SDL automata carry out the ISDN communication according to protocol D of standard Vn-1 on the terminal side or the standard Vn for the network. The automata deal with the ISDN messages and the information elements using SDL internally, but exchange them in binary form. In order to do that the automata use codec C.

The codecs C change the SDL information elements and the ISDN messages transmitted by the automata into binary code and decode the binary messages received by the automata into SDL information elements and ISDN messages. The codecs C have been developed in C language.

The operator starts the simulation by activating the SDL answering systems. The aim of the simulation is to detect interworking problems between the terminal Vn-1 and the network Vn in relation to exchanges of ISDN messages and ncoding / decoding of messages and information elements.

The answering systems also have to verify the reliability of the scenario (the establishment or release of a call), as well as verifying the compatibility of the value interpretation. In fact, a value could have different interpretations depending on the protocol version in use.

2.2. Development of an answering system

2.2.1. Aim of the answering system

The answering systems (one terminal side and one network side) must allow the activation of the overall SDL system and the automatic operation of the simulation. They allow the modelling of the behaviour of the application oriented layer of an actual terminal (to make a call, to answer a call, to release a call). The answering systems are designed under the form of SDL automata.

The aim of the answering systems is to generate simulation scenarios which must link the establishment and release of the calls, using, if necessary, all the parameter values associated with the layer 3 primitives, i.e. all the fields of information elements which make up the message exchanges between the layer 3 automata.

The answering system is composed of two distinct parts:

- One part consists of a module of predefined scenarios, i.e. a module of sequences of the layer 3 service primitives; the progress of a scenario chosen by the operator supposes the execution of a scenario Vn-1 on the terminal side and the execution of a scenario Vn on the network side (a sub-set of predefined scenarios exists on both sides).
- The other part is for the activation and parameterization of these scenarios:
  - the activation involves selecting a scenario and linking different scenarios if necessary; it also involves synchronising the two scenarios which are carried out on both the terminal side and the network side.
  - the parameterization is carried out either by fixing the parameter values of the layer 3 service primitives, or by scanning these values, i.e. by repeating the same scenario many times with different parameters; the simulation can be done completely automatic.

2.2.2 Presentation of the functional modules of the answering system

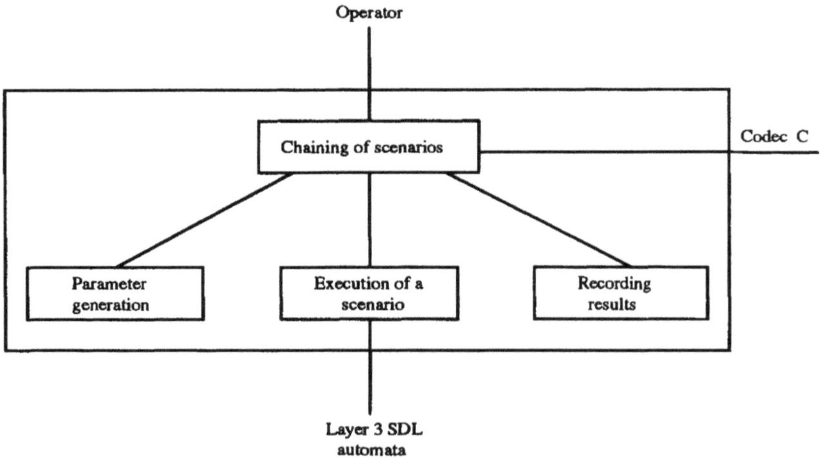

Operator
Chaining of scenarios
Codec C
Parameter generation
Execution of a scenario
Recording results
Layer 3 SDL automata

Figure 2 Link description between each function.

The answering system offers the following functions :

- *Linking of scenarios :* activating and synchronising the scenarios, with a scanning of parameter values if necessary, and the initialisation/termination of the codecs.
- *Execution of a scenario :* activation of the layer 3 automaton concerned, in accordance with the linking of the anticipated service primitives in the scenario and their parameterization. This could be scenario Vn-1 on the terminal side or scenario Vn on the network side.
- *Generation of parameters :* generation of layer 3 service primitives in order to give all possible combinations of information elements for each layer 3 message and for all the possible values of each of the fields of information elements. This could be service primitives Vn-1 on the terminal side or Vn on the network side.
- *Recording of results :* recording of the scenarios and the parameterization which were carried out, as well as the results (OK or NOK) in a trace file.

2.3. Development of a codec system

2.3.1 Aim of the codec system C

The aim of the decoder is to decode the binary messages according to a given specification and to detect non-conformity in relation to this specification.

The aim of the encoder is to encode the SDL message according to a given specification.

2.3.2 Presentation of the functional modules of the codec system

The diagram below shows the functions of the codec system.

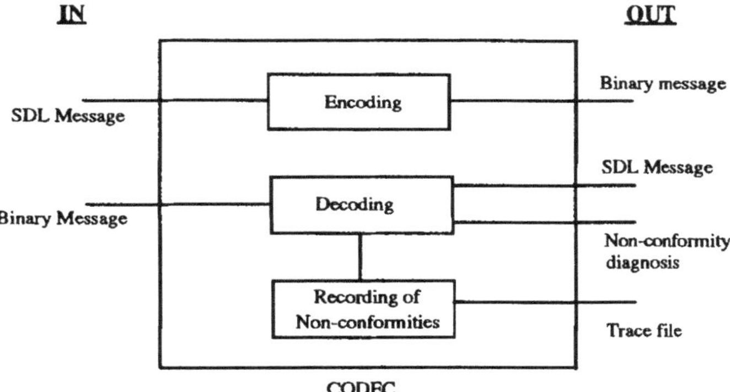

Figure 3 Description of the input/output information.

The codec system offers the following functions :

- Encoding process Vn-1
- Decoding process Vn-1
- Recording of non-conformities Vn-1
- Encoding process Vn
- Decoding process Vn
- Recording of non-conformities Vn

2.4 Implementation of the simulations

The simulation is started when the operator activates the SDL answering systems. Its aim is to detect the non-conformities between the terminal Vn-1 and the network Vn in relation to exchanges of ISDN messages (using the answering systems and layer 3 automata) and encoding / decoding of messages and information elements (using the answering systems, layer 3 automata and codecs).

The simulator actually uses C language to encode the information elements. In future it would be more judicious to use ASN.1, but this will require a considerable amount of work.

The ultimate stage will involve the analysis of the incompatibilities. This will make it possible to detect those which will be likely to prevent the interworking of the new version of the network with a given terminal.

3. CONCLUSION

The results of the simulation were very useful because the analysis of them shows us some new problems which were not identified "manually". Hopefully these "new" incompatibilities were not so serious because error management procedures described in the ISDN layer 3 protocol [1] are able to handle them.

This first experiment was very useful because it gave us the following information :
- "Manual" checking of protocol compatibility cannot be exhaustive.
- Automatic checking is much faster (1 month instead of 5 months).
- Automatic checking has to be improved in order to include help for results analysis.
- Automatic checking has to be improved in order to use terminal profiles representing exactly the behaviour of one terminal instead of using a general profile (complete specification).

Then we need to work into 3 directions :
- improvement of the result function in order to help the operator to identify the incompatibilities,
- use of ASN.1 [4] for the description of messages and information elements,
- development of a tool able to generate a specific terminal profile from the layer 3 specification [3] and from the PICS (Protocol Implementation Conformance Statement) and the PIXIT (Protocol Implementation eXtra Information for Testing) values.

4. DEFINITIONS - REFERENCES

4.1. Glossary

ISDN : Integrated Services Digital Network
Vn : Version n

4.2. References

[1] ETS 300 403-1 : ISDN; DSS1; Signalling network layer for circuit-mode basic call control; part 1 : protocol specification
[2] ETS 300 403-2 : ISDN; DSS1; Signalling network layer for circuit-mode basic call control; part 1 : SDL
[3] CSE P 22-30 : Layer 3 of call control protocol D for access to ISDN
[4] ASN.1 : Abstact Syntax Notation One (IS 8824 & IS 8825 1988)
[5] SDL : Specification and Description Language (Z.100 1992)

7

The analysis of integration test result for ATM switching systems

Seung-kook Cheong
Kyeong-ho Lee, Taeg-won Jeong

Electronics and Telecommunications Research Institute
161 Kajong-Dong, Yusong-Gu, Taejon, 305-600, KOREA
FAX : +82-42-860-5410
<e-mail address>
skjeong@nice.etri.re.kr/khlee@nice.etri.re.kr/twjeong@nice.etri.re.kr

Abstract

This paper presents a new analysis method of integration test result for ATM switching systems. The new analysis method, called XTAM(eXchange system Test result Analysis Method), will provide numerical parameters from test results which are vague, qualitative and descriptive. The analysis method is based on Fuzzy theory. This method has been successfully applied to the ATM switching system which is under development by ETRI in Korea.

Keywords
Switching System, Fuzzy Set, Test Analysis, XTAM, Integration Test, Regression Test

1 INTRODUCTION

System integration test is one of the most important activities during the development of complex systems such as ATM, N-ISDN, PSTN switching systems.[1] The test is valuable only if the result is correctly analyzed and notified to the related developers. Therefore, correct analysis of test results is indispensable for the improvement of system. Usually test results are used just as data for debugging rather than as data for influences of the errors found during the test and for the consideration needed to implement more functions.[2-3]

The limited use of test results can be considered as the waste of resources.

In this paper a new analysis is proposed to give quantitative description about the influences of faults on the system, extent of implementation for each function, stability of the system, etc. This analysis is based on the Fuzzy theory.

2 ANALYSIS METHODOLOGY

System integration tests are carried out for the total system rather than for each part of the system during the development of system. Errors, caused by faults in hardware, software, firmware, or combination of several factors, are debugged according to the information found during system tests[3].

The test of systems and the analysis of test results are usually done at the source level[4-7]. Hence there are not much study on the analysis of test results at system level. In our approach test results are analyzed to provide various parameters which are helpful for various R&D stages. The analysis system, called XTAM(eXchange system Test result Analysis Method) has the following characteristics :

- Define probable(expected) errors
- Integration and analysis of errors found during tests
- Numerical representation of errors which are descriptive and qualitative
- Transform numerical representation value into possibility distribution function
- Extract raw data needed for the analysis
- Analyze and find distribution for each error type.

At first, probable errors are defined and categorized. Probable errors depend on R&D environments. In this paper, 7 types of errors are defined as in Table 1. The error types should be defined such that an error may be contained in a unique error type.

Table 1 Types of probable errors

Error Type	*Function*	*Cause of error*
a	Errors by Hardware Failure	Hardware malfunction
b	Firmware Logic Error	Firmware
c	Operating System Error	Operating System
d	Data Base Error	DBMS or incorrect data
e	Administration Application Error	Administration software
f	Maintenance Application Error	Maintenance software
g	Call Processing Application Error	Call processing software

Next stage is detailed classification and numerical representation of found errors by system testing execution. This should be done carefully by system engineering experts with abundant experiment, since the numerical value is the raw data for analysis. Basically the numerical representation procedure of errors which are descriptive and qualitative are based on discussion with system function developers. Also, found errors during system tests may belong to several predefined error types(refer. Table 1). The numerical value is defined as the membership function of fuzzy set. The following describes how to transform descriptive and qualitative data into numerical value(membership function in fuzzy set).

- Each found error is considered as independent and must decide probable error types. For example, if found error belongs error types a, b, and e, then the membership function of each error type(a, b, and e) will be determined some value from 0 to 1 ($0 \le \mu_x(x) \le 1$).
- The membership function for each determined error type takes any value from 0 to 1. The value is determined according to the following relation. $\mu_x(x) = 0$ no have relation, $0 < \mu_x(x) < 0.3$ have weak relation, $0.3 \le \mu_x(x) < 0.7$ have middle relation, and $0.7 \le \mu_x(x) \le 1$ have strong relation with predefined error types.
- The exact value of this membership function is determined as the mean by several experts. For example, if 3 experts assign 0.3, 0.4, 0.2 for error respectively, it is decided to 0.3.

After the numerical representation of errors, errors are categorized as in Table 2. using the possibility distribution function in fuzzy theory. Raw data, possibility distribution function, are acquired from membership function of fuzzy set.

Table 2 Possibility distribution of found errors

Found	*Error Types'*							
Error	*a*	*b*	*c*	*d*	*e*	*f*	*g*	*Remark*
$v=E_1$	$\pi_\upsilon(a)$	$\pi_\upsilon(b)$	$\pi_\upsilon(c)$	$\pi_\upsilon(d)$	$\pi_\upsilon(e)$	$\pi_\upsilon(f)$	$\pi_\upsilon(g)$	Error types are defined in Table 1.

After the preparation of possibility distribution function, errors are analyzed to give useful parameters using the data in possibility distribution function. Parameters are defined and found according to the following definitions.

$$|I| = \sum_{y \in Y} \mu_I(y), \quad |T| = \sum_{x \in X} \mu_T(x), \quad \|I\| = \frac{|I|}{|Y|}, \quad \|T\| = \frac{|T|}{|X|} \tag{1}$$

where, |I| is data for the improvement of system as the scalar cardinality of fuzzy set for each of the found error. |T| is data for the influence of found errors on the system as the scalar cardinality of fuzzy set for each of the error. Y is the set of probable errors and X is the set of found errors. ||I|| is the relative size of Fuzzy set for each of the found error and ||T|| is the relative size of Fuzzy set for each of the error.

$$\alpha = \frac{\sum \|T\|}{|Y|}, \quad \beta = \frac{\sum \|I\|}{|X|}, \quad \gamma = \frac{\|T\|}{FN} \tag{2}$$

where, α ,α-cut set in fuzzy set, is data for error types and functions which should be considered carefully for improvement. β is data for system stability as the relative cardinality of fuzzy set that found errors are elements of predefined error types. γ is the ratio of application software errors(error types e, f, and g) to functions of the system as the relative size of fuzzy set for error types e, f and g. FN is the total number of functions of the system.

The parameters found in the previous step can be used in various analyses. In this paper, the parameters are used in the following 5 kinds of analyses :

- Analysis of scalar cardinality distribution for each error type. => This shows the degree of improvement for each version upgrade.
- Analysis of relative cardinality distribution for each error type. => This shows the functions that should be considered carefully in the next improvement and the error types that have strong influence on the system.
- Analysis of found error distribution for each improvement or software version. => This can be used to tell the influence of errors on the degree of improvement.
- Analysis of found error distribution for each function of application software. => This tells not only the degree of implementation for each function but also is very important because the improvements of software are continued after the freezing of hardware
- Analysis of system stability for each software version. => System stability is analyzed using relative cardinality of fuzzy set comprised of fuzzy values for error types. The result can also be used as the basis for the evaluation of R&D results.

3 APPLICATION

The analysis method, implemented as XTAM, is applied to the ATM switching system which is under development by ETRI, Korea. Raw data are gathered from tests of several software versions for the switching system. Software versions are named as V1, V2, and V3 for convenience.

Raw data, needed for possibility distribution function, are acquired with the experience and the help of system engineering expert. The analyses of raw data give the results in Figures 1, 2, and 3. Figure 1 represents the degree of improvement for each error type with software versions. In the Figure, error types a, b, c, d, and e are less frequent with software versions while other error types are not improved. The trend of error type g may be explained the instability of the software. Figure 3 (b) shows the decrease of total errors. However, this does not mean that error of every type is less frequent as shown by Figure 1.

Figure 2 shows the relative cardinality distribution which shows the possibility for each error type. α value represents the what should be carefully debugged for improvement as mentioned before. Since α values are 0.23, 0.25 and 0.24 for V1, V2, and V3, respectively, efforts should be given to the error types of hardware, administration and maintenance for V1 and V2 and to the error types of hardware, firmware, administration, and maintenance for V3. From this it can be said that errors in maintenance application part are most influential.

Figure 3 (a) shows error distribution for each function group. The value is the ratio of scalar cardinality to the number of needed functions. The figure shows highest error ratio in the implementation of administration functions for V1 and V2 while shows highest error ratio in the implementation of administration functions for V3. This fact is a very important information in the improvement of software versions.

Figure 3 (b) shows the influence of errors on the system stability. The value remains nearly constant irrespective of error numbers and extent of improvement for each error type compared with that in Figures 1 and 2. This shows the fact that the errors are critical or appears in several types of errors. Therefore it may be said that the software package is not stable. The stability of a system depends on the degree of influence caused by errors.

The above results can be summarized that most errors come from hardware, maintenance and administration functions and the number and types of error decreases as the upgrade of system.

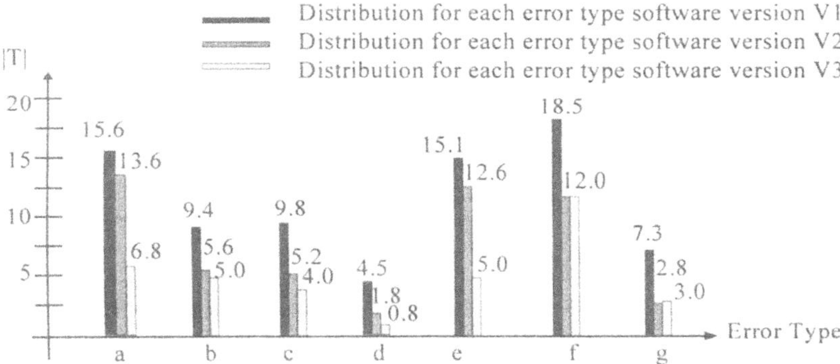

Figure 1 Scalar cardinality distribution for each error type

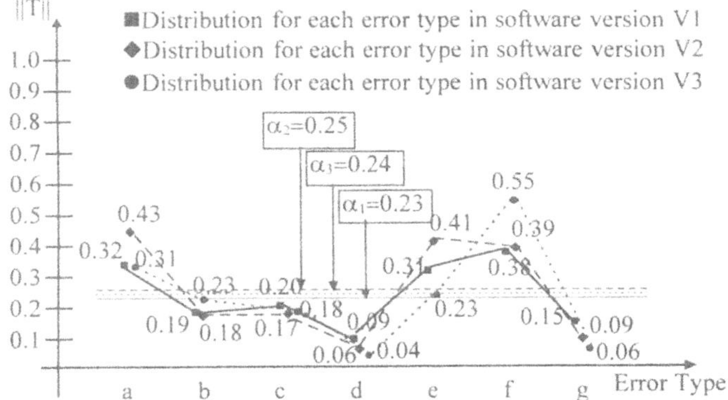

Figure 2 Relative cardinality distribution for each error type

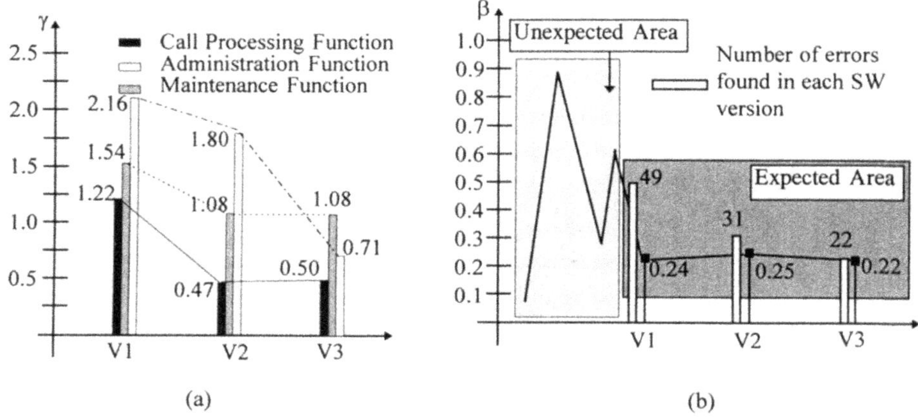

Figure 3 Error distribution for each application software of ATM Exchange and the relation with between number of errors found and system instability distribution in software version

4 CONCLUSION

This paper proposes a new method, called XTAM(eXchange system Test result Analysis Method) to analyze system test result. The new method has been applied for the analysis of test result for ATM switching system. By the application of new analysis method, it is possible to know the degree of system improvement, to determine parts that need careful consideration in the improvement of the system, to know the extent of implementation of functions, and to know the influence of errors for each type on the system and the stability of the system. Thus, the application of the new analysis method, XTAM, will make it easier to assure the quality of switching systems.

Since XTAM is based on possibility distribution function, correct raw data is required for correct analysis. After the correct analysis of errors found during tests by experts, it is possible to get possibility distribution function.

[References]

[1] Martin L. Katcher, "Integration and System Testing: a Methodology," ISS Session 14 C Paper 1, pp1-5, 7-11 May 1984.

[2] S.K.Cheong, S.K.Noh, and T.W.Jeong, "Testing Evaluation Method of ATM Switching System,"Analysis of Trend in Electronics and Telecom. vol.9 no.4, pp1-7, Jan.1995, ETRI

[3] S.K. Cheong, K.H. Lee, and T.W.Jeong, "On the testing of TDX-10 SSP System," Analysis of Trend in Electronics and Telecom. vol.14 no.2, pp1-7, July 1992, ETRI

[4] William E. Howden, "Errors, Design Properties and Functional Program Tests," B.Chandrasekaran and S.Radicchi, *COMPUTER PROGRAM TESTING*, North Holland,

pp115-127,1981.

[5] Lee J.White, "Basic Mathematical Definitions and Results in Testing," B.Chandrasekaran and S.Radicchi, *COMPUTER PROGRAM TESTING*, North Holland, pp.13-24,1981.

[6] Timothy A. Budd, "Mutation Analysis : Ideas, Examples, Problems and Prospects," B.Chandrasekaran and S.Radicchi, *COMPUTER PROGRAM TESTING*, North Holland, pp129-148,1981.

[7] Leon Osterweil, "A STRATEGY FOR INTEGRATING PROGRAM TESTING AND ANALYSIS", B.Chandrasekaran and S.Radicchi, *COMPUTER PROGRAM TESTING*, North Holland, pp187-229,1981.

[BIOGRAPHY]

Seung-Kook Cheong was born in Korea in 1959. He received the B.S in electronic engineering from Taejeon Open National University, Korea, in 1982, and M.S in electronic engineering from Hannam University, Korea, in 1992. Since 1985, he has worked for Switching Technology Division of ETRI, as a senior researcher, pursuing R&D on the switching systems for PSTN, N-ISDN, IN, and B-ISDN. His interested research parts are switching systems engineering for public network, ATM network design, and its evaluation.

Kyeong-Ho Lee* was born in Korea in 1956. He received the B.S and M.S in electronic engineering from Kwangwun University, Korea, in 1980, and 1982, respectively. Since 1982, he has worked for Switching Technology Division of ETRI, pursuing R&D on the switching systems for PSTN, N-ISDN, IN, and B-ISDN. Currently, he is the head of Quality Assurance Section. His research interests include network design, telecommunication modeling and its performance evaluation.

Taeg-Won Jeong* was born in Korea in 1955. He received the B.S and M.S in electrical engineering from Seoul National University, Korea, in 1979, and 1981, respectively, and Ph.D. in electrical engineering from University of Florida, USA, in 1991. Since 1983, he has worked for ETRI, pursuing R&D on the switching systems for PSTN, N-ISDN, IN, and B-ISDN. Currently, he is the head of ATM Service Section.

PART FOUR

ODP Testing

8
The specification and testing of conformance in ODP systems

P.F. Linington, J. Derrick and H. Bowman
University of Kent
Computing Laboratory, University of Kent, Canterbury, CT2 7NF, UK
Phone: + 44 1227 764000, Fax: + 44 1227 762811,
Email: {P.F.Linington,J.Derrick,H.Bowman}@ukc.ac.uk.

Abstract

Open Distributed Processing (ODP) is a joint standardisation activity of the ISO and ITU. A reference model has been defined which describes an architecture for building open distributed systems. This paper introduces the key aspects of the reference model of open distributed processing, including the ODP conformance framework. We discuss how specific formal techniques are used in the ODP viewpoints, along with the implications for conformance assessment using such techniques. Particular attention is given to the role of consistency in the conformance assessment process. Finally, we review the current work on an ODP conformance testing methodology.

Keywords: Open Distributed Processing; Conformance; Formal Description Techniques.

1 THE OPEN DISTRIBUTED PROCESSING MODEL

Open Distributed Processing (ODP) [29] is a joint standardisation activity of the ISO and ITU. A reference model has been defined which describes an architecture for building *open* distributed systems [36, 33]. Central to this architecture is a *viewpoints* model. This enables distributed systems to be described from a number of different perspectives. There are five viewpoints: *enterprise*, *information*, *computational*, *engineering* and *technology*. Requirements and specifications of an ODP system can be made from any of these viewpoints.

The framework for the standardization of Open Distributed Processing is provided by the Reference Model for Open Distributed Processing (RM-ODP), the essential, normative parts of which were completed early in 1995. The RM-ODP gives a structure for a family of standards for common interfaces, reusable components and supporting notations needed for the establishment of a wide range of distributed systems.

The reference model is published as ISO 10746, and is in four parts [29]:

Part 1: Overview. This provides an introduction to the reference model, explaining its objectives and giving background to the concepts defined. Part 1 also includes a number of larger examples of ODP use. This part was completed somewhat after the main normative parts (2 and 3), in May 1996.

Part 2: Foundations. This part defines the basic concepts which make up the ODP Object Model, and supporting concepts needed to describe the specification techniques and architectures which are based on it. It also defines the ODP approach to conformance.

Part 3: Architecture. This part defines five viewpoints focussed on various major concerns in the creation of distributed systems, and identifies a set of functions needed to construct such systems.

Part 4: Architectural Semantics. This part provides a formal basis for the reference model. It does this by giving interpretations of the main ODP concepts in a number of standardized formal description techniques. Further detail is given later in section 3.2 of this paper. Part 4 is currently undergoing its final phase of balloting as an International Standard.

This paper introduces the key aspects of the reference model of open distributed processing, including the ODP conformance framework, in section one. Section two discusses the requirements on formal techniques that ODP makes. The use of specific techniques within the ODP viewpoints is considered in section three, along with the implications for conformance assessment using such techniques. Particular attention is given to the role of consistency in the conformance assessment process. We review the current work on an ODP conformance testing methodology in section four, and we conclude in section five.

1.1 The ODP Object Model

ODP is object based; the initial set of concepts used in the reference model define objects which are encapsulated and which interact only at well-defined interfaces. It is not object oriented, because the strong emphasis on implementation inheritance normally found in object oriented programming languages is not appropriate when defining objects which are to form the parts of a distributed system.

An object in ODP can have any number of interfaces, which may be of the same type or of different types. This ability to define objects with multiple interfaces, and to define the dynamic creation and deletion of interfaces, gives a powerful tool for describing the evolution of object behaviour and the development of system configuration. In particular, the ability to define the interaction of objects at an interface which is subsequently hidden forms the basis of many forms of object composition.

The behaviour at an interface is expressed in terms of a set of interactions, considered as atomic at the level of abstraction being used. The general object model does not place constraints on the nature of these interactions, but the more detailed architectural part concentrates on two kinds of interaction - operation invocation and stream flow - leading to two major categories of interface.

In addition to the basic object model, concepts are defined which relate to the specification languages which are used in defining object based systems. These include ideas of behaviour specification, type, class, instantiation and template. The definitions given are generic, capturing concepts found in slightly differing forms in a number of different specification languages. They indicate points where language specific rules are likely to be found associated with a particular notation, but the representation in the various languages of interest is left as part of the architectural semantics.

1.2 The ODP Viewpoints

One of the most important structuring principles in ODP is the definition of a set of viewpoints. The idea is to divide the complete system specification, which may be very large and complex, into a number of areas of concern appropriate to the different stakeholders in the design process. If the different viewpoints are well chosen, they can be developed with a fair degree of independence, simplifying the specification activity. Of course, the viewpoints remain interlocking views of a single system, and so they are not fully independent. Correspondences between them must be declared, even if only by identifying elements which are referred to in more than one viewpoint as equivalent.

The ODP architecture defines five viewpoints, as defined below. The rules associated with each viewpoint effectively define the grammar of a corresponding viewpoint language, and the viewpoint specifications produced for a system must be valid utterances in these languages.

The enterprise viewpoint

The starting point for a system design is the identification of the roles played by entities influencing or influenced by the system, the form of the agreement or contract relating these roles, and the policies which are to direct the detailed system design. In many respects the enterprise viewpoint establishes and documents the boundary conditions constraining the remainder of the design activities.

The information viewpoint

The interaction of different objects depends on the sharing between them of a sufficient interpretation of the parameters of the interaction for there to be meaningful communication. The objective of the information viewpoint is to provide a definition of the shared model used to interpret the information communicated. There will either be a single information model or a small number of overlapping information models, reflecting the domain structure in which the system is to operate.

The enterprise and information viewpoints are concerned with the environment in which the system is to operate, rather than with the distribution process itself. The individual elements which participate in the system's behaviour are not yet visible.

The computational viewpoint

The functional decomposition of the system into a collection of interacting objects forms the basis for distribution. Once a computational model of the system has been defined, the way is open for a statement of the mapping from this configuration to the available physical resources, and the communications requirements can then be deduced from the computational statement of the information flows needed.

The specific details of the object model implicit in the computational language effectively define a virtual machine for the execution of the computational specification. It is this virtual machine which forms the basis for portability and process migration.

The engineering viewpoint

The engineering viewpoint defines the mechanisms which support the various actions and interactions required in the computational specification. These engineering specifications can be seen as the templates which would be used when creating an infrastructure on which to interpret the computational specification.

A large part of the engineering specification is concerned with the definition of channels capable of supporting the various kinds of computational interaction at operational and stream interfaces. This includes the specification of recipes to achieve various transparencies - solving some of the common problems found in distributed systems, such as location or failure independence. Other aspects are concerned with the local behaviour of objects and with the management of resources needed to support it.

The technology viewpoint

The technology viewpoint adds little to the general specification of system behaviour. It is concerned with the collection together of any necessary pointers to pre-existing specifications, such as the specification of supporting protocol stacks or collections of data elements. The linkages thus established form an important step in the interpretation of system behaviour from test results, but the detailed conformance requirements will generally be found in the specifications referenced.

The tool chain

One of the aims of ODP is to provide a coordinated family of standards which encourages the creation of a powerful tool chain able to construct implementations from the set of viewpoint specifications. For any particular design, the computational viewpoint provides the application structure, the enterprise and information viewpoints provide common definitions, policies and constraints, and the engineering viewpoint provides standard templates for the reuse of established communication and resource management techniques. Such tool chains will play an important role in making the development of distributed systems a cost-effective process.

1.3 The ODP Functions

Having established the viewpoint languages, the architecture next defines a set of common functions needed to support distribution. Many of these are concerned with aspects of channel support, either directly or by provision of the additional information necessary to support channel establishment.

Examples of ODP functions are the trading function [10], used in locating suitable services by type when building configurations, and the type repository function [27], used in managing and federating the different type systems shared between many ODP components. Components performing these two functions are already being standardized, with the trader currently nearing publication.

The ODP functions are most closely related to the engineering viewpoint, but the whole set of viewpoints is used when standardizing them.

1.4 The ODP Conformance Framework

ODP defines a basic framework for conformance as part of the reference model, rather than it being retrofitted later, as in OSI. The framework identifies the three key roles as being those of the standardizer, implementor and tester.

The foundation concepts include the idea of a *reference point* as being a specific location at which one or more interfaces can be localized, so that interactions at those interfaces can be observed. A standardizer can then declare some subset of the possible reference points in the system configuration for the component being defined as *conformance points*, stating the set of conformance requirements to be met at each of these points.

There are four different kinds of reference point, corresponding to different interfacing techniques and different testing technologies. They are:

1. interworking reference points, at which interactions take place on an observable communication medium, so that communication events can be deduced from observation of the medium. Most OSI conformance testing is concerned with interworking reference points.

2. programmatic reference points, which represent boundaries between software components within a system. The boundary, for example, can be between an application and its library or middleware, or at the protection boundary of the system kernel. Events in such an interface can be observed by using a suitable software test harness, either to provide an artificial environment or to gain access to the interface within the operational system.

3. perceptual reference points, which represent interactions between the system and the physical world which thus need to be checked by direct observation. A perceptual reference point may be at a human to computer interface, or it may be at a remote sensor or physical actuator in a process control system or for a robot.

4. interchange reference points, which are points at which interchangeable media are read or written. Examples here might be the checking of the correctness of discs passed between otherwise unconnected systems, or testing of the correctness of physical audit or archive media.

Once the conformance requirements have been established by the standardizer, the implementor must create an implementation and make a claim of its conformance to the standard. This involves the implementor naming the standard and declaring which physical locations and test procedures correspond to the various reference points required. In general, this declaration will involve the implementor in declaring a suitable test and interpretation procedure which can be carried out to identify the primitive actions required by the standard.

This is no different in principle from the situation in traditional protocol testing, where the communication medium and the form of any supporting lower layers must be declared. However, it may be considerably more complex in practice because of the wider range of supporting mechanisms which may be available. Thus, for example, the testing of a network protocol requires the identification of data link service events, but there are only a limited number of commonly used network and datalink profiles. On the other hand, the checking of a subtype relationship in a federated type repository may require the testing of steps in an algorithm supported in different locations by a variety of standardized and proprietary interaction mechanisms, depending on the engineering domains involved.

From these requirements comes the idea of a general Implementation Conformance Statement (ICS) and a much fuller set of extra information for testing (IXIT) than in previous conformance testing activities.

Finally, the tester must use this collection of information to plan and execute a testing programme. The testing observations must be interpreted using the information in the IXIT to recognize (or not) the observations as being the events referenced in the standard, and then the behaviour, in terms of these events, must be checked for correctness. The broader scope and size of distributed systems makes exhaustive testing completely impractical. At the same time, however, the growing reliance on the software tool chain offers some hope of simplifications by certifying key components, such as Interface Definition Language [26] compilers and stub generators. It is becoming possible to validate that these tools generate correct engineering viewpoint channel components from correct IDL, thus eliminating a large body of potential run-time testing.

2 FORMAL TECHNIQUES IN ODP

The use and choice of formal description techniques (FDTs) within ODP has implications for both the development and conformance assessment cycles. A number of different techniques and combinations of techniques have been considered, and part 4 of the reference model provides a detailed interpretation of the basic concepts from the descriptive model in a number of formal description techniques. Within ODP, formal description is viewed as enabling precise, unambiguous, and abstract definition and interpretation of ODP standards, and FDTs have been widely used in a number of different contexts [16, 7, 39, 35].

Techniques being considered include those arising from considerations of communication and concurrency (in particular from protocol engineering) such as LOTOS [2], Estelle [21] and SDL [6], together with those arising from general software engineering concerns, for example Z [40], VDM [30] and Raise [18]. The architectural semantics currently considers the languages LOTOS, Estelle, SDL and Z.

Although ODP can be viewed as a natural progression from OSI, ODP includes a wider range of modelling concepts which are needed in order to encompass the concerns of ODP; these range from enterprise policy specification to the description of engineering infrastructures. These differing concerns place a number of requirements on FDTs in order that they may support specification and testing of modern distributed systems. Current languages (both informal and formal) do not support all of these requirements, and it is likely that combinations of languages will be used for the specification of ODP systems and standards.

One major departure from OSI is that ODP modelling is object-based: objects are encapsulated and they interact via a number of interfaces (potentially more than one). In addition, there is a requirement within ODP to be able to specify the composition of objects and for incremental specification using inheritance. The growth of object orientation as a modelling paradigm has led to the definition of a number of object-oriented flavours of FDTs, all of which are able to support the required concepts to various degrees. These flavours include SDL-92 which provides an upwardly compatible object-oriented version of the extended finite state machine notation SDL; Object-Z [11], an object-oriented version of the state based language Z and proposals on how to use LOTOS in an object-oriented style.

However, of these variants only SDL-92 is as yet standardized, and clearly languages or methods advocated in a standard must be sufficiently mature and stable, and therefore should be standardized. Thus it is likely that consideration of FDTs in the ODP reference model will be restricted to LOTOS, SDL-92, Estelle and Z as the standardised or near standardised languages.

Two further ODP modelling requirements which are not in general supported by standardised formal languages are dynamic re-configuration and the ability to express non-functional requirements.

Because ODP systems can be modified and extended during their lifetime, ODP offers a flexible model of configuration. For example, faulty components may need to be replaced or it may be desirable to enable components to migrate in order to enhance performance and availability. The majority of semantic models of distribution and concurrency, e.g. labelled transition systems, finite state machines, event structures or petri nets, only allow static configuration. As a result the ability of LOTOS to model dynamic reconfiguration properly is limited and rather clumsy, [34, 31] discuss the support for dynamic reconfiguration in LOTOS, and the extensions needed to support it fully. On the other hand SDL and Estelle lack a formal semantics, a necessary pre-requisite for convincing modelling of dynamic reconfiguration.

The term non-functional refers to properties not identifiable in terms of a sequence of interactions between communicating objects; for example, quality of service and security issues are usually deemed non-functional. The provision of support for multi-media in ODP means that such requirements are important for the specification and testing of ODP systems. For example, the expression of real-time quality of service constraints, such as latency, throughput and jitter, is significant. [3] discuss the demands made upon

ODP modelling notations by the use of stream bindings and real-time synchronisation in the specification of continuous media.

Of the available notations, SDL, SDL-92 and Estelle all support a model of quantitative time, but it is far from clear that these are appropriate for definition of quality of service constraints [3]. LOTOS supports the expression of the temporal ordering of actions, but not the expression of real-time which is needed for quality of service constraints. A number of real-time extensions to LOTOS have been proposed, and one such variant, E-LOTOS, is currently undergoing standardisation. In addition to the real-time aspects, the new standard contains improved means to model mobility and reconfiguration [20]. The specification of non-functional requirements for an ODP system raises significant issues concerning automatic test generation from FDTs which involve real-time.

As a general purpose language, Z has the ability to support the specification of real time systems without extending the syntax. Work on real time specification using Z includes [12, 14, 32, 37]. However, as time is not built into the language as a pre-defined modelling concept, any real time specification using Z has to model time explicitly in a rather ugly way by using clocks and ticks etc.

As can be seen all four languages LOTOS, Estelle, SDL and Z provide support for modelling in ODP in a number of ways. However, no one language has all the facilities required for the specification of modern distributed systems.

This consideration, and others, point towards the use of multiple languages for the specification of a single ODP system. Single languages do not have the generality or expressiveness to support the full range of ODP specifications across all the viewpoints. Even wide spectrum FDTs such as Raise [18] are not able to embrace all needs (Raise, for example, has no support for real-time modelling). Thus, it is now accepted that a multiple language specification paradigm must be employed and that mechanisms must be provided in order to enable these FDTs to co-exist. This has implications for both the consistency checking and testing aspects of ODP conformance assessment which we describe below.

3 VIEWPOINT LANGUAGES

As indicated earlier, viewpoint languages are used to express specifications in each of the ODP viewpoints. The RM-ODP defines the concepts and rules of the viewpoint languages. However, the reference model is not prescriptive in the choice of specification notation; rather, the intention is that particular existing notations will be instantiated for each of the viewpoint languages, by supporting the concepts and rules defined in the RM-ODP.

Because the modelling concepts and abstraction levels of particular viewpoints vary, different FDTs are applicable to different viewpoints. A testing methodology for ODP then has to encompass the testing of a system potentially specified from a number of different viewpoints each employing a different notation at a different level of abstraction. We review here the specific requirements made by each viewpoint, and discuss the consequences for conformance assessment within ODP.

Enterprise

Enterprise modelling entails statements of policy, of organizational objectives and obligations which must be discharged. None of the concurrency based FDTs are thus seen as suitable candidates for the enterprise viewpoint language. Current enterprise modelling is performed in informal diagrammatic notations [17]; however, the semantics of the informal diagrammatic notations is usually not precisely specified, and thus not suitable for automatic test generation. A logical approach may be applicable to the type of abstract statements of system constraints that are required in the enterprise viewpoint. Therefore a combination of temporal logic and Z is a possibility for a formally based approach.

Because the enterprise viewpoint specifies organizational objectives and policy constraints it is not clear to what extent current automatic test suite generation techniques are applicable to it. However, since this viewpoint might specify key enterprise objectives, it would be instructive to see to what extent enterprise specifications can be used to offer guidance in the selection of test cases.

Information

Z is recognised as highly appropriate for information modelling, e.g. [38, 15]. Z is able to specify the format of information and operations to access and manipulate information without prescribing a particular implementation. Furthermore, many of the information viewpoint concepts have a natural interpretation in Z, which facilitates the formalisation of the information viewpoint in Z in the architectural semantics (see below).

The abstract data typing (ADT) languages incorporated into LOTOS and SDL are also possible vehicles for information specification. However, the correspondence between such ADT notations and the information language concepts is not as natural as it is for Z. This is because the definition of information in ODP centres on the notion that information must be exchangeable amongst users, which relates more directly to the behavioural part.

In LOTOS, information is modelled in ACT ONE along with operations that can act upon them. The behaviour expression part of the specification may then use and manipulate instances of these items. However, it is not the behaviour expressions themselves that change the information items, rather it is the ACT ONE expressions that are associated with the action denotations of the behaviour expressions that manipulate the information. As an example, consider the event offer $g!push(val, empty\_queue)$ in a behaviour expression. It is not the event offer that manipulates the information, but the ACT ONE expression $push(val, empty\_queue)$. Thus, only ACT ONE manipulates the information; however, ACT ONE can only be used within the behaviour expression of LOTOS.

This has implications for both the location of conformance points in an information viewpoint and automatic test suite generation from this viewpoint. Current automatic test suite generation techniques based upon LOTOS (and other process algebras) generate tests from the process algebraic part of the language, and are not capable of independent ACT ONE generated tests. Hence these techniques could only be used in conjunction with other viewpoints (for example, the computational viewpoint), and therefore the viewpoints could not be assessed separately for conformance against an implementation.

Work on test generation from Z specifications includes [43, 8, 5, 42, 19], however, little of this work is specifically targeted towards distributed systems or ODP in particular. Exceptions to this include [8, 42]. [8] generalises the CO-OP method of extracting test suites from LOTOS specifications and applies this to Z specifications. Work done under the Prost project in the UK on the testability of managed object specifications is also relevant to the information viewpoint. In [42] an object-oriented variant of Z is used to specify managed objects, and an inheritance hierarchy is constructed which facilitates the construction of a sound and complete test suite. Importantly, though, the test generation aims to supply heuristics and is not automatic. The heuristics provide a collection of tests together with a residual component which makes explicit the functionality not covered by the test suite. The tests generate form an independent and orthogonal collection of tests.

Because of the inheritance hierarchy, the reuse of tests between related specifications is maximised. A prototype tool-set developed by Logica provides organisational support for the collection of test specifications as they are generated.

The use of Pascal as the Estelle data language prevents Estelle from being an appropriate vehicle for information modelling.

Computational

The computational objects and their interfaces become visible in this viewpoint. The computational viewpoint also identifies the candidates for distribution (the choice being resolved in the engineering viewpoint); hence languages used in this viewpoint need to support the specification of interaction and synchronisation. Languages such as LOTOS, SDL and Estelle all offer considerable support in this respect; however, the encapsulation into objects is not directly supported (except in SDL-92). The work on automatic test suite generation in the languages LOTOS, SDL and Estelle is directly relevant to this viewpoint, where the concerns are similar to those found in OSI. Relevant work is surveyed in [24].

Z lacks explicit support for interaction and synchronisation, although one of the object-oriented dialects of the language, e.g. [7], may be more applicable. The object-oriented versions of the language offer encapsulation and define mechanisms to specify interaction and communication, either by adopting conventions or by definition of appropriate operators. Test generation methods for Z outlined above are again relevant to conformance in this viewpoint. However, aspects of concurrency and hiding have traditionally had less emphasis in Z, and this is reflected in the automatic test generation research using the language.

Engineering

The requirements for engineering viewpoint specification have many similarities to those for the computational viewpoint. Thus, from the potential candidate languages it is reasonable to consider LOTOS, SDL and Estelle as suitable choices and Z as less appropriate.

Technology

Specification in this viewpoint is primarily concerned with referencing appropriate standards and technologies to use in order to realise the specifications of the other viewpoints. This could involve referencing conformance standards and choices and combinations of standards as specified in a profile. Thus, extensive FDT specification is not a major requirement of this viewpoint, although it should be noted that the appropriate standards and technologies are not always rigourously specified, so FDTs may be useful for this purpose as well.

3.1 Consistency

The consistency of multiple viewpoint specifications is an aspect of conformance assessment that is new in ODP. The separation of concerns provided by the viewpoints can simplify the conformance assessment process, but it adds an additional obligation to be discharged, namely to show that all the viewpoints are consistent. Providing mechanisms to demonstrate that multiple viewpoint specifications are consistent is seen as essential to ODP. Furthermore, to use FDTs effectively within ODP, different FDTs are applicable to different viewpoints. This raises the issue of how to ensure consistency between different languages with different underlying semantics. This issue of consistency also arises outside ODP. For example, within OSI two formal descriptions of communication protocols can co-exist and there is no guarantee that, when the two protocols are implemented on the basis of these specifications, processes which use these two protocols can communicate correctly [13].

Consistency checking is relevant to conformance assessment, because even though the check is applied before any implementation is produced, clearly no implementation can hope to conform to all the viewpoints if they are inconsistent. Thus consistency checking can be viewed as part of ODP conformance assessment which is necessary before product testing can begin.

There are a number of different interpretations of the meaning of consistency [4]: one interpretation is to view consistency in terms of whether specifications impose contradictory requirements. Another interpretation is in terms of finding a common implementation, and as such is based on a notion of conformance. The final interpretation is in terms of behavioural compatibility of specifications. We adopt the most general of these possible interpretations, and define that:

> "A collection of viewpoints is consistent if and only if it is possible for at least one example of an implementation to exist that can conform to all viewpoints."

This definition of consistency hinges on the notion of conformance. Therefore, as usual, we divide conformance testing into two parts.

Firstly, we consider formal conformance up to implementation specifications (a relation *conf* between specifications is used for this purpose) and then we consider conformance testing of implementation specifications (essentially a very detailed specification

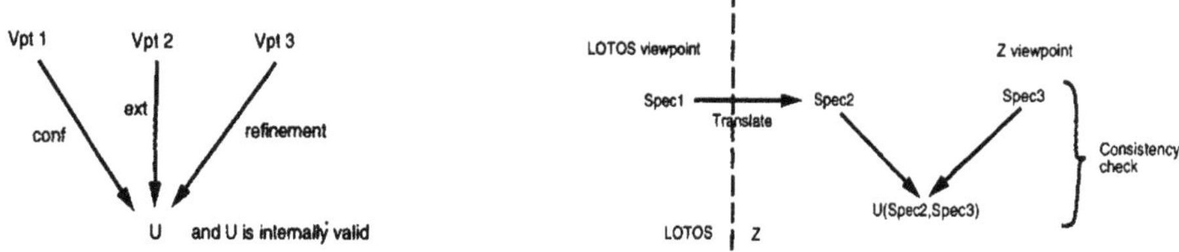

Figure 1: (a) Consistency checking in general, and (b) between LOTOS and Z.

that won't be refined further) to real implementations. The latter is needed because implementation specifications relate to real implementations in different ways for different FDTs and, in particular, for some FDTs not all implementation specifications are implementable. For example, a Z specification that contains an operation $[n! : \mathbf{N} | n! = 5 \wedge n! = 3]$ has no real implementation. Since the implementability of a specification is a property that depends on the FDT used, we will capture this property in our model by an assertion Ψ, which we call *internal validity*.

Then we can view a collection of viewpoints as consistent whenever an internally valid implementation specification exists which conforms to each of the viewpoints. Notice that because different (formal) languages have different notions of conformance, different relations are likely to be used with respect to the different viewpoints (e.g. we might use conformance in one viewpoint and functionality extension in another), see Figure 1 (a).

This methodological framework is compatible with current standardisation work on formal methods in conformance testing [25]. [25] defines a framework for the use of formal methods in conformance testing; in particular, it defines the meaning of conformance if formal methods are used for the specification of a communication protocol or service. In the framework, $MODS$ is used to represent all internally valid implementation specifications under the *testing assumption* that any real implementation can be modelled by an element of $MODS$ (which might be a labelled transition system, finite state machine etc). Conformance is asserted by means of a relation $\underline{\text{imp}}$ between $MODS$ and the set of specifications $SPECS$ (i.e. $\underline{\text{imp}} \subseteq MODS \times SPECS$). Then consistency is defined in [25] as finding a common element of $MODS$ conforming to each viewpoint. The work at Kent diverges from [25] because to define consistency across viewpoints we consider $MODS$ to include possibly unimplementable specifications, and therefore place an additional check on the implementable specification, that it be internally valid.

With this addition in place, a framework such as [25] can be extended to provide a methodology for consistency checking of multiple ODP viewpoint specifications. Mechanisms needed to support consistency checking can then be defined in terms of *refinement*. One specification is said to be a refinement of another if it restricts the set of conformant implementation specifications.

One way of determining whether two specifications are consistent is to unify them. A *unification* of two specifications is their least common refinement. We then check for consistency by determining whether the unification is internally valid, which guarantees that a conformant implementation of the unification exists.

The definition of consistency allows the differing aspects of FDTs to surface within the mechanism in appropriate ways because it involves two distinct parts: firstly the construction of a unification, and secondly verification of internal validity. Which part is appropriate depends on whether the behavioural or logical aspects are dominant in the FDT used. For example, consistency checking in Z and in LOTOS have a very different character. With LOTOS the central issue is finding a unification, while with Z the central issue is demonstrating that a unification does not contain any contradictions and can thus be implemented (assuming the specifications to be unified were themselves implementable).

Consistency checking between viewpoint specifications in LOTOS can involve a number of different refinement relations (**red, ext** etc). It is possible to obtain syntactic definitions of various kinds of unifications, some of these are guaranteed to be least unifications, and classified consistency induced by different refinement relations, [41].

In Z, finding a least unification of two viewpoints is an almost syntactical operation. For any two viewpoints, it is possible to construct a candidate unification, which is the least common refinement if one exists, [1]. Two conditions characterise whether the candidate is a refinement.

Unification combined with verifying internal validity for the unification forms a suitable method of consistency checking in a single FDT environment. However, since specifications in different FDTs cannot be unified, a translation mechanism is needed to transform a specification in one language to a specification in another language. In order to support consistency checking between viewpoints written in Z and LOTOS, a translation between the two languages has been defined in [9]. This can be combined with the Z unification mechanisms to support consistency checking between LOTOS and Z, as shown in Figure 1 (b).

An important aspect in deriving unifications is the correspondence between the viewpoints. Naming alone is insufficient to determine which parts of which viewpoint refer to the same object. Correspondence relations (which are similar to retrieve or abstraction relations [40]) document the dependency between the viewpoints, and the unification is relative to this correspondence relation.

A framework for consistency checking such as that just described has important implications for test case generation. In fact, a major consequence of the move from Open Systems Interconnection to Open Distributed Processing is the influence that viewpoints modelling has on generating tests. Specifically, the tests for an ODP system must be appropriate with regard to all the ODP viewpoints. There seem to be two alternative approaches to enhancing test generation strategies in order to fulfil the requirements of ODP viewpoints modelling.

- **Generation from Multiple Viewpoints.** In this approach, classic methods of deriving tests from a single specification are enhanced to embrace multiple viewpoints. For example, test cases could be derived from each viewpoint in turn and then composed in order to reduce the tests to only those that are appropriate for all viewpoints. Clearly, an explosion in the number of test cases when tests are generated from each viewpoint would be an important issue here.

- **Generation from a Unification Specification.** Central to the framework for consistency checking is the concept of a unification. Such a unification can be viewed as an "implementation specification" for the five viewpoints, i.e. a single specification that reflects all the viewpoints (by being a refinement of them all) and is as developed a description of the system under consideration as can be expressed in the specification domain. So, an alternative approach is to first unify the viewpoint specifications and then derive tests from this unification using classic test case generation techniques. It should be noted thought that it is still not clear that a manageable unification can always be derived from an ODP specification.

There is a clear trade-off between these two approaches. In the first approach the specific process of generating tests is hard and must handle the multiple viewpoints problem, but no explicit unification of the ODP viewpoints is required. In contrast, in the second approach the process of generating tests is unchanged from that traditionally considered, but an actual unification must be derived and this may be difficult. Thus, the former approach puts the emphasis on extending test case generation technology, while the latter emphasizes extending software development technology for multiple viewpoints architectures.

3.2 The ODP Architectural Semantics

The need for an architectural semantics was recognised from the start of the work on the ODP reference model and is reflected in the inclusion of the architectural semantics as Part 4 of the standard. This provides an interpretation of the ODP modelling and specification concepts in LOTOS, Estelle, SDL and Z. The aim of this work is to enable formal description of standards for ODP systems to be developed in a sound and uniform way by providing a link between the ODP modelling concepts and each of the different FDTs (with differing semantics).

Architectural semantics grew out of work on formal description of the protocol layers of the OSI reference model, where it was realised that specifications of protocol entities in different FDTs could not easily be combined. The problem had arisen because OSI concepts were being given totally different interpretations in different FDTs. By defining an architectural semantics, each architectural concept (interaction point, service access point etc) is given a fixed interpretation in different FDTs.

The formalization of the ODP modelling concepts in its architectural semantics consists of formalizing parts 2 and 3 of the reference model (i.e. the prescriptive parts). The RM-ODP defines a set of basic modelling concepts which define an interpretation for a vocabulary together with a collection of viewpoints which specify different aspects of a distributed system. The architectural semantics respects this distinction and comprises an interpretation of modelling concepts in a number of different FDTs together with a formalization of the viewpoint languages (directly and in particular FDTs). Conformance assessment of an ODP system written using a formal technique begins with the architectural semantics, both as a means to interpret the specification, and hence to provide for meaningful test generation, and to define the location of conformance and reference points, i.e. at which locations the testing will take place. As an illustration we discuss the approach taken in the semantics using LOTOS and Z.

Modelling concepts

Rather than provide a mapping of all the ODP modelling concepts, the architectural semantics focusses on the most basic, leaving interpretation of higher level architectural concepts to be made indirectly through their definition in terms of the most basic ones. The focus on object-based modelling means that many of the basic concepts centre around these definitions, and provide interpretations of Object, Interface, Composition of objects, Type, Class etc. In addition, there are interpretations of notions of Action, Behaviour, Communication, Refinement, Pre- and Post- condition etc.

The architectural semantics also defines which relation in a language is to be used to assert conformance. For example, in LOTOS conformance and refinement are then identified with the relations **conf** for the former, and **red** or **ext** for the latter.

Refinement has a well defined interpretation in Z, although importantly there is no general non-transitive relation that corresponds to conformance. Thus Z supports a series of refinements from abstract to more concrete specifications, but not the ability to determine conformance to an implementation specification (that might be written in another formalizations, e.g. labelled transition systems). The definition of a suitable conformance relation is an area that needs further study.

Formalization of Viewpoint Languages

Concepts from the viewpoint languages can be represented directly in particular formal techniques, we illustrate this by describing the information viewpoint in Z, and part of the computational viewpoint in LOTOS.

Formalization of information viewpoint in Z

An information specification corresponds to a Z specification, with the data being represented by the types and the information processing activities given by the possible behaviours as described by the operations in the specification. A static schema then corresponds to the bindings of the variables within the state schema at any point in time. An invariant schema corresponds to predicates given in the state schema and axiomatic descriptions within a specification (i.e. the predicates that the specification always satisfies).

Finally, dynamic schemas correspond to the possible behaviours of a Z specification. Since behaviour is given by the performance of the Z operation schemas, a dynamic schema corresponds to one or more operation schemas in the Z specification. Note that within Z the representation of invariant schemas and dynamic schemas are combined. Any

Z operation must satisfy the invariants within state schemas and axiomatic descriptions, thus any behaviour represented by the Z operations is constrained by the invariant schemas by default. The initialization schema of a Z specification is an example of a information viewpoint static schema, as such an initialization describes the values of variables (i.e. a binding) at a particular point in time, namely initially.

Because behaviour is represented by the Z operation schemas, a conformance statement in an information specification corresponds to one or more operation schemas. If a conformance point in the engineering or computational viewpoints has behaviour that uses information, a check on the use of the information is required to ensure conformance. This behaviour is said to conform if the post-condition and invariant predicates of this information manipulation are satisfied in the associated Z schemas.

A reference point will occur at an interface where tests can be applied to check for conformance. In Z, interfaces are associated with collections of operation schemas (the rest being hidden as described above), so reference points can be considered to reside at the pre-conditions of the operation schemas. However, a Z specification will not in itself identify which reference points are programmatic, perceptual, interworking or interchange. Such identification would accompany the specification as informal commentary.

Formalization of computational viewpoint in LOTOS

The mapping between computational viewpoint concepts and LOTOS is less complete than the Z formalization of the information viewpoint. This is because the computational viewpoint contains a number of modelling concepts for which a direct representation in LOTOS is not obviously available. For example, modelling streams and environmental contracts in LOTOS is problematic because of the lack of real-time support in LOTOS.

However, a subset of the computational viewpoint modelling concepts can be reflected. For example, a computational object is represented by an instantiation of a LOTOS process with the behaviour of the object modelled as the behaviour of the LOTOS process. ODP operations are modelled by LOTOS actions and the interface of an ODP object is the behaviour of a LOTOS process after gates have been hidden. Subtyping relationships between computational interfaces are viewed in ODP in terms of a concept called behavioural compatibility. This concept has in turn been identified with the LOTOS *conf* relation. However, there is now some doubt over this choice.

The exact relationship between conformance concepts and LOTOS constructs is not precisely specified in the current version of the architectural semantics. However, in a general sense, reference points will be identified with interfaces at which tests can be applied. The behaviour of specific actions from amongst those offered by an interface would be monitored during testing. Again, identification of the nature of the identified reference points, i.e. programmatic, perceptual, interworking or interchange, would have to be added as informal commentary.

4 AN ODP CONFORMANCE TESTING METHODOLOGY - CURRENT WORK

The RM-ODP provides a basic framework for describing conformance requirements and testing procedures for distributed systems. However, considerable detail needs to be added before workable conformance requirements can be specified. At the same time there is a need for a consistent approach across a number of areas of standardization, including ODP. In particular, there are closely related requirements arising in work originating from the programming language community to define open system environments. This has lead to requirements for OSE profile definition in part 3 of ISO TR 10000 [28].

At the same time, there is increasing emphasis on development of cost effective standards and support for product testing within ISO. A recent JTC1 Interoperability Policy Statement indicated that individual subcommittees should be given authority and responsibility for the development of an assessment methodology for their areas of activity. The ITU-T is expected to define a question on OSE conformance assessment in the next study period.

On a practical level, the final editing of the ODP Trader standard has indicated an urgent need for a consistent ODP style for the drafting of conformance statements for system components, and for a common understanding of the way middleware requirements and assumptions are to be expressed. More compact representation of test requirements and test purposes are needed to avoid the combinatorial explosion.

These considerations have lead to a general agreement within the ODP community that work on an ODP Conformance Testing Methodology is needed. SC21/WG7 has responded to this need by using the well established SC21 question procedures to clarify scope and direction of work before starting any necessary standardization activity. The question was formulated in the SC21 meeting in Ottawa in July 1995, and work started on it in the meeting in Kansas City in May 1996.

4.1 The WG7 Conformance Testing Question

The question posed was as follows (the text of the question quoted here and a summary of discussion to date can be found in [23]):

"Are any new standardization activities needed to support the conformance testing of ODP systems, components and supporting tools? To what extent can ODP requirements be met by standards which already exist or are under development, and are new liaison activities needed to ensure ODP requirements are taken into account in other work?"

Issues to be addressed:

1. JTC1 standards should avoid unnecessary constraint and proliferation of options; are there any features of the ODP architecture which lead to unnecessary complexity in the conformance testing process?
2. Both the RM-ODP and the work on OSE profiling define four kinds of reference point - in ODP terms, programmatic, interworking, perceptual and interchange

reference points. In OSE, these are considered as interfaces to systems, while in ODP they are interfaces to objects. The implication of this fine grain structure and the associated object encapsulation for testing methodology requires study.

3. Current work on conformance testing has concentrated on the testing of single interfaces. In ODP, there is a need for testing which involves multiple interfaces:
   (a) ODP objects can have multiple interfaces , and assessment of conformance to the object's behaviour may require correlation of tests at a number of interfaces.
   (b) One ODP computational interface can correspond to a number of engineering interfaces of different kinds within the supporting channel (e.g. an interworking reference point and one or more programmatic reference points).
4. The ODP computational model includes stream interfaces; a methodology is required for testing the conformance of stream interfaces.
5. The ODP concept of viewpoint is intended to support separation of concerns, and it is expected that the specifications in the various viewpoints will be combined by suitable tools (as, for example, when an information or computational declaration of need for a transparency results in incorporation of mechanisms based on an engineering specification). The conformance testing methodology may need to distinguish validation of this tool chain from detailed testing of the individual implementations produced.
6. The ODP concept of environment contract allows performance constraints to be placed on implementations. There is a need to study the methodology for testing conformance to these performance requirements.
7. The behaviour of an ODP computational object may require that, in defined circumstances, the object should initiate an invocation at one of its client interfaces. Current conformance testing methodology has concentrated on the testing of objects acting as servers. Study is needed to determine if extensions to the methodology are required to cover testing of client behaviour.

4.2 Constructing an answer

The initial discussions on this question have focussed on removing any architectural uncertainty and on establishing priorities for the work.

One of the key aspects seems to be the need to clarify what kind of standard will include statements of conformance requirements. Consideration needs to be given to component standards, component composition standards and notational standards. These different classes of standard will have quite different forms of assessment, with corresponding style of ICS and IXIT. ODP conformance will place emphasis on aspects of cooperative, multiparty operation, and dynamic configuration not addressed in protocol testing. On the other hand, some traditional concerns can be de-emphasised as a result of the introduction of more powerful notations; static conformance virtually disappears as a result of the widespread use of notations such as IDL.

On the other hand, there is now a level of agreement that apparent conflict in (2) above is not real; the system is, at some level of abstraction, an object, and there is only a problem if the system is seen as an absolute, rather than relative, structuring concept.

Other inputs to date have primarily been concerned with clarifying particular issues in the question text, and further input is being sought. The Kansas City meeting defined priorities for the work, identifying three areas as the most urgent. These were (again, quoting from [23]):

1. the creation of a model conformance statement, using the ODP Trader as an example, which can be used as a pattern for other component standards being developed. The Implementation Conformance Statement and Test Cases for the ODP Trader are to be published as a separate part [22], and so can be developed further without delaying the main part of the technical specification. It was noted that much duplication might be avoided by the definition of implied requirements from an IDL specification.

2. the study of certification of the tool chain as a way to simplify the assessment of conformance of complex systems, rather than exhaustive testing of the actual system.

3. initial development of a methodology for the testing of conformance of stream implementations. There is, as yet, little consensus in this area and thus development of an initial draft is urgent.

5 FUTURE ACTIVITY

This paper has outlined a wide area of work in which there is currently considerable activity. It has also indicated that there is now a well-established body of theory associated with specification and conformance testing. Over the next year or two a draft answer to the ODP question should be agreed and published, and work started on the standardization of the standard for an ODP conformance methodology.

It is to be hoped that conformance testing experts and ODP experts will work together to establish the necessary framework for the creation of reliable and cost effective distributed systems in the future.

References

[1] E. Boiten, J. Derrick, H. Bowman, and M.Steen. Consistency and refinement for partial specification in Z. In M.-C. Gaudel and J. Woodcock, editors, *FME'96: Industrial Benefit of Formal Methods, Third International Symposium of Formal Methods Europe*, volume 1051 of *Lecture Notes in Computer Science*, pages 287–306. Springer-Verlag, March 1996.

[2] T. Bolognesi and E. Brinksma. Introduction to the ISO Specification Language LOTOS. *Computer Networks and ISDN Systems*, 14(1):25–59, 1988.

[3] H. Bowman, L. Blair, G. S. Blair, and A. Chetwynd. Formal description of distributed multimedia systems; an assessment of potential techniques. *Computer Communications*, 18(12):964–977, 1995.

[4] H. Bowman, E.A.Boiten, J. Derrick, and M. Steen. Viewpoint consistency in ODP, a general interpretation. In *First IFIP International workshop on Formal Methods for Open Object-based Distributed Systems*, Paris, March 1996. Chapman & Hall. To appear.

[5] D. Carrington and P. Stocks. A tale of two paradigms: Formal methods and software testing. In J.P. Bowen and J.A. Hall, editors, *ZUM'94, Z User Workshop*, pages 51–68, Cambridge, United Kingdom, June 1994.

[6] CCITT Z.100. *Specification and Description Language SDL*, 1988.

[7] E. Cusack. Object oriented modelling in Z for Open Distributed Systems. In J. de Meer, V. Heymer, and R. Roth, editors, *IFIP TC6 International Workshop on Open Distributed Processing*, pages 167–178, Berlin, Germany, September 1991. North-Holland.

[8] E. Cusack and C. Wezeman. Deriving tests for objects specified in Z. In J. P. Bowen and J. E. Nicholls, editors, *Seventh Annual Z User Workshop*, pages 180–195, London, December 1992. Springer-Verlag.

[9] J. Derrick, E.A.Boiten, H. Bowman, and M. Steen. Supporting ODP - translating LOTOS to Z. In *First IFIP International workshop on Formal Methods for Open Object-based Distributed Systems*, Paris, March 1996. Chapman & Hall. To appear.

[10] Draft Rec. X.950 — ISO/IEC DIS 13235-1. *Open Distributed Processing - Trading Function - Part 1: Specification*, May 1996.

[11] R. Duke, G. Rose, and G. Smith. Object-Z: A specification language advocated for the description of standards. *Computer Standards and Interfaces*, 17:511–533, September 1995.

[12] M. Engel. Specifying real-time systems with Z and the duration calculus. In J.P. Bowen and J.A. Hall, editors, *ZUM'94, Z User Workshop*, pages 282–294, Cambridge, United Kingdom, June 1994.

[13] A. Fantechi, S. Gnesi, and C. Laneve. Two standards means problems : A case study on formal protocol descriptions. *Computer Standards and Interfaces*, 9:11–19, 1989.

[14] C.J. Fidge. Specification and verification of real-time behaviour using Z and RTL. In J. Vytopil, editor, *Formal Techniques in Real-Time and Fault-Tolerant Systems*, Lecture Notes in Computer Science, pages 393–410. Springer-Verlag, 1992.

[15] J. Fischer, A. Prinz, and A. Vogel. Different FDT's confronted with different ODP-viewpoints of the trader. In J. C. P. Woodcock and P. G. Larsen, editors, *FME'93: Industrial Strength Formal Methods*, LNCS 670, pages 332–350. Springer-Verlag, 1993.

[16] R. Gotzhein and F. H. Vogt. The design of a temporal logic for Open Distributed Systems. In J. de Meer, V. Heymer, and R. Roth, editors, *IFIP TC6 International Workshop on Open Distributed Processing*, pages 229–240, Berlin, Germany, September 1991. North-Holland.

[17] J. J. Van Griethuysen. Enterprise modelling, A necessary basis for modern information systems. In J. de Meer, V. Heymer, and R. Roth, editors, *IFIP TC6 International Workshop on Open Distributed Processing*, pages 29–68, Berlin, Germany, September 1991. North-Holland.

[18] The RAISE Language Group. *The RAISE Specification Language.* Prentice Hall, 1992.

[19] H-M. Horcher. Improving software tests using Z specifications. In J. P. Bowen and M. G. Hinchey, editors, *Ninth Annual Z User Workshop*, LNCS 967, pages 152–166, Limerick, September 1995. Springer-Verlag.

[20] ISO, Kansas City. *Minutes of the ISO/IEC JTC1/SC21/WG7/E-LOTOS meeting*, May 1996.

[21] ISO 9074. *Estelle, a Formal Description Technique based on an extended state transition model*, June 1987.

[22] ISO/IEC DIS 13235-2. *Open Distributed Processing - Trading Function - Part 2: Implementation Conformance Statements and Test Cases*, May 1996.

[23] ISO/IEC JTC1 SC21/WG7 N1163. *Working Document on Question 7/003 on an ODP Conformance Testing Methodology*, 1996.

[24] ISO/IEC JTC1/SC21. *FMCT guidelines on "Test Generation methods from Formal descriptions"*, December 1995. ISO/IEC ITU-T Interim meeting on FMCT.

[25] ISO/IEC JTC1/SC21. *Formal methods in conformance testing - Part 1: Framework*, December 1995. ISO/IEC ITU-T Interim meeting on FMCT CD 13245-1.

[26] ISO/IEC JTC1/SC21 N10389. *Open Distributed Processing - Interface Definition Language*, April 1996.

[27] ISO/IEC JTC1/SC21 N10389. *Open Distributed Processing - Type Repository Function*, May 1996.

[28] ISO/IEC TR 10000-3. *Information Technology - Framework and Taxonomy of International Standardized Profiles - Part 3: Principles and Taxonomy for Open System Environment Profiles*, 1995.

[29] ITU Recommendation X.901-904 — ISO/IEC 10746 1-4. *Open Distributed Processing - Reference Model - Parts 1-4*, July 1995.

[30] C. B. Jones. *Systematic Software Development using VDM.* Prentice Hall, 1989.

[31] T. Koch, B. Kramer, and N. Volker. Modelling dynamic ODP-configurations with LOTOS. In J. de Meer, B. Mahr, and O. Spaniol, editors, *2nd International IFIP TC6 Conference on Open Distributed Processing*, pages 346–351, Berlin, Germany, September 1993.

[32] L. Lamport. TLZ. In J.P. Bowen and J.A. Hall, editors, *ZUM'94, Z User Workshop*, pages 267–268, Cambridge, United Kingdom, June 1994.

[33] P. F. Linington. RM-ODP: The Architecture. In K. Raymond and L. Armstrong, editors, *IFIP TC6 International Conference on Open Distributed Processing*, pages 15–33, Brisbane, Australia, February 1995. Chapman and Hall.

[34] E. Najm, J-B. Stefani, and A. Fevrier. *Introducing Mobility in LOTOS.* ISO/IEC JTC1/SC21/WG1 approved AFNOR contribution, July 1994.

[35] P. F. Pinto and P. F. Linington. A language for the specification of interactive and distributed multimedia applications. In B. Mahr J. de Meer and O. Spaniol, editors, *IFIP International Conference on Open Distributed Processing*, pages 217-234, Berlin, Germany, September 1993. North-Holland.

[36] K. Raymond. Reference model of open distributed processing (RM-ODP): Introduction. In K. Raymond and L. Armstrong, editors, *IFIP TC6 International Conference on Open Distributed Processing*, pages 3–14, Brisbane, Australia, February 1995. Chapman and Hall.

[37] A.R. Ruddle. Formal methods in the specification of real-time, safety-critical control systems. In J. P. Bowen and J. E. Nicholls, editors, *Seventh Annual Z User Workshop*, pages 131-146, London, December 1992. Springer-Verlag.

[38] S. Rudkin. Modelling information objects in Z. In J. de Meer, V. Heymer, and R. Roth, editors, *IFIP TC6 International Workshop on Open Distributed Processing*, pages 267–280, Berlin, Germany, September 1991. North-Holland.

[39] M. Van Sinderen and J. Schot. An engineering approach to ODP system design. In J. de Meer, V. Heymer, and R. Roth, editors, *IFIP TC6 International Workshop on Open Distributed Processing*, pages 301–312, Berlin, Germany, September 1991. North-Holland.

[40] J. M. Spivey. *The Z notation: A reference manual.* Prentice Hall, 1989.

[41] M. W. A. Steen, H. Bowman, and J. Derrick. Composition of LOTOS specifications. In P. Dembinski and M. Sredniawa, editors, *Protocol Specification, Testing and Verification, XV*, pages 73-88, Warsaw, Poland, 1995. Chapman & Hall.

[42] S. Stepney. Testing as Abstraction. In J. P. Bowen and M. G. Hinchey, editors, *Ninth Annual Z User Workshop*, LNCS 967, pages 137–151, Limerick, September 1995. Springer-Verlag.

[43] P. Stocks and D. Carrington. Deriving software test cases from formal specifications. In *6th Australian Software Engineering Conference*, pages 327–340, July 1991.

9

An approach to the validation of open object-based distributed applications[1]

S. Pickin, C. Sánchez, J. Sánchez, J.C. Yelmo,[a] J.J. Gil, E. Rodríguez[b]
[a] *DIT, ETSI-Telecomunicación, Universidad Politécnica de Madrid. Cuidad Universitaria, 28040 Madrid, Spain. pickin@dit.upm.es*
[b] *Telefónica Investigación y Desarrollo. Emilio Vargas 6, 28043 Madrid, Spain. err@tid.es*

Abstract

The validation of object-based distributed applications is a difficult and as yet largely unexplored problem. We describe here a practical approach over the OMG's CORBA platform and architecture. Our approach hinges on specifying interfaces in CORBA Interface Definition Language (IDL) at other levels besides the object level and adjoining two types of complementary semantic information - then the basis of the validation methods. We then indicate how standardized FDTs such as SDL and LOTOS can be used in such a validation framework.

Keywords

ODP system validation, practical testing methodology, conformance testing, formal notations.

1 INTRODUCTION

In the ISO Open Distributed Processing (ODP) standards, it is recognized that object-based technologies constitute the most promising framework for distributed applications operating in an open heterogeneous environment. While still in the draft stage, these standards were adopted as the basis of the future telecommunications architecture by the Telecommunications Information Networking Architecture consortium (TINA-C). At the same time the Object Management Group (OMG) was defining their Common Object Request Broker Architecture (CORBA) with similar, if less far-reaching and more implementation-oriented, objectives to those of ODP. Though much of the ODP corpus has now reached the stage of international standard and numerous implementations of the CORBA standards are commercially available, validation in this setting has not yet received much attention.

---

[1] This work has been partly sponsored by the following Spanish research bodies: the Centre for Industrial and Technological Development (CDTI) and the Plan for Electronics and Computer Science (PEIN) as part of the OTELSO Project (EUREKA 1001) ; the Interministerial Commission for Science and Technology (CICYT) as part of the EMEDAS project.

This article discusses, in the context of our methodology, issues of particular relevance to object-based distributed system validation. We take a pragmatic view and draw on work of both the software and protocol validation communities. In order to carry out work with industrial relevance and in keeping with our objective of employing standards (even if *de facto*) where possible, we have worked with CORBA (and therefore deal only with operational interfaces and do not consider group communication). Our objective is to define and implement a validation environment for CORBA applications, using the Orbix[1] CORBA implementation.

2 VALIDATION ASPECTS OF THE DEVELOPMENT METHODOLOGY

Here we present the aspects of our methodology relevant to validation.

Our development methodology structures the application into components, with interfaces described in CORBA IDL, at different levels. Associated to the components are two types of *invocation scenarios*: external and internal. The former concern communications between the component and its peers and the second concern communications between its subcomponents. The latter will therefore include projections and decompositions of the former. Three categories of scenarios, describing obligatory, optional and forbidden behaviour, can be used. The scenarios are not intended to furnish a complete description of behaviour. Thus, a conformant implementation may legally execute traces which are not described by any of the scenarios. These may include traces with a prefix in common with a trace so described. This has obvious consequences for the discriminating power of invocation-trace testing based on these scenarios. Projections of external scenarios onto individual interfaces of a component (*interface scenarios*) are of particular importance, as explained below. The language chosen for describing the invocation scenarios is the standardized Message Sequence Chart (MSC) notation.

In this methodology, *assertions*, relating input to output, are associated to some of the external scenarios. The input (resp. output) can refer to both input (resp. output) parameters of the scenario invocations as well as to accessible aspects of the state of the IUT prior to (resp. following) the execution of the trace described in the scenario. Commonly, assertions are placed on scenarios describing single RPC calls, i.e. on CORBA two-way operations.

A set of interface scenarios and associated assertions can be bundled together with the signature of the interface (in CORBA IDL) and the joint signature of all partners appearing in the interface scenarios (also in CORBA IDL) into a development *contract*[2] [Pickin et al, 1996]. This extension of the usual notion of interface can be particularly useful for large-scale components, in particular for organizing a division of work during the design phase, for facilitating prototyping activities and for validating the service offered through a component interface.

Note that the scenarios and assertions specified in earlier phases may need to be evolved to accommodate design changes, while taking into account that modification of contractual features should involve renegotiation. Moreover, the addition of more scenarios or assertions is permitted in all phases, and in particular the testing phase, with the proviso that the contractual ones are clearly distinguished as such.

The difficulties of using current FDTs to specify object-based distributed systems, together with the extensions and modifications which are needed in order to make such specifications possible, have been treated in [Stefani, 1991], and are also discussed in [Pickin et al, 1996]. In

---

[1] Orbix is a trademark of IONA Technologies Inc.

[2] note that this notion of contract is not identical to that of ODP, TINA or other work.

view of these difficulties, our methodology only provides for FDT-based development of critical components of object-based distributed applications. This is done through the use of a CORBA-style of specification and the notion of an "FDT-wrapper" [Pickin et al, 1996].

3 VALIDATION OPTIONS FOR CORBA APPLICATIONS

3.1 Automatic (active) testing

Preambles and postambles: In the case of a non-FDT based component, pre-/post- amble specification is necessarily an informal procedure which may or may not use other scenarios.

In the case where the IUT is a distributed component, ensuring that it is placed in a particular state in the pre-/post- amble is a major difficulty. In such a case, we have to content ourselves with an abstract, high-level notion of state.

The distinction between the initialization of the test environment and the test preamble may not always be obvious. Generally speaking, the preparation of the environment of the IUT (setting environment permissions/variables, server registrations, dummy server activations, etc.) is part of the initialization procedure whereas the preparation of the IUT itself constitutes the preamble. However, consider IUT permissions or the case discussed in Section 4, where the IUT is to activate the test controller itself.

In the framework of the methodology discussed earlier, two types of active test can be derived:

- **Invocation-based tests**: From the invocation scenarios, tests checking that the CORBA invocations were executed in the correct order can be generated by using the MSCs expressing the obligatory and optional invocation scenarios as test purposes. These tests are evidently less discriminating than tests derived from complete specifications of behaviour; in the general case there can be no *fail* verdict, only *pass* and *inconclusive*. Nevertheless, in cases where the size and the branching of the behaviour graph are not excessive, exploration of the principal execution paths denoted by the scenarios can be of great value. Interpretation of an inconclusive verdict requires an idea of the completeness of the set of scenarios, knowledge of the problem domain and, in many cases, knowledge of the implementation.
  In the case where the IUT is a critical component whose implementation results from an FDT-based development, more discriminating tests can be derived. In this case, each message of the scenario corresponds, for example to a LOTOS action or SDL input/output. Relatively efficient tools are available for finding occurrences of the scenario in the behaviour graph of the specification[1]; after finding all the suitable ones, one reachable via the shortest path is selected. The location of this occurrence in the behaviour graph then enables a suitable preamble and postamble to be derived. A verdict of fail can then be assigned if, during the execution of the test body, a transition not foreseen by the formal specification occurs as in, for example, [Grabowski et al, 1993]. The details depend on the type of FDT wrapper.
- **Assertion-based tests** From the assertions associated to a scenario, black-box tests checking that they hold on execution of that scenario can be derived. Evidently, for true black box component testing, only those aspects of the state of the component which are verifiable using externally visible operations of that same component (where this includes reading

[1] though not always with the standardized causal semantics of the MSCs!

attribute values) should be used in the assertions. For true black box service testing, any other operations used to verify an assertion on a given operation must be visible in the same interface as that operation.

If the same scenario exists at different structural levels, the assertions associated to it at these different levels are not necessarily identical.

We derive assertion-based tests using a procedure which adapts the public domain tools of the Assertion Definition Language (ADL) project [Sankar et al, 1994] to use with CORBA.

3.2 Automatic passive testing or observation

We concentrate here on passive testing which takes advantage of the specifications used in our methodology. Many other dynamic properties, see the framework of [Babaoglu et al, 1995], could only be checked by ensuring causal delivery to the tester. In general, this requires the introduction of extra fields in system messages. As there are currently no OMG-standardized mechanisms for this (see [Maffeis, 1995]), such introduction would compromise the openness of the system. Two types of passive test can be derived in our methodology:

- **Invocation-based tests**: In run-time checking, the behaviour of the IUT is not originated and rigorously controlled by the tester. As the testing is based on an incomplete behaviour specification, the lack of success of a non-failure test only indicates a possible problem. To avoid being too intrusive, such testing should not signal possible problems too often. The "success" of a failure test, on the other hand, always indicates a problem. Therefore, failure tests are better suited to use in run-time checking of invocation orderings.
  The principal difficulty to be treated in invocation-based passive testing is the interleaving of the traces described in the scenarios, a phenomenon which increases with the scale of the component under test and with the number of messages in the test scenario. Though the complexity engendered by concurrency increases for larger scale components, the use of threads leads to similar problems even for the smallest scale components. Abstract interpretation of MSCs (see Section 4) reduces but does not obviate the problem. The interleaving of scenarios with a non-empty message alphabet intersection, in particular different instances of the same scenario, could still lead to failures going unnoticed by the failure tests or to a possible problem being signalled too often by other tests.
- **Assertion-based tests**: The difficulties engendered by the interleaving of the execution traces described in the scenarios mean that the assertions which are the most suited to run-time checking are those which are associated to the simplest scenarios describing single two-way operations. For these, the underlying system takes responsibility for pairing up the request and the reply. Each time such an operation is used then, its functionality is checked using the assertions. Including such tests as a permanent part of the interface facilitates implementation changes.

3.3 Monitoring and visualization

Here, though any verdict is completely manual, information considered to be of particular importance for determining correct behaviour can be presented so as to reflect this importance. In the framework of our methodology, two types of visualization have been defined; in both cases, user-defined multiple visualizers (implemented as CORBA servers) can show the same or different parts of the executing application (and can be executing on any host):

- **dynamic visualization of execution traces** (as MSCs): This diagram shows the invocations which have been performed between the different monitored entities, in the simplest case, CORBA servers. The sending of a local event counter value with each monitored event[1] is sufficient to ensure that the MSCs displayed respect causality (except for identical messages between the same entities which will not be a problem).
- **dynamic visualization of the application architecture** (as DARTS[2]): This diagram shows the bindings established between the stubs and interfaces of the monitored entities (currently CORBA servers). This diagram complements the execution trace visualization and can also be used as the vehicle for modifications to the monitoring session.

3.4 Monitoring, visualization and interactive testing

Though relatively informal, this type of debugging or operational validation is highly valued by developers. To the monitoring and visualization facilities, we can add the following:

- A facility for automatically checking a previously executed trace against a set of scenarios. Here, the boundaries of the test body are defined interactively on the MSC representation of the trace. The user is then informed whether the MSC of the executed trace corresponds fully, partially or not at all to one or more of a selected set of scenarios.
- A facility for automatically checking the current dynamic architecture against a model. We provide the capability to dynamically check that entities have been activated/instantiated and/or that bindings have been established in accordance with a previously defined model.

4 SOME ISSUES IN THE VALIDATION OF CORBA APPLICATIONS

Use of MSCs: The scenarios are described using the MSC notation, whose interpretation is as a causality diagram. However, use in object-based distributed system validation requires some extensions to the current MSC standard, in particular, the event-oriented syntax and a notational convention for denoting RPC-like calls. The first already is, and the second almost certainly will be, part of the MSC96 standard. In addition, the HMSCs of MSC96 could be used to express relations between scenarios associated to a given interface/component.

In using MSCs to specify execution traces (rather than to reflect execution traces as in monitoring), it must be decided whether they are to be interpreted as abstractions of trace segments or as complete trace segments, i.e. whether or not a conformant trace may contain other messages imbetween those specified, provided these extra messages are not members of that MSC's message alphabet[3]. We advocate use of the concrete interpretation for active testing, the abstract interpretation for passive testing and user-choice for interactive testing.

Test architecture considerations: Apart from the visualizers used in monitoring, the validation software is composed of a controller and a series of probes and/or dummies. The former are used in the case where we use the actual implementation of a service needed by the IUT from its environment, while the latter are used in the case where we substitute such an implementation by one with the minimal functionality needed to test the IUT. That is, whenever the

---

[1] along with: the identity of the entity in which the event occurs, the message identity, the event type (send or receive) and for a receive/send event, the source/destination identity respectively.

[2] DARTS is a visual specification language which is presented in [Sánchez et al, 1995].

[3] analogous to completing an FSM with loop transitions to the same state or with transitions to a fail state.

IUT behaves as a CORBA client in active or passive testing, either the real IUT environment (so probes) or the test software (so dummies) must offer the corresponding interfaces. The probes are implemented using the Orbix filters and send event notifications to the controller and/or visualizers. As mentioned previously, our purposes require that the probes insert local event counter values in the event notification messages but do not require addition of timestamps in system messages.

Location transparency in object-based distributed systems means that the controller need never itself be distributed and that the dummies can also be located at the same host, thus alleviating the sychronization problems of distributed system testing. To avoid inter-process communication within the test software, it is preferable to incorporate the dummies and the controller into a single CORBA server (Unix process), though this many not always be possible, e.g. when the IUT expects to activate different dummies in different CORBA servers. Note that in the case where the IUT expects to activate a dummy server which is implemented in the same server as the test controller, this will mean that the test will proceed with the IUT activating the test controller itself!

5 CONCLUSIONS

To date, validation of object-based distributed systems has not received much attention. This is in spite of the fact that, due to the efforts of the ISO ODP group and the OMG, such systems are beginning to gain widespread acceptance. The approach to the validation of CORBA applications presented here centres on the identification of components in the design phase, the description of their interface signatures in CORBA IDL, and the association of two types of semantic constraints to these components and their interfaces. This information is then used to drive the validation of the resulting implementation. In the case where FDTs are used to specify critical CORBA components, in accordance with the approach we have developed using a specification style and FDT wrappers, we point out how more satisfactory test suites for these components can be generated semi-automatically from the MSCs viewed as test purposes.

6 BIBLIOGRAPHY

Babaoglu, Ö., Fromentin, E., Raynal, M. (1995). *A Unified Framework for the Specification and Run-time Detection of Dynamic Properties in Distributed Computations.* Department of Computer Science, University of Bologna, Technical Report UBLCS-95-3. Jan. 1995, revised Feb 1995.

Grabowski, J., Hogrefe, D. and Nahm, R. (1993). Test case generation with test purpose specification by MSCs, in *SDL'93: Using Objects* (ed. O. Færgemand and A. Sarma), proc. SDL'93. North Holland.

Maffeis, S. (1995). Adding group communication and fault tolerance to CORBA, in Proc. of *USENIX Conference on Object-Oriented Technologies*, Monterey, CA, USA. Jun. 1995.

Pickin, S., Sánchez, C., Yelmo, J.-C., Gil, J., Rodriguez, E. (1996). Introducing formal notations in the development of object-based distributed applications, in Proc. of *First IFIP International Workshop on Formal Methods for Open Object-based Distributed Systems* (ed. Stefani, J.-B., Najm, E.), Chapman & Hall 1996.

Sánchez, J., León G., and Sánchez, C. (1995). Visual Analysis of Specifications, in Proc. of *21st EUROMICRO Conference* (ed. Kavanaugh, M.). IEEE Computer Society Press, 1995

Sankar S. and Hayes, R. (1994). ADL— an interface definition language for specifying and testing software. *ACM SIGPLAN notices*, 29(8):13–21, Jan. 1994.

Stefani, J.-B. (1991). ODP: the next target for the application of FDTs, in *Formal Description Techniques III* (ed. J. Quemada, J. Mañas and E. Vázquez), proc. FORTE'90. North Holland.

Simon Pickin received a Master's degree in Computer Science from Imperial College, London in 1988 having previously studied Mathematics and Theoretical Physics at Sussex University (B.Sc.), Cambridge University (Maths Tripos, Part III) and King's College, London (M.Sc.).

He spent over five years at the CNET, France Télécom's research centre, working on formal specification, protocol and telecom service engineering and particularly on verification and validation. Since 1995 he has worked as a collaborating researcher at the Departamento de Ingeniería de Sistemas Telemáticos (DIT) of the Universidad Politécnica de Madrid (UPM).

His research interests include verification and validation, formal methods and object-based distributed systems

Carlos Antonio Sánchez Tarnawiecki received the Electronics Engineering degree from the Universidad Nacional de Ingeniería (UNI) in Lima, Peru in 1980 and obtained his Masters degree in Communication Networks and Systems from the ETSIT of the UPM in 1990.

From 1981 to 1990 he worked as researcher and lecturer in the UNI Electrical and Electronic Engineering faculty. Since 1991 he has worked as a collaborating researcher in the software engineering group of the Telematic Systems Engineering Department (DIT) of the UPM.

His research interests include formal methods and their application in distributed systems, protocol engineering and telecommunications software.

Jesús Sánchez Allende received his PhD in Telecommunications Engineering from the Technical School of Telecommunications Engineering (ETSIT) of the UPM in 1994.

Since 1990 he has been working as a resercher in the software engineering group of the Telecommunications System Engineeriong Department (DIT) of the UPM. At the DIT he has been involved in R&D tasks in the fields of formal specification languages, visual formal languages, concurrent engineering, prototyping and has participated in several research projects. In 1994 he obtained a lectureship at the University Alfonso X El Sabio.

His research interests include software development environments, formal methods, open distributed processing and graphical fomalims for software design, development, analysis and testing.

Juan-Carlos Yelmo García received the Telecommunications Engineering degree from the Technical School of Telecommunications Engineering (ETSIT) of the UPM in 1990.

From 1988 to 1990 he worked at Telefónica I+D in the design and implementation of TMN communications software. In 1991 he joined the staff of the ETSIT in the DIT. He has partipated in several research projects at both national and international level and has published several papers mainly in the fields of FDT-based software development and technology transfer of software development methods.

His research interests include software development environments, software process modelling and enaction, open distributed platforms and architectures and formal description techniques.

Emilio Rodriguez Rodriguez received the Telecommunications Engineering degree in 1987 and the Specialization qualification in Switching Architectures and Data Communication in 1992 from the ETSIT of the UPM.

Since 1988 he has worked at Telefónica I+D in the development of TMN software, in the creation and maintenance of the company's software development methodology and in the provision of methodological support to software development projects. He has been active in the introduction of OO techniques in Telefonica I+D and has participated in several european collaborative projects.

His research interests include distributed object-based systems, rapid prototyping, groupware technology, workflow techniques, and the software development life-cycle.

Juan José Gil Ríos received the Telecommunications Engineering degree from the Escuela Técnica Superior de Ingenieros de Telecomunicación (ETSIT) of the UPM in 1989.

Since 1989 he has worked in Telefónica I+D in software development methodologies and in the use of formal techniques in telecommunications systems: their application in software engineering, verification and validation, their relation with object-oriented techniques and derivation of software prototypes from formal specifications.

His research interests include distributed object-based systems, formal techniques, rapid prototyping and validation.

PART FIVE

Test Generation 1

Testing deterministic implementations from nondeterministic FSM specifications

A. Petrenko [1], *N. Yevtushenko* [2], *and G. v. Bochmann* [3]

1 *CRIM, Centre de Recherche Informatique de Montréal,*
1801 Avenue McGill College, Montréal, H3A 2N4, Canada,
Phone: (514) 398-3054, Fax: (514) 398-1244
petrenko@crim.ca

2 *Tomsk State University, 36 Lenin str., Tomsk, 634050, Russia,*
yevtushenko@elephot.tsu.tomsk.su

3 *Université de Montréal,*
CP. 6128, Succ. Centre-Ville, Montréal, H3C 3J7, Canada,
Phone: (514) 343-7535, Fax: (514) 343-5834,
bochmann@iro.umontreal.ca

Abstract

In this paper, conformance testing of protocols specified as nondeterministic finite state machines is considered. Protocol implementations are assumed to be deterministic. In this testing scenario, the conformance relation becomes a preorder, so-called reduction relation between FSMs. The reduction relation requires that an implementation machine produces a (sub)set of output sequences that can be produced by its specification machine in response to every input sequence. A method for deriving tests with respect to the reduction relation with full fault coverage for deterministic implementations is proposed based on certain properties of the product of specification and implementation machines.

Keywords

Conformance testing, test derivation, fault detection, I/O nondeterministic FSMs, equivalence and reduction relations

1 INTRODUCTION

Conformance testing of protocol implementations is often formalized as the FSM equivalence problem (Moore, 1956) and (Hennie, 1964). In particular, we are given two machines defined over the same input alphabet, one is referred to as the specification machine, the other is referred to as the implementation machine. The latter is treated as a black-box, so little is usually known about the implementation machine prior to testing; yet one typically assumes an upper bound on the number of its states (Gill, 1962). It is required to determine by testing whether the two are equivalent. A corresponding test suite is said to be complete with respect to equivalence relation in the class of implementation machines within the assumed bound on the number of states. The problem of deriving such a test suite for a given deterministic specification machine has recently attracted close attention in the literature (Vasilevski, 1973), (Chow, 1978), (Sidhu and Leung, 1989), (Fujiwara, Bochmann et al. 1991), (Ural, 1992), (Bochmann and Petrenko, 1994), (Yannakakis and Lee, 1995), and (Petrenko and Bochmann, 1996). Here we take a step further addressing a more general problem of testing a so-called reduction relation between FSMs (Petrenko, Bochmann, and Dssouli, 1993) and (Petrenko, Yevtushenko, and Bochmann, 1994). Specifically, we assume that an implementation machine is deterministic, but its specification machine is not necessarily deterministic. In this case, the implementation to be conforming is required to satisfy the reduction relation, i.e. the inclusion of output traces must hold for every input trace. The classical FSM equivalence problem becomes then a special case of the FSM reduction problem. A nondeterministic machine is evidently a more versatile paradigm for describing the protocol behavior than a deterministic one. A nondeterministic machine can represent, for example, a 'loose' description of the required behavior which contains options left for the protocol implementation. Most existing protocols allow these options. The nondeterministic machine paradigm is also useful for embedded testing. As shown in (Petrenko, Yevtushenko, and Dssouli, 1994) and (Petrenko, Yevtushenko, Bochmann, and Dssouli, 1996), testing a deterministic FSM embedded within a given system of communicating FSMs can be reduced to that of an appropriate nondeterministic FSM. Thus, the test derivation problem for the reduction relation is of both, theoretical and practical interests.

Not much work, however, has been done to solve this problem. In (Petrenko, Yevtushenko, Lebedev, and Das, 1993), it is demonstrated that the problem can be solved at least for a narrow subclass of nondeterministic FSMs. In this paper, we present a refined method for test derivation based on that work and analysis of properties of the product of the specification and implementation machines. The method is now extended to cover more general FSMs. Called the 'State-Counting method' (as in the previous work), it handles an arbitrary observable FSM which can be deterministic or not, completely or partially specified, and guarantees complete fault coverage within a given class of deterministic implementations with respect to the reduction relation. We also undertake a more profound study on state distinguishability in the context of the reduction relation.

This paper is organized as follows. Section 2 contains basic notions and definitions related to the model of a nondeterministic FSM. In Section 3, we present the SC-method for deriving test suites complete w.r.t. the reduction relation. The method is then extended in Section 4 to partially specified machines. In the concluding section, we discuss further research problems.

2 PRELIMINARIES

A finite state machine (FSM), often simply called a machine throughout this paper, is an initialized observable (possibly nondeterministic) Mealy machine which can be formally defined as follows. A *finite state machine* A is a 5-tuple (S, X, Y, h, s_0), where S is a finite set of n states with s_0 as the initial state; X - a finite set of input symbols; Y - a finite set of output symbols; and h - a behavior function $h: S\times X\rightarrow \mathbb{P}(S\times Y)$, where $\mathbb{P}(S\times Y)$ is the set of all nonempty subsets of $S\times Y$, such that $|\{s' \mid (s',y) \in h(s,x)\}|\leq 1$ for all $(s,x)\in S\times X$ and all $y\in Y$. (Starke, 1972). The machine A becomes deterministic when $|h(s,x)|=1$ for all $(s,x)\in S\times X$. In a deterministic FSM, instead of the behavior function which is required for expressing a nondeterministic behavior, we use two functions: the next state function δ, and the output function λ.

We extend the behavior function to a function on the set X^* of all input sequences containing the empty sequence ε, i.e., $h: S\times X^* \rightarrow \mathbb{P}(S\times Y^*)$. For convenience, we use the same notation h for the extended function, as well, since in our discussions, this does not imply any contradiction. Assume $h(s,\varepsilon) = \{(s,\varepsilon)\}$ for all $s\in S$, and suppose that $h(s,\beta)$ is already specified. Then $h(s,\beta x) = \{ (s',\gamma y) \mid \exists s''\in S [(s'',\gamma) \in h(s,\beta) \;\&\; (s',y)\in h(s'',x)] \}$.

The function h has two projections: the first projection h^1 and the second projection h^2, where $h^1(s,\alpha) = \{ s' \mid \exists \beta\in Y^* [(s',\beta) \in h(s,\alpha)] \}$ and $h^2(s,\alpha) = \{\beta \mid \exists s' \in S [(s',\beta) \in h(s,\alpha)] \}$, for all $\alpha\in X^*$. Given $s\in S$, $\alpha\in X^*$, $\delta\in Y^*$, we use $h^1_{\delta}(s,\alpha)$ to denote a state (if exists) where the input sequence α takes the FSM A from s producing the output sequence δ.

Let $s\in S$, $\alpha\in X^*$ and $\delta\in h^2(s,\alpha)$. We say that the I/O sequence α/δ *visits* state s' from the state s if there exists a nonempty prefix β/γ of α/δ such that $h^1_{\gamma}(s,\beta) = s'$. In the case of a deterministic FSM A, we say that an input sequence α applied at the state s visits state s' of A if a non-empty prefix of β is a transfer sequence from the state s to state s'. Given states s, $s'\in S$, a sequence $\alpha\in X^*$ such that $h^1(s,\alpha)\ni s'$ is a *transfer* sequence from s to s'. If $h^1(s,\alpha)=\{s'\}$ then α is said to be a *deterministic transfer* sequence from s to s'. A state s is called *d-reachable* if there exits a deterministic transfer sequence from s_0 to s. For any state s, the empty sequence ε is a deterministic transfer sequence from s to s, therefore any FSM has at least one d-reachable state, namely the initial state. A set of (deterministic) transfer sequences from the initial state s_0 to all the (d-reachable) states of A is a (deterministic) *cover* of the FSM A. We consider here only connected machines. Given an FSM $A = (S, X, Y, h, s_0)$, A is said to be *connected* if for any state $s\in S$, there exists a transfer sequence $\alpha\in X^*$ from s_0 to s. In this paper, we use two types of covers for test derivation from an FSM. We use a traditional cover (often called a state cover set), defined as a set of transfer sequences used to take the machine A from the initial state to every its state. Note that, due to possible nondeterminism of the given machine, a single input sequence of a cover may serve as a transfer sequence for a number of states. We also use a deterministic cover V_A for the given FSM A. Obviously, in the class of deterministic machines, the two notions of a cover coincide.

To construct deterministic transfer sequences, we delete outputs from the FSM A and apply a standard technique for determinizing of the obtained automaton (Hopcroft and Ullman, 1979). A state s is a d-reachable state in A if and only if there exists a set $\{s\}$ among the states of the deterministic automaton. An input sequence that takes the automaton to the state $\{s\}$ is a deterministic transfer sequence from the initial state to the state s in the FSM A.

Given two states s of the FSM A and t of the FSM $B= (T, X, Y, H, t_0)$, state t is said to be a *reduction* of s, written $t\leq s$, if, for all input sequences $\alpha \in X^*$, the condition $H^2(t,\alpha) \subseteq h^2(s,\alpha)$ holds; otherwise t is not a reduction of s, written $t\not\leq s$. States s and t are *equivalent*, $s\cong t$, iff $s\leq t$ and $t\leq s$. On the class of deterministic machines, the relations coincide. We will also use weaker versions of equivalence and reduction relations, namely the E-equivalence and E-reduction, as well as their negations w.r.t. a given set E of input sequences, $E\subseteq X^*$; we use $\cong_E$, $\leq_E$, $\not\cong_E$ and $\not\leq_E$, to denote these relations, respectively. Given two machines, A and B, B is a reduction of A, written $B\leq A$, if the initial state of B is a reduction of the initial state of A. Similarly, the equivalence relation between machines is defined, $B\cong A$, iff $B\leq A$ and $A\leq B$.

Theorem 2.1. Given an FSM A, let B be a deterministic FSM over the same input alphabet. If $B \leq A$ then, for each state t of B, there exists a state s of A such that $t \leq s$ and, for each d-reachable state s of A, there exists a state t of B such that $t \leq s$.

Unlike to the case of deterministic FSMs, not every state of the nondeterministic FSM should correspond to some state of its reduction. However, as is established in Theorem 2.1, each d-reachable state of the FSM must have a corresponding state. Moreover, different states of the nondeterministic FSM may correspond to the same state of its reduction. The situation is similar to the case of compatible states of a partial deterministic FSM in context of state minimization (Grasselli and Luccio, 1965). We now establish necessary and sufficient conditions when two states of the FSM cannot correspond to a single state in any deterministic reduction of the FSM.

Given an FSM A, states s and r of A are said to be *r(1)-distinguishable* if there exists an input $x \in X$ such that $h^2(s,x) \cap h^2(r,x) = \varnothing$. Suppose we have determined all the pairs of r(j)-distinguishable states for $j=1,\ldots,k-1$; $k>1$. States s and r of A are said to be *r(k)-distinguishable* if they are r(j)-distinguishable, $j<k$, or there exists an input $x \in X$ such that for every output $y \in h^2(s,x) \cap h^2(r,x)$ states $h^1_y(s,x)$ and $h^1_y(r,x)$ are r(j)-distinguishable, $j<k$. Two states are said to be *r-distinguishable* if there exists an integer k such that the states are r(k)-distinguishable. Since the set S of states of A is finite there exists $k \leq C^2_n$ such that the sets of pairs of r(k)-distinguishable and r(k+1)-distinguishable states coincide. By definition, any two separable states of the FSM A, i.e. states $s,r \in S$ for which there exists an input sequence $\alpha \in X^*$ such that $h^2(s,\alpha) \cap h^2(r,\alpha) = \varnothing$ (Starke, 1972), are r-distinguishable.

Theorem 2.2. Given an FSM A and states s and r of A, let B be a deterministic FSM over the same input alphabet as A. If there exists a state of B that is a reduction of the states s and r then the states s and r are not r-distinguishable.

Proof. Let $B = (T, X, Y, \Delta, \Lambda, t_0)$. 1. If there exists a state t of B, $t \leq s$, $t \leq r$, then the states s and r are not r(1)-distinguishable.

2. Assumption of induction. Let the statement of Theorem hold for all integers less than k, $k>1$, i.e. if a state of B is a reduction of the states s and r then the states s *and* r are not r(j)-distinguishable, for each j, $j<k$.

3. Suppose now that the states s and r are r(k)-distinguishable and there exists a state t of B, $t \leq s$, $t \leq r$. Then there exists an input $x \in X$ such that for every output $y \in h^2(s,x) \cap h^2(r,x)$ states $h^1_y(s,x)$ and $h^1_y(r,x)$ are r(j)-distinguishable, $j<k$. If t is a reduction of s and r then the state $t'=\Delta(t,x)$ of B should be a reduction of the states $h^1_y(s,x)$ and $h^1_y(r,x)$, where $y=\Lambda(t,x) \in h^2(s,x) \cap h^2(r,x)$. The latter contradicts the assumption of induction. Thus, if t is a reduction of states s and r then they are not r(k)-distinguishable for any k, i.e. they are not r-distinguishable.

❑

Combining Theorem 2.2 with the results of (Damiani, 1994), we have the following fact. A state of a deterministic FSM is not a reduction of two states of the FSM A if and only if these states are r-distinguishable.

The definition of r-distinguishable states implies an inductive procedure for constructing a set of input sequences r-distinguishing two given states s and r of the FSM A. We use $W(s,r)$ to denote this set. For any deterministic FSM B over the same input alphabet as A and any state t of B, the state t is not a reduction of both states s and r w.r.t. the set $W(s,r)$. The procedure for constructing $W(s,r)$ resembles that for determining the compatibility of states in a partial deterministic FSM (Grasselli and Luccio, 1965). We omit details, due to space constraints.

Based on the determined sets $W(s,r)$ for all pairs of r-distinguishable states, we define a so-called 'r-identifier' of a state of the FSM A as a set of sequences that r-distinguish the given state from any other r-distinguishable state of A. The union of r-distinguishing sets $W(s_i,s_j)$ over all states s_j of the FSM A that are r-distinguishable with s_i is said to be a *(harmonized) r-identifier* W_i of state s_i. The case $|W_i|=1$ resembles the notion of a UIO-sequence used in literature for deterministic FSMs. The set W_i becomes empty when state s_i cannot be r-distinguished from any other state. We define a *family* of harmonized r-identifiers as the set $\{W_i| s_i \in S\}$ and further call it simply a family of r-identifiers of the FSM A. The union of r-identifiers over all states of the FSM A is said to be an *r-characterization set* W of A. It is a generalization of a classical notion of a characterization set of a deterministic machine (Kohavi, 1978).

The equivalence and reduction relations serve the conformance relations between implementations and their FSM specifications for deriving test suites. Let a specification FSM A be defined over an input alphabet X. We assume that all potential implementations are represented by a set $\Im(X,Y')$ of deterministic FSMs defined over alphabets X and Y' (sometimes called a fault domain). A universal set of all deterministic FSMs with at most m states over input alphabet X is denoted by $\Im_m(X)$.

A test suite is a finite set E of finite input sequences of the FSM A. A test suite E is said to be *complete* for A w.r.t. the reduction relation in the class $\Im(X,Y')$ iff, for all $B \in \Im(X,Y')$, $B \not\leq A$ implies $B \not\leq_E A$. A test suite is said to be m-complete for A if it is complete in the fault domain $\Im_m(X)$.

Theorem 2.3. Given a specification FSM A over alphabets X and Y, a fault domain $\Im(X,Y)$ and a complete test suite E w.r.t. the reduction relation in the class $\Im(X,Y)$, let $\alpha\beta \in E$ be an input sequence where β is an sequence of length L, such that, for each output sequence γ of A to α, the set of output sequences of A to β at the state $h^1_\gamma(s_0,\alpha)$ contains each sequence of Y^* of length L. The complete test suite E reduced by replacing the sequence $\alpha\beta$ with α is complete in $\Im(X,Y)$.

Thus, in the case where implementations are known to preserve the output alphabet of their specification, a test suite can be reduced. We will, however, consider a more general case where the fault domain is the set $\Im_m(X)$.

3 CHECKING THE REDUCTION RELATION

In this section, we give a refined version of the method for test derivation based on an early version outlined in (Petrenko, Yevtushenko, Lebedev, and Das, 1993) for a rather narrow subclass of FSMs where each state is a d-reachable and the relation of r-distinguishabilility only includes pairs of separable states. The method is now extended to cover FSMs with states which are not d-reachable. We preserve the name 'State-Counting method' (SC-method for short); the name reflects the fact that test derivation for reduction relation relies upon counting appropriate states rather than upon checking individual transitions in conventional methods for equivalence relation. Another new feature of the SC-method is that state identification is now based on a more subtle distinguishability of states which may be nonseparable. We think to have also found a more appropriate technique for presenting the main ideas of the method which helps us find new avenues for further optimizing tests with guaranteed fault coverage. Our technique is based upon properties of the product of given specification and implementation machines.

3.1 Product of FSMs

Let A be a given (possibly nondeterministic) specification FSM and B be a deterministic implementation FSM over the input alphabet of A. Suppose that the FSM B is known. Then, to verify whether or not the FSM B is a reduction of A, we construct the product $A \times B$. Its initial state is the pair of initial states of the two machines A and B, the remaining states are determined by performing a reachability analysis. For a conforming implementation machine B that is a reduction of A, the output of the FSM B belongs to a set of outputs of A for any reachable state of the product and any input. The two machines, B and $A \times B$ are equivalent. If, however, B is not a reduction of A then there exists a reachable state of $A \times B$ and an input x such that the output of B is not in the set of outputs of A. In this situation, the machines cannot agree on any common output and the product is said to produce a special output *'fail'*. If the product at state (s,t) produces the output *'fail'* to an input x then we assume that the input x takes $A \times B$ from the state (s,t) to a designated state '*Fail*', called the *fail-state*. A sequence distinguishing B from A is a transfer sequence taking the product $A \times B$ from the initial state to the fail-state.

Formally, we define a product as follows. Let $A = (S, X, Y, h, s_0)$ be a given (possibly nondeterministic) specification FSM and $B = (T, X, Y', \Delta, \Lambda, t_0)$ be a deterministic implementation FSM. We define a machine $(S \times T \cup \{Fail\}, X, Y \cup \{fail\}, \psi, \varphi, s_0t_0)$, where for all $(s,t) \in S \times T$, $x \in X$,

$\psi(st,x) = [h^1_{\Lambda(s,x)}(s,x), \Delta(s,x)]$ and $\varphi(st,x) = \Lambda(t,x)$ if $h^2(s,x) \ni \Lambda(t,x)$;

otherwise $\psi(st,x) = Fail$ and $\varphi(st,x) = fail$.

$\psi(Fail,x) = Fail$ and $\varphi(Fail, x) = fail$, for all $x \in X$.

We use Q to denote the set of all states of this machine reachable from the initial state. Then $(Q, X, Y \cup \{fail\}, \psi, \varphi, q_0)$, where $q_0 = s_0t_0$, is called the *product* $A \times B$.

Assume now that we are required to test an unknown implementation FSM B against a given specification FSM A. We only know that the FSM B belongs to a given fault domain. Suppose that we could enumerate all machines in a given fault domain. Then a test suite for the FSM A complete in the fault domain could be obtained in a straightforward manner. In particular, for each FSM B, we construct the product $A \times B$ and determine at least one input sequence that takes the product from the initial state to the fail-state, whenever B is a nonconforming implementation machine. The union of such sequences for all machines in the fault domain gives a desired test suite. Such a solution can be costly, moreover, all the machines of the fault domain are simply not possible to enumerate in a realistic situation. There is a need for another approach that does not require each possible implementation machine separately.

The idea behind such an approach is based on the existence of certain properties shared by all input sequences causing, at least once, the output *'fail'* in the product $A \times B$ for each non-conforming FSM B of the given fault domain. As we shall show, based on these properties, a complete test suite can be derived without explicitly enumerating machines of a fault domain.

3.2 *M*-complete cover of an FSM

Given an FSM $A = (S, X, Y, h, s_0)$, a set of input sequences of A is said to be an *m-complete cover* for the FSM A if it is a cover of the product $A \times B$ for any FSM $B \in \Im_m(X)$. We use C_m to denote an m-complete cover for A.

Lemma 3.1. Given an FSM A and an m-complete cover C_m for A, the set C_m is an m-complete test suite for A.

Given the FSM A with n states and any $B\in\Im_m(X)$, the number of states in the product $A\times B$ does not exceed $mn+1$ and any state of this machine is reachable from its initial state by an input sequence whose length does not exceed mn. Thus, the set X^{mn} of all input sequences of length up to mn, is an m-complete cover for the FSM A with n states and, according to Lemma 3.1, an m-complete test suite for the FSM A (Petrenko, Yevtushenko, Lebedev, and Das, 1993).

Given a set of states $P\subseteq Q$ of the product $A\times B$ and state q', a sequence α is a *transfer sequence from P to q'*, if there exists a state $q\in P$ such that α is a transfer sequence from q to q'. If the length of the transfer sequence α from P to q' does not exceed that of any other sequence from P to q' then α is said to be a *minimal* transfer sequence from P to q'.

Let $\hat{S}$ be the set of all d-reachable states of A and V_A be a deterministic cover of the FSM A such that $|V_A|=|\hat{S}|$. We use $P(V_A)$ to denote the set of states where the sequences of V_A take the product $A\times B$ from the initial state q_0. The set $P(V_A)$ contains $|\hat{S}|$ states. Let also $\alpha_i\in V_A$ denote a deterministic transfer sequence of the FSM A from the initial state to a d-reachable state s_i. Since the product has at most $mn+1$ states, length of a minimal transfer sequence from the set $P(V_A)$ to any reachable state of the product does not exceed $mn-|P(V_A)|+1 = mn-|\hat{S}|+1$. Therefore, the union of the sets $\alpha_i X^{mn-|\hat{S}|+1}$ over all sequences $\alpha_i\in V_A$ is a cover of the product machine $A\times B$. It is also an m-complete cover for the FSM A, since B is an arbitrary FSM of the set $\Im_m(X)$.

Theorem 3.2. Given an FSM A, let V_A be a deterministic cover, and $\hat{S}$ be the set of all d-reachable states of A. Then the set $V_A X^{mn-|\hat{S}|+1}$ is an m-complete test suite for the FSM A.

A test suite of Theorem 3.2 can often be reduced by deleting certain sequences from the set $X^{mn-|\hat{S}|+1}$. Let $B\in\Im_m(X)$. Given an input sequence β, if among the states, visited by the transfer sequence β from a certain state of the set $P(V_A)$ to the fail-state of the product, either a state of the set $P(V_A)$ occurs or one state appears more than once, then the sequence β is not a minimal transfer sequence from $P(V_A)$ to the fail-state and can therefore be reduced. Based on this property of minimal transfer sequences from the set $P(V_A)$, we can construct an m-complete test suite for A as follows.

For any d-reachable state $s_j\in\hat{S}$, we determine the traversal set $C_m(s_j)$ of input sequences as follows. An input sequence β is included into the set $C_m(s_j)$ if, for each sequence $\gamma\in h^2(s_j,\beta)$, there exists a d-reachable state $s\in\hat{S}$ visited by β/γ exactly m times from the state s_j or there exists a state $s\in S\backslash\hat{S}$ visited by β/γ exactly $m+1$ times while, for any proper prefix β' of β, there exists $\gamma'\in h^2(s_j,\beta')$ such that the property does not hold for the sequence β'/γ'.

We use $\alpha_j C_m(s_j)$ to denote the result of concatenation of a sequence $\alpha_j\in V_A$ that takes the FSM A from the initial state to the state $s_j\in\hat{S}$, with all sequences of the set $C_m(s_j)$.

Theorem 3.3. Given an FSM A, a deterministic cover V_A of A, the union E of sets $\alpha_j C_m(s_j)$ over all $\alpha_j\in V_A$ is an m-complete test suite for A.

Proof. Let $B\in\Im_m(X)$ be a deterministic FSM and $P(V_A)$ be a set of states where the sequences from the set V_A take the product $A\times B$ from the initial state. If the state $Fail\in P(V_A)$ then an appropriate sequence $\alpha\in V_A$ is a transfer sequence from the initial state to the state

Fail, and by construction, the state *Fail* is visited by an appropriate sequence of the set E. Assume then that $Fail \notin P(V_A)$. Let an input sequence βx applied at some state $(s_j,t_j) \in P(V_A)$ be a minimal transfer sequence from $P(V_A)$ to *Fail*, i.e. the sequence $\alpha_j \beta x$ takes the product $A \times B$ from the initial state to the fail-state and γ be the output sequence of B to β at the state t_j. Since βx is a minimal transfer sequence from $P(V_A)$ to the state *Fail*, the pair β/γ is an I/O sequence of A at the state s_j. Moreover, the sequence β applied at the state $(s_j,t_j) \in P(V_A)$ is a minimal transfer sequence from $P(V_A)$ to the state q of $A \times B$, where β takes the product machine from the state (s_j,t_j).

Suppose that the sequence β/γ applied at the state s_j visits l times a state $s \in \hat{S}$ of the FSM A and $l \geq m$. In this case, the sequence β applied at the state (s_j,t_j) visits l states $(s,t_1),\ldots,(s,t_l)$ of the product $A \times B$. Since the FSM B has at most m states and the set $P(V_A)$ contains a pair (s,t) for an appropriate state t of the FSM B, among these states either a state from the set $P(V_A)$ occurs or at least one state appears more than once. In the both cases, the sequence β is not a minimal transfer sequence from $P(V_A)$ to q.

Similar to this, β is not a minimal transfer sequence from $P(V_A)$ to q if β visits l times a state $s \in S \setminus \hat{S}$ of the FSM A and $l \geq m+1$. Thus, the sequence β is a proper prefix of an appropriate sequence of the set $C_m(s_j)$ and there exists a sequence $\alpha_j \alpha \in E$ with a prefix $\alpha_j \beta x$ that visits the state *Fail* from the initial state of the product $A \times B$.

□

Compared to Theorem 3.2, Theorem 3.3 offers a more economical way of constructing an m-complete test suite. To assure that a set of input sequences of the specification FSM A visits the state *Fail* of the product $A \times B$ for any $B \in \Im_m(X)$ we include in the traversal set $C_m(s_j)$ each input sequence if there may exist a product machine $A \times B$, $B \in \Im_m(X)$, such that the sequence may turn to be a minimal transfer sequence from $P(V_A)$ to *Fail*. A sequence $\beta \in C_m(s_j)$ is expanded by appending all inputs until it visits an appropriate state s of the FSM A m or $m+1$ times for each output sequence of A to β at the state s_j. The size of the traversal sets $C_m(s_j)$ is exponential and it is, therefore, important to determine cases where their sequences can be terminated as early as possible. For a specification FSM such that none of its states are r-distinguishable and no state is a reduction of any other state, it seems nearly impossible to reduce the size of the traversal sets. Certain sufficient conditions enforcing an earlier termination of sequences of the traversal sets $C_m(s_j)$ can be established provided that a given specification FSM A has r-distinguishable states.

3.3 Reducing traversal sets

Let an FSM A have states, say, s_1 and s_2, r-distinguished by a set $W(s_1,s_2)$ of input sequences. Given an FSM B, let the product $A \times B$ have states (s_1,t) and (s_2,t) for an appropriate state t of the FSM B. Then we refer to these states as to *conflicting* states. Due to the properties of the set $W(s_1,s_2)$, there exists an input sequence $\alpha \in W(s_1,s_2)$ such that the output response of B to the input sequence α at the state t is not in the set of output sequences of the FSM A to α at least at one of states s_1 and s_2. Thus, the input sequence α takes the product $A \times B$ at least from one of the states (s_1,t) and (s_2,t) to the fail-state. In other words, if a transfer sequence β applied at some state of the product visits the two conflicting states, the sequence α applied at (s_1,t) or (s_2,t) could be used as a shortcut to reach the fail-state in this product.

We now analyze a string of states visited by a minimal transfer sequence β from the set of states $P(V_A)$ to the fail-state of the product $A \times B$ and establish sufficient conditions when the set of states visited by the β, along with states of the set $P(V_A)$, contains conflicting states.

Lemma 3.4. Given FSMs A and B, $B \in \Im_m(X)$, a set D of pairwise r-distinguishable states of A together with its subset $\hat{D}$ of d-reachable states, let an input sequence βv, $v \neq \varepsilon$, applied at some state $(s,t) \in P(V_A)$ be a minimal transfer sequence from $P(V_A)$ to the fail-state of $A \times B$ and γ be the output response of B to the sequence β applied at the state t. If the I/O sequence β/γ, applied at the state s of A, visits states of D more than $m-|\hat{D}|$ times then the set of states visited by the β, applied at the state (s,t) of $A \times B$, together with the states of $P(V_A)$, contains conflicting states (s_1,t') and (s_2,t'), $s_1,s_2 \in D$.

Proof. Let $B=(T, X, Y, \Delta, \Lambda, t_0) \in \Im_m(X)$, and the sequence βv applied at a state $(s,t) \in P(V_A)$ be a minimal transfer sequence from $P(V_A)$ to the fail-state of $A \times B$ and $\gamma = \Lambda(t,\beta)$. Then the pair β/γ is an I/O sequence of the FSM A. Suppose that the I/O sequence β/γ applied at the state s of A visits l times states of D and $l > m-|\hat{D}|$. In this case, the sequence β applied at the state (s,t) traverses l states $(s_1,t_1),\ldots,(s_l,t_l)$ of the product, where $s_i \in D$, $i=1,\ldots,l$. Since the FSM B has at most m states, among the states $(s_1,t_1),\ldots,(s_{m-|\hat{D}|+1},t_{m-|\hat{D}|+1})$ combined with states of the set $\{(s',t')|\ s' \in \hat{D}\} \subseteq P(V_A)$, there can be at most m states with pairwise distinct states of the FSM B. The set $P(V_A)$ contains at least $|\hat{D}|$ distinct pairs (s',t'), $s' \in D$. Thus, at least two states in the union of the sets $\{(s_i,t_i)|\ i=1,\ldots,m-|\hat{D}|+1\}$ and $\{(s',t'),\ s' \in \hat{D}\} \subseteq P(V_A)$ have the same state of B. Because of β being a minimal transfer sequence from $P(V_A)$ to the fail-state, the corresponding two states of the product cannot coincide. Thus, among the states (s_1,t_1), ..., $(s_{m-|\hat{D}|+1},t_{m-|\hat{D}|+1})$ visited by β and states of the set $P(V_A)$, there exist two distinct states (s_1,t') and (s_2,t'), $s_1,s_2 \in D$, for an appropriate state t' of the FSM B.

❑

Let β be an input sequence. We denote $\Im_m(\alpha_j\beta)$, where $\alpha_j \in V_A$, the set of all implementation FSMs $B \in \Im_m(X)$ such that an input sequence with the prefix β applied at the state $(s_j,t_j) \in P(V_A)$ is a minimal transfer sequence from $P(V_A)$ to the fail-state of the product $A \times B$. Based on Lemma 3.4, the following statement can be established.

Lemma 3.5. Given FSM A, an input sequence β, an r-characterization set W and a d-reachable state s_j of A, let for each $\gamma \in h^2(s_j,\beta)$, there exists a set D of pairwise r-distinguishable states of A such that the I/O sequence β/γ, applied at the state s_j of A, visits states of D more than $m-|\hat{D}|$ times, where $\hat{D}$ is the subset of d-reachable states of D. Then the union of the sets $\alpha_i W$ over all $\alpha_i \in V_A$, and the sets $\alpha_j\beta' W$ over all nonempty prefixes β' of β is a test suite complete for A in the class $\Im_m(\alpha_j\beta)$.

The lemma states that replacing an exponential expansion (Theorem 3.3.) of an input sequence for which the conditions of Lemma 3.5 hold, by a certain polynomial set of input sequences preserves the fault coverage. Thus, an m-complete test suite can now be obtained as a union of corresponding sets over all d-reachable states of A and input sequences satisfying Lemma 3.5. However, a test suite complete in the class $\Im_m(\alpha_j\beta)$ can often be reduced if we use a family of state r-identifiers instead of an r-characterization set W. The procedure for deriving a test suite $T_m(\alpha_j\beta)$ complete for A in the class $\Im_m(\alpha_j\beta)$ for the sequence β satisfying Lemma 3.5, includes the following steps.

1. Find a deterministic cover V_A of the FSM A.
2. For each $\alpha_i \in V_A$, that takes the FSM A from the initial state to a d-reachable state s_i, concatenate α_i with every sequence of the set W_i. Let E be the union of obtained sets over all $\alpha_i \in V_A$.
3. For each nonempty prefix β' of the sequence β determine $h^1(s_0,\alpha_j\beta')$. Then concatenate $\alpha_j\beta'$ with all sequences of every W_i, $s_i \in h^1(s_j,\alpha_j\beta')$. Let $\alpha_j(\beta @\{W_i | s_i \in S\})$ denote the result of this concatenation.
4. Find the union $T_m(\alpha_j\beta)$ of E and $\alpha_j(\beta @\{W_i | s_i \in S\}$.

Theorem 3.6. Given an FSM A, let $T_m(\alpha_j\beta)$ be the set of input sequences derived from A by the above procedure. Then the set $T_m(\alpha_j\beta)$ is a complete test suite for the FSM A in the class $\Im_m(\alpha_j\beta)$.

Proof. Let $B=(T, X, Y, \Delta, \Lambda, t_0) \in \Im_m(\alpha_j\beta)$, the sequence $\beta\nu$ applied at a state $(s_j,t_j) \in P(V_A)$ be a minimal transfer sequence from $P(V_A)$ to the fail-state of $A \times B$ and $\gamma = \Lambda(t,\beta)$. If $\nu = \varepsilon$ then $Fail \in P(V_A)$. If the state $Fail \in P(V_A)$ then an appropriate sequence $\alpha \in V_A$ is a transfer sequence from the initial state to the state $Fail$. Let then $\nu \neq \varepsilon$. Then $\beta\gamma$ is an I/O sequence of A. The sequence β satisfies the conditions of Lemma 3.5; therefore, there exists a set D of pairwise r-distinguishable states of A such that the I/O sequence β/γ, applied at the state s_j of A, visits states of D more than $m-|\hat{D}|$ times. Due to Lemma 3.4, among the states visited by β applied at state (s_j,t_j) and states of the set $P(V_A)$, there exist two distinct states (s_1,t) and (s_2,t), $s_1,s_2 \in D$, for an appropriate state t of the FSM B. Thus, among sequences V_A and sequences $\alpha_j\beta'$, where β' is a nonempty prefix of β, there exist sequences α' and α'' that take the product to the states (s_1,t) and (s_2,t), where $s_1,s_2 \in D$. The states s_1 and s_2 of A are r-distinguished by an appropriate sequence $\delta \in W_1 \cap W_2$ and, by construction, $\alpha'\delta, \alpha''\delta \in T_m(\alpha_j\beta)$. Thus, at least one of the two sequences, $\alpha'\delta$ or $\alpha''\delta$, takes the product to the fail-state.

❑

3.4 The SC-Method

Based on Theorem 3.6, an m-complete test suite can now be derived as the union of test suites complete in classes $\Im_m(\alpha_j\beta)$ over all sequences $\alpha_j \in V_A$ and all input sequences β such that, for d-reachable state s_j of A and each sequence $\gamma \in h^2(s_j,\beta)$, there exists a set D of pairwise r-distinguishable states of A such that the I/O sequence β/γ, applied at the state s_j of A, visits states of D more than $m-|\hat{D}|$ times. The SC-method for constructing an m-complete test suite includes the following steps.

1. Find a deterministic cover V_A of the FSM A.

2. Find all pairs of r-distinguishable states of A and determine all maximal sets $D_1,...,D_k$ of pairwise r-distinguishable states. For each D_r, $r=1,...,k$, find a maximal subset $\hat{D}_r$ of d-reachable states.
3. Construct a family $\{W_i | s_i \in S\}$ of r-identifiers of A.
4. For any d-reachable state s_j, derive the traversal set $Tr_m(s_j)$ as follows. An input sequence β is included into the set $Tr_m(s_j)$ if, for each sequence $\gamma \in h^2(s_j,\beta)$, there exists a set D_r such that its states are visited by β/γ exactly $(m-|\hat{D}_r|+1)$ times from the state s_j.
5. For each traversal set $Tr_m(s_j)$ and every sequence $\beta \in Tr_m(s_j)$, construct the test suite $T_m(\alpha_j\beta)$ complete in the class $\Im_m(\alpha_j\beta)$, by use of the above given procedure (Section3.3).
6. Find the union E of $T_m(\alpha_j\beta)$ for all $\alpha_j \in V_A$ and $\beta \in Tr_m(s_j)$ (note that each sequence that is a prefix of another sequence can be deleted from E to simplify the result).

Theorem 3.7. Given an FSM A, let E be the set of input sequences derived from A by the SC-method. Then the set E is an m-complete test suite for the FSM A.

Proof. Consider an m-complete test suite $V_A X^{mn-|\hat{S}|+1}$ from Lemma 3.2. Let $B \in \Im_m(X)$, and the sequence v, $v \in X^{mn-|\hat{S}|+1}$, applied at some state (s_j,t_j) be a minimal transfer sequence from $P(V_A)$ to the fail-state of $A\times B$. Determine a minimal prefix β of v such that $\beta \in Tr_m(s_j)$. Due to Theorem 3.6, a test suite $T_m(\alpha_j\beta) \subseteq E$ contains an input sequence that takes the product $A\times B$ to the fail-state.

❑

Example. We consider the FSM A shown in Figure 1. State 3 cannot be deterministically reached from the initial state 1, all the other states are d-reachable. We choose a minimal d-reachable state cover set $V_A=\{\varepsilon, a, ab\}$, the empty sequence ε serves a transfer sequence for the initial state, a for state 2, and ab for state 4. Next, we check whether the states are r-distinguishable. The sequence a r-distinguishes states 2 and 3; the sequence aa r-distinguishes states 1 and 2; aaa - states 1 and 3. States 2 and 4 are r-distinguished by the sequence b; the states 3 and 4 - by the sequence bb. States 1 and 4 are not separable but they are r-distinguished by the set $\{aaa, ab\}$ of input sequences. In fact, there are two common output responses x and y of A to the input a at the states 1 and 4. The I/O sequence a/x takes the FSM A from the states 1 and 4 to the states 2 and 4 which are separated by the input sequence b while the I/O sequence a/y takes the FSM A from the states 1 and 4 to the states 2 and 1, respectively, which are r-distinguished by the input sequence aa.

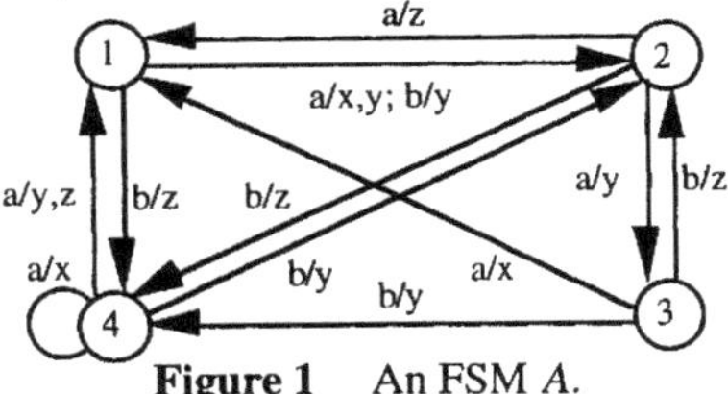

Figure 1 An FSM A.

States 1,2, 3, and 4 form a single maximal set D of pairwise r-distinguishable states and $W_1 = \{aaa, ab\}$, $W_2 = \{aa, b\}$, $W_3 = \{a, bb\}$, and $W_4 = \{aaa, ab, bb\}$. State 3 is not a d-reachable state, so the subset of pairwise r-distinguishable d-reachable states is $\hat{D} = \{1,2,4\}$. We assume that the number of states in any implementation is at most four (m=4) and proceed by determining traversal sets for d-reachable states. The termination rule for expanding input sequences becomes m-$|\hat{D}|$+1 = 4-3+1=2, in other words, states from D should be visited twice before any input sequence can be terminated. Since D contains all the states of the FSM A, it is required to apply X^2 at each d-reachable state, thus $Tr_4(i)=\{a,b\}^2$, for each i=1,2,4. The union of complete test suites $T_4(\alpha_i\beta)$ over all sequences $\alpha_j\beta$, $\alpha_j \in V_A$, $\beta \in Tr_4(j)$ is an m-complete test suite (m=4). As an example, consider the sequence $aab \in a\{a,b\}^2$. One can assure that $h^1(1,a)$ = {2}, and so the sequence a should be concatenated with W_2; $h^1(1,aa)$ = {1,3} and the sequence aa is concatenated with $W_1 \cup W_3$; $h^1(1,aab)$ = {1,2}, thus, the sequence aab is concatenated with $W_1 \cup W_2$.

4 EXTENSION TO PARTIALLY SPECIFIED MACHINES

The model of partially specified finite state machines is useful for describing the behavior of systems where transitions out of certain states on some inputs are not defined, these are 'don't care' transitions. Implementation machines are usually assumed to be completely specified. Implementing a partial specification amounts to completing it in a certain way. The model defined in Section 2 is, in fact, a completely specified (complete) finite state machine. Now we formally define a partial FSM (PFSM) and generalize the reduction and equivalence relations.

A *partial* finite state machine A is an observable partial FSM, i.e. 6-tuple (S, X, Y, h, s_0, D_A), where S is a set of states with s_0 as the initial state; X - a finite set of input symbols; Y - a finite set of output symbols; D_A - a specification domain, $D_A \subset S \times X$; and h - a behavior function h: $D_A \rightarrow \mathbb{P}(S \times Y)$ such that $|\{s' \mid (s',y) \in h(s,x)\}| \leq 1$ for all $(s,x) \in D_A$ and all $y \in Y$. Replacing D_A by $S \times X$, we obtain a complete FSM.

Any I/O sequence specified in an observable machine takes it from its initial state to a unique state. However, in a nondeterministic machine, a specified input sequence may lead to several states. Generally speaking, these states may have different unspecified inputs. Here, we restrict ourselves to a class of machines with so-called harmonized traces (Petrenko, Yevtushenko, Lebedev, and Das, 1993). States of such a machine once reached from the initial state with the same specified transfer sequence have the same set of specified (unspecified) inputs. Figure 2 shows an example of a partial machine with harmonized traces. Each input sequences specified at the initial state of such machine does not execute any 'don't care' transition. We use X_A^* to denote the set of all sequences specified at the initial state.

To test a machine against its PFSM specification, we have to compare the I/O behaviors of a complete implementation FSM $B = (T, X, Y, H, t_0)$ and a partial specification FSM A (S, X, Y, h, s_0, D_A).

An FSM B is a *quasi-reduction* of a PFSM A, written $B \leq_{quasi} A$, iff for all input sequences $\alpha \in X_A^*$ the condition $H^2(t_0,\alpha) \subseteq h^2(s_0,\alpha)$ holds; otherwise $B \not\leq_{quasi} A$.

An FSM B is *quasi-equivalent* to a PFSM A, written $B \equiv_{quasi} A$, iff for all input sequences $\alpha \in X_A^*$ the condition $H^2(t_0,\alpha) = h^2(s_0,\alpha)$ holds; otherwise $B \not\equiv_{quasi} A$. This relation originates from the quasi-equivalence relation introduced in (Gill, 1962) for deterministic machines which corresponds to a so-called weak conformance (Sidhu and Leung, 1989), (Yannakakis and Lee, 1995). On the class of deterministic machines, quasi-reduction and quasi-equivalence coincide.

According to definitions of quasi-equivalence and quasi-reduction relations, deriving test suites, we should omit input sequences on which the behavior of the specification machine is not defined. Thus, all complete test suites can be determined as subsets of the set X_A^*. With this exception, the definitions of complete test suites for partial machines repeat that for complete machines.

A partial machine A with harmonized traces can often be treated as a special complete nondeterministic machine $\tilde{A}$ by treating its transitions on unspecified inputs as 'don't care' transitions to a *trap* state (Unger, 1969). Such transitions are labeled with an input not specified at the current state of A and all outputs of some superset of Y. The superset Y' of Y represents all outputs in the class of implementation machines. The trap state has looping transitions labeled with all inputs in X and all outputs. Input sequences leading $\tilde{A}$ into the trap state are sequences not specified in A, they constitute the set $X^*\backslash X_A^*$. The machine $\tilde{A}$ is said to be the *completed form* of A. The completed form of a PFSM reflects a rather general completeness assumption, namely 'undefined by default', used in protocol testing (Petrenko, Bochmann, and Dssouli, 1993). Figure 2 shows an example. Here 'any' stands for an arbitrary output in Y', a 'black hole' represents a trap state. The completed form is necessary nondeterministic even when a given machine is deterministic.

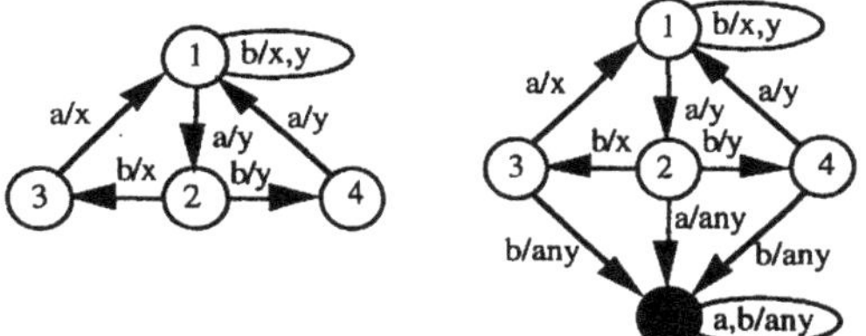

Figure 2 A PFSM A with harmonized traces and its completed form.

By construction, the completed form $\tilde{A}$ of a PFSM A with harmonized traces is a machine that is quasi-equivalent to A. In fact, for every input sequence specified in A, the sets of output sequences produced by their initial states coincide, i.e. $\tilde{A}\cong_{quasi}A$. (It is not necessarily true for an arbitrary partial nondeterministic machine). Every deterministic machine has harmonized traces, therefore, the problem of test derivation from a partial deterministic FSM w.r.t. the quasi-equivalence relation and that for a complete nondeterministic FSM w.r.t. the reduction relation become equivalent problems. We have an even more general fact.

Theorem 4.1. Let $\tilde{A}$ be the completed form of a partial machine A with harmonized traces. Assume that an FSM B over the same input and output alphabets is complete and deterministic. Then $B\leq_{quasi}A$ if and only if $B\leq\tilde{A}$.

Proof. I. $B\leq_{quasi}A \Rightarrow B\leq\tilde{A}$. Let $A = (S, X, Y, h, s_0, D_A)$ be a partial nondeterministic machine, $\tilde{A} = (S, X, Y, \tilde{H}, s_0, D_A)$ be its completed form, and $B = (T, X, Y, \delta, \lambda, t_0)$ be a complete deterministic machine. Assume that $B\leq_{quasi}A$, but $B\not\leq\tilde{A}$. In this case, there exists an input sequence α such that $\lambda(t_0,\alpha) \notin \tilde{H}^2(s_0,\alpha)$. By virtue of definition of the completed form, $\tilde{A}\cong_{quasi}A$, that is $\tilde{H}^2(s_0,\beta) = h^2(s_0,\beta)$ for all input sequences $\beta\in X_A^*$ Assuming $\alpha\in X_A^*$ leads us to a contradiction, as $\lambda(t_0,\alpha) \notin h^2(s_0,\alpha)$ and B is not a quasi-reduction of A. Suppose therefore that $\alpha\notin X_A^*$ and $\alpha= \beta\gamma$, where $\beta\in X_A^*$ and $\gamma\in X^*$. A has harmonized traces, then its completed form $\tilde{A}$ in each state of the set $\tilde{H}^1(s_0,\alpha)$ produces in response to γ all output sequences of the length of γ. Then $\lambda(t_0,\beta) \in \tilde{H}^2(s_0, \beta)$, and $\lambda(\delta(t_0,\beta),\gamma) \in \tilde{H}^2(s,\gamma)$ for any $s\in \tilde{H}^1(s_0,\beta)$. This again leads us to a contradiction.

II. $B \le \tilde{A} \Rightarrow B \le_{quasi} A$. Assume that $B \le \tilde{A}$, but $B \not\le_{quasi} A$. If B is not a quasi-reduction of A then there exists an input sequence $\alpha \in X_A^*$ such that $\lambda(t_0,\alpha) \not\in h^2(s_0,\alpha)$. $\tilde{A} \cong_{quasi} A$, it means that for all input sequences $\alpha \in X_A^*$ the condition $h^2(s_0,\alpha) = \tilde{H}^2(s_0,\alpha)$ holds. Thus, $\lambda(t_0,\alpha) \not\in \tilde{H}^2(s_0,\alpha)$. A contradiction.

❑

Based on this theorem, the problem of test derivation from a partial FSM with harmonized traces w.r.t. the quasi-reduction relation can be reduced to that from its completed form w.r.t. reduction relation. The SC-method serves this purpose. It follows, however, from our discussions that the trap state does not require any identification (anyway, every other state is a reduction of the trap state) neither should transitions into the trap state be covered by a test suite. In other words, we have the following fact as a corollary to the above theorem.

Corollary. Let E be a complete test suite for the completed form $\tilde{A}$ of a partial machine A with harmonized traces for the reduction relation in the class of deterministic implementation machines. Then $E \cap X_A^*$ is a complete test suite for A w.r.t. the quasi-reduction relation in the same class of implementations.

Note that constructing a complete test suite exclusively from the set X_A^* of specified input sequences becomes essential in situations where undefined transitions are treated as 'forbidden' transitions, as explained in (Yevtushenko and Petrenko, 1990), (Petrenko, 1991), (Petrenko and Yevtushenko, 1992), (Luo, Petrenko, and Bochmann, 1994), and (Yannakakis and Lee, 1995). The difference from the latter work is that we consider here a wider class of partial machines that are not necessarily reduced. (Yannakakis and Lee, 1995) gives no solution for partial machines with compatible, i.e. indistinguishable states, but our method is fully applicable to such machines.

5 CONCLUSION

We have presented a refined version of the test derivation method (SC-method) which, for a given FSM, generates a test suite in the context of the reduction relation. The SC-method is proven to provide full fault coverage on the pre-determined class of deterministic implementations. It can be applied to various classes of specification FSMs, including partially specified machines with compatible states provided that they are observable. This limitation is by no means prohibitive, as any FSM with harmonized traces has an equivalent observable form. Our method follows a new principle of constructing test sequences, namely counting appropriate states visited by test sequences, unlike conventional methods that strictly follow the transition checking principle.

Next step in this direction would be to further elaborate the proposed approach taking into account, for example, that the reduction relation may hold between a number of states in a given specification machine, all these states can correspond to a single state of an implementation FSM. It is also interesting to establish which states of a specification machine (along with d-reachable states) should have a corresponding state in an implementation FSM.

Acknowledgments

This research was partly supported by an NSERC strategic research grant and by the Russian Found for Fundamental Research. The authors would like to thank the anonymous referees for their comments that helped improve the presentation of this paper.

6 REFERENCES

Bochmann, v.G. and Petrenko, A. (1994) Protocol testing: review of methods and relevance for software testing. *ISSTA'94 ACM International Symposium on Software Testing and Analysis*. pp. 109-124

Chow, T. S. (1978) Testing software design modeled by finite-state machines. *IEEE Transactions on Software Engineering*, Vol. SE-4, No 3, pp. 178-187.

Damiani, M. (1994) Nondeterministic finite-state machines and sequential don't cares. *Proceedings of the European Conference on Design&Test,* pp. 192-198.

Fujiwara, S., Bochmann, v.G., Khendek, F., Amalou, M., and Ghedamsi, A. (1991) Test selection based on finite state models. *IEEE Transactions on Software Engineering,* Vol. SE-17, No. 6, pp. 591-603.

Gill, A. (1962) *Introduction to the theory of finite-state machines,* NY, McGraw-Hill, 270p.

Grasselli, A. and Luccio, F. (1965) A method for minimizing the number of internal states in incompletely specified sequential networks. *IEEE Transactions on Electronic Computers,* No. 6, pp. 350-359.

Hennie, F. C. (1964) Fault detecting experiments for sequential circuits. *Proceedings of the IEEE 5th Ann. Symp. on Switching Circuits Theory and Logical Design,* pp. 95-110.

Hopcroft, J.E., and Ullman, J.D. (1979) Introduction to automata theory. languages and computation. Addison-Wesley Publishing Company, Inc., 418 p.

Kohavi, Z. (1978) *Switching and finite automata theory,* N.Y., McGraw-Hill.

Luo, G., Petrenko, A., and Bochmann, v.G. (1994) Selecting test sequences for partially-specified nondeterministic finite state machines. *Protocol Test Systems VII (the Proceedings of IFIP WG 6.1 International Workshop on Protocol Test Systems 1994),* Chapman & Hall, 1995, pp. 95-110.

Moore, E. F. (1956) Gedanken-experiments on sequential machines, *Automata Studies,* Princeton University Press, Princeton, NJ, pp. 129-153.

Petrenko, A. (1991) Checking experiments with protocol machines. *IFIP Transactions, Protocol Test Systems, IV (the Proceedings of IFIP TC6 Fourth International Workshop on Protocol Test Systems, 1991),* 1992, North-Holland, pp. 83-94.

Petrenko, A. and Yevtushenko, N. (1992) Test suite generation for a fsm with a given type of implementation errors. *IFIP Transactions Protocol Specification, Testing, and Verification XII (the Proceedings of IFIP TC6 12th International Symposium on Protocol Specification, Testing, and Verification 1992),* pp. 229-243.

Petrenko, A., Bochmann, v.G., and Dssouli, R. (1993) Conformance relations and test derivation. *IFIP Transactions Protocol Test Systems VI (the Proceedings of IFIP TC6 Fifth International Workshop on Protocol Test Systems,1993)* North-Holland, 1994, pp. 157-178.

Petrenko, A., Yevtushenko, N., Lebedev, A., and Das, A. (1993) Nondeterministic state machines in protocol conformance testing. *IFIP Transactions Protocol Test Systems VI (the Proceedings of IFIP TC6 Fifth International Workshop on Protocol Test Systems 1993),* North-Holland, 1994, pp. 363-378.

Petrenko, A., Yevtushenko, N., and Dssouli, R. (1994) Testing strategies for communicating fsms. *Protocol Test Systems VII (the Proceedings of IFIP WG 6.1 International Workshop on Protocol Test Systems, 1994)* Chapman & Hall, 1995, pp. 193-208.

Petrenko, A., Yevtushenko, N., and Bochmann, v.G. (1994) Experiments on nondeterministic systems for the reduction relation. *Technical Report 932, Université de Montréal*, 23p.

Petrenko, A. and Bochmann, v.G. (1996) On fault coverage of tests for finite state specifications. To appear in a special issue on Protocol Testing of *Computer Networks and ISDN Systems*.

Petrenko, A., Yevtushenko, N., Bochmann, v.G., and Dssouli, R. (1996) Testing in context: framework and test derivation. *Technical Report 1011, Université de Montréal,* To appear in a special issue on Protocol Engineering of *Computer Communications Journal*.

Starke, P. H. (1972) *Abstract automata.* North-Holland/American Elsevier, 419p.
Sidhu, D. P. and Leung, T. K. (1989) Formal methods for protocol testing: a detailed study. *IEEE Transactions on Software Engineering*, Vol SE-15, No 4, pp. 413-426.
Vasilevski, M. P. (1973) Failure diagnosis of automata. *Cybernetics*, Plenum Publishing Corporation, New York, No 4, pp. 653-665.
Unger, S. H. (1969) *Asynchronous sequential switching circuits* Wiley-Interscience.
Ural, H. (1992) Formal methods for test sequence generation. *Computer Communications*, Vol. 15, No. 5, pp. 311-325.
Yannakakis, M. and Lee, D. (1995) Testing finite state machines: fault detection. *Journal of Computer and System Sciences*, 50, pp. 209-227.
Yevtushenko, N. and Petrenko, A. (1989) Fault-detection capability of multiple experiments. *Automatic Control and Computer Sciences*, Allerton Press, Inc., N.Y., Vol. 23, No. 3, pp. 7-11.
Yevtushenko, N. and Petrenko, A. (1990) A method of constructing a test experiment for an arbitrary deterministic automaton. *Automatic Control and Computer Sciences*, Allerton Press, Inc., N.Y., Vol. 24, No. 5, pp. 65-68.

7 BIOGRAPHY

Alexandre Petrenko received the Diploma degree in electrical and computer engineering from Riga Polytechnic Institute in 1970 and the Ph.D. in computer science from the Institute of Electronics and Computer Science, Riga, USSR, in 1974. In 1996, he has joined CRIM, Centre de Recherche Informatique de Montréal, Canada. From 1992 to 1996, he was a visiting professor/researcher of the Université de Montréal. From 1982 to 1992, he was the head of a research department of the Institute of Electronics and Computer Science in Riga. From 1979 to 1982, he was with the Networking Task Force of the International Institute for Applied Systems Analysis (IIASA), Vienna, Austria. From 1969 to 1979, he was a researcher and the head of a research department of the Institute of Electronics and Computer Science in Riga. His current research interests include high-speed networks, communication software engineering, formal methods, conformance testing, and testability.

Nina Yevtushenko received the Diploma degree in radio-physics in 1971 and Ph. D. in computer science in 1983, both from the Tomsk State University, Russia. She is currently a Professor at that University. Her research interests include the automata and FSM theory and testing problems.

Gregor v. Bochmann received the Diploma degree in physics from the University of Munich, Munich, West Germany, in 1968 and the Ph.D. degree from McGill University, Montreal, P.Q., Canada, in 1971. He has worked in the areas of programming languages, compiler design, communication protocols, and software engineering and has published many papers in these areas. He holds the Hewlett-Packard-NSERC-CITI chair of industrial research on communication protocols in the Département d'informatique et de recherche opérationnelle, Université de Montréal. His present work is aimed at design methods for communication protocols and distributed systems. He has been actively involved in the standardization of formal description techniques for OSI. He is presently one of the scientific directors of the Centre de Recherche Informatique de Montréal (CRIM).

11

An approach for testing asynchronous communicating systems

Myungchul Kim[*], *Samuel T. Chanson*[**], *Sungwon Kang*[*], *Jaehwi Shin*[*]

[*] *Korea Telecom R & D Group*
Sochogu, Seoul, Korea 137-140
{mckim,kangsw,jhshin}@sava.kotel.co.kr

[**] *Hong Kong Univ. of Science & Technology*
Clear Water Bay, Kowloon, Hong Kong
chanson@cs.ust.hk

Abstract

This paper studies the problem of testing concurrent systems as blackboxes described in *asynchronous* Communicating Finite State Machines. We present an approach to derive test cases in a succinct and formal way for the concurrent systems. The approach avoids the state space explosion problem by introducing a causality relation model to express true concurrency. A new test architecture for executing the test suite is also described.

Keywords

Protocol Testing, Concurrent System, Conformance Testing, Logical Time.

1 INTRODUCTION

Concurrent behavior of a distributed system is difficult to specify and analyze. In order to specify the concurrent behavior of communication protocols and services in a distributed system, formal description techniques such as Communicating Finite State Machines (CFSMs), Estelle, SDL, LOTOS, and others have been developed. These techniques allow interactions among multiple machines (or modules or processes) to be specified. In order to increase confidence in protocol products made according to international standards, various protocol (conformance and interoperability) testing methods have been developed and applied. In the ISO conformance testing methodology, an Implementation Under Test (IUT) is viewed as a blackbox for testing purposes (ISO, 1994). Up until now, most work on automatic test case generation has been focused on single machine specifications even though the implementation may consist of multiple machines. Typically, a preprocessing stage is used to translate the multiple machines into a single system in order to obtain a specification with sequential behavior because we have a variety of testing methods for sequential systems and understand sequential behavior better. In this stage, the "state space explosion" problem often occurs.

In this paper, we present a succinct and formal methodology using true concurrency for generating test cases from a concurrent system described as asynchronous CFSMs and then applying the test cases to the concurrent system. Briefly speaking, the approach is to adopt techniques for distributed systems and apply them into protocol conformance testing on a concurrent system. The difficulty in testing a concurrent system is occurred by intentional nondeterminism existing in the specification and unintentional nondeterminism either arising from the concurrent environment or due to the blackbox approach (Kim, 1995). Our work is to deal with the unintentional nondeterminism in asynchronous CFSMs.

Related papers exist in the software engineering field such as (Young, 1988, Taylor, 1992, Tai, 1995). In the protocol engineering area, there are a few papers such as (Araka ,1991, Lee, 1993, Caval, 1994, Luo 1994, Ulric 1995). Our work is different from the existing work from the fact that our work aims at testing *asynchronous* CFSMs as treating them *blackboxes* and is a *non-heuristic* approach using a *true concurrency* model. It is an effort for coping with inherently difficult issues in testing a concurrent system.

The rest of the paper is organized as follows; Section 2 surveys some related work and Section 3 summarizes logical clocks (Lampo, 1978, Rayna, 1992a) for maximizing concurrency which is a crucial step in testing concurrent systems. Section 4 presents the model we propose. In addition, test case generation as well as application of the model is described. The section also gives definitions such as Observability Rate and Controllability Rate. Finally Section 5 concludes the paper.

2 RELATED WORK

Our work aims at testing asynchronous CFSMs which are treated as blackboxes. We provide a non-heuristic approach which uses a true concurrency model. By comparison, most of the existing work in software engineering such as (Young, 1988, Taylor, 1992, Tai, 1995) is based on synchronous communicating programs where the internal codes are accessible (whitebox approach).

The area of protocol testing has more similarity to our work, thus the following is a brief survey of some related papers on testing concurrent systems in protocol testing. (Araka, 1991) proposes a test case generation method for concurrent programs as a blackbox. They generate the control flow graph representing the global behavior of a given concurrent program and then apply conventional test case generation methods on the control flow graph. To avoid state space explosion, blackbox equivalence between system behavior is introduced. The approach is a modified technique from (Kajiw, 1985), and works for asynchronous CFSMs as blackboxes. However, the method is not formally presented.

Lee et. al (Lee, 1993) present an approach for conformance testing of protocols specified as a collection of synchronous CFSMs. The approach has two parts: pruning and a guided random walk procedure. This allows some parts of the behavior space to be removed from consideration. Also, the model is based on interleaving in order to explore the state space.

(Caval, 1994) presents a test sequence generation method based on Unique Input/Output sequences and signatures in a framework which allows conformance test sequences to be obtained from LOTOS specifications. The model is for synchronous communicating systems.

Luo et al. (Luo, 1994) present a model of generating test sequences for concurrent systems and communication protocol model as communicating nondeterministic finite state machines (CNFSMs). After CNFSMs are transformed into a single nondeterministic finite state machine (NFSM) by reachability analysis, test sequences are generated from the resulting NFSM using the generalized Wp-method. The approach meets the state space explosion problem easily because it adopts the reachability analysis in the transformation phase.

In (Ulric, 1995), an approach to deriving a concurrent transition tour as a test suite from multi-module specifications described by a restricted class of Calculus of Communicating Systems is presented. This approach avoids the state space explosion problem by introducing a true concurrency model. Thus this model is for synchronous communicating systems and uses some heuristics in terms of the order of composition.

The work in (Verha, 1992) shows that test cases generated using algorithms for synchronous communication can not be used for asynchronous systems. Thus, our work is different from all of work listed above in a sense that our work aims at testing *asynchronous* CFSMs as treating them *blackboxes* and is a *non-heuristic* approach using a *true concurrency* model.

3 BACKGROUND: LOGICAL CLOCKS AND MAXIMIZING CONCURRENCY

Since Lamport's pioneering work (Lampo, 1978), there has been considerable research on concurrency, logical clocks and global states for distributed systems (Fidge, 1991, Matte, 1988, Charr, 1989, Rayna, 1992a, Rayna, 1992b, Kim, 1993). The logical clock is used to determine whether or not two events are concurrent. The concurrency relation is then used to compute global states.

Given a complete specification, causality relationship of events may be obtained readily using a concurrency model (Kim, 1993). Here, we adopt logical clocks (Lampo, 1978, Rayna, 1992a) to incorporate the duration of an operation in the model even though the causality relation is known and does not have to be computed on-the-fly. The purpose is to handle unintentional nondeterminism either arising in the concurrent environment or due to the blackbox approach in asynchronous CFSMs.

3.1 Logical clocks

Lamport showed that distributed asynchronous executions may be characterized by a partial order relation on the events produced. The relation, called the *causality* relation (or *happened before* relation) and denoted by $\rightarrow$, is defined in the following way (Lampo, 1978). Let E be a set of events produced by the execution of a distributed system. For e, e′, e″$\in$ E, e $\rightarrow$ e′ holds if

1. e and e′ are events in the same process and e precedes e′,
2. e is a send event and e′ is the corresponding receive event, or
3. there exists e″ such that e $\rightarrow$ e″ and e″$\rightarrow$ e′.

Two events e and e′ are said to be *causally related* if e → e′ or e′ → e holds. If neither e → e′ nor e′ → e holds, these events are *concurrent* (or *mutually independent*), denoted by e || e′.

According to the above principles, a number of timestamping mechanisms are devised such as linear time, vector time, and matrix time (Rayna, 92a). Linear time c_i associated with process P_i, for example, can be computed as follows:

- When P_i executes an internal event or a send event, the time is computed by $c_i = c_i + d$. In case of a send event, the message carries the updated c_i value as the timestamp.
- When P_i executes a receive event where the message contains timestamp *ts*, the time is computed by $c_i = max(c_i, ts) + d$.

The value *d* represents the duration of the corresponding operation.

In case of vector time (Fidge, 91, Matte, 1988), the logical global time represented by an n-dimensional vector $vt_i[1..n]$ is associated with each process P_i where n is the number of counters which is equal to the number of processes. The vector $vt_i[j]$ denotes the j-th counter maintained by process P_i. All values in $vt_i[1..n]$ are initialized to zero. The principle of the technique is the following:

- When P_i executes an internal event, vt_i is advanced by setting $vt_i[i] := vt_i[i] + d$.
- When P_i executes a send event, vt_i is advanced by setting $vt_i[i] := vt_i[i] + d$. The message carries the updated vt_i value.
- When P_i executes a receive event and reads a message containing vt_j, vt_i is advanced by setting $vt_i[k] := max(vt_i[k], vt_j[k])$, $1 \leq k \leq n$, and then $vt_i[i] := vt_i[i] + d$.

If event *x* at site *i* and event *y* at site *j* are timestamped respectively by *vh* and *vk*, we have:

$x \rightarrow y$ if and only if $(vh[i] \leq vk[i]$ and $vh[j] < vk[j])$ ----- (1)

$x \parallel y$ if and only if $vh[i] > vk[i]$ and $vh[j] < vk[j]$ ----- (2)

A computation may be graphically displayed in a *time-event sequence* diagram, such as the one shown in Figure 1. The origin of an arrow represents a *send* event and the head of an arrow indicates a *receive* event. Figure 1 shows an example of vector clocks progress with the value d = 1. The values of *vt* are represented by (i_1, i_2, i_3).

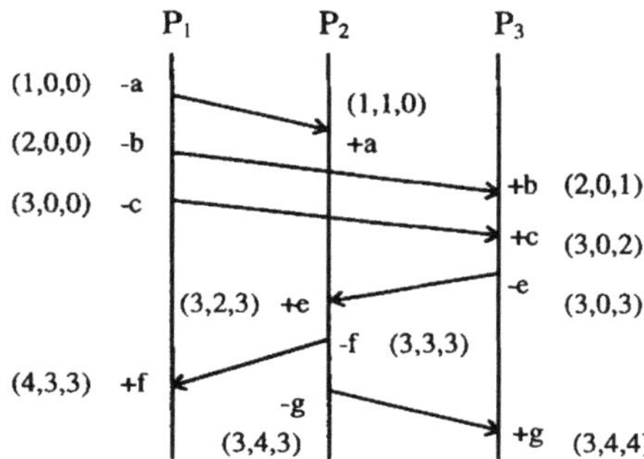

Figure 1 A computation.

Using (1) and (2) given above, we can decide whether two events are concurrent or not. For example, event '-b' precedes event '+b' and '+c' because of (1) and event '-b' and '+a' are concurrent because of (2). Also note that not only are '-b' and '+a' concurrent because of (2) but '+a' and '-c' are also concurrent by (2) event though '-b → -c'.

3.2 Maximizing concurrency

Any analysis technique should be independent of real time effects, such as system load and processor speed. To abstract away from any particular machine executing a distributed program, we assume that each event consumes approximately the multiple amount of a time unit. By the assumption, we can set durations of events to any integer numbers. These assumptions can be achieved by adjusting the granularity of the events. Such an approach allows us to precisely characterize the causality constraints inherent within a distributed computation without being bothered by the perturbations caused by a given implementation.

In the model given in (Rayna, 1992b), the execution time based on time units can be computed using linear logical time described in the previous section. For simplicity, we further assume that the duration of each event is a unit time. Based on this, we draw a time-event sequence diagram by aligning all events with the same logical time represented by the same dotted row, as shown in Figure 2. Those events are concurrent ones which can be represented by '||' relation defined in previous section. Then, the time-event sequence diagram for a given computation is unique and minimal in terms of the number of time units (Rayna, 1992b). In the figure, synchronization delays represented by circles occur where events are not produced by processes.

For example, the linear logical time of event '-e' is '5'. In Figure 2, the logical time of every event is increased by '1' and every event is aligned with respect to the logical time for later usage.

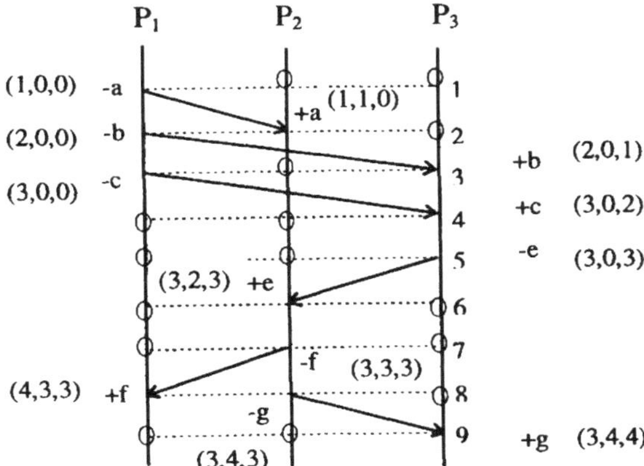

Figure 2 A computation with linear logical clock.

4 FORMAL MODEL

In this section, we propose a model to derive test cases from a concurrent system in a succinct and formal way. The concurrent system is represented in asynchronous CFSMs and treated as blackboxes.

4.1 System modeling

A concurrent system consists of a number of sequential systems. Each sequential system is modeled as a Finite State Machine (FSM). The environment (e.g., operators, testers, and/or subsystems which are not under test) of a concurrent system is also modeled in term of FSMs. It is assumed that there is no global physical clock; that is, the system model is *asynchronous*. FSMs exchange messages to communicate and synchronize with each other. The underlying message passing system provides asynchronous communication and is assumed to be reliable; that is, there is no message loss, distortion or duplication.

An event is the atomic execution of an operation within a system. There are three types of events:

- Internal event which models the execution of an internal operation of an FSM.
- Send event which models a message send operation of an FSM.
- Receive event which models a message receive operation of an FSM.

We assume that a concurrent system consists of subsystems with no access to global variables. Thus, there are no dataraces in this model. Also each subsystem is assumed to be a blackbox. Therefore, only communication events such as message send and receive events are considered. Also a send (or receive) event takes a logical unit time as we assumed in Section 3.

Definition 1: A Global Event (GE) is defined as a set of events occurred at the same time in all FSMs of a concurrent system. Namely, GE = $(e_1, \ldots, e_i, \ldots, e_n)$ where $1 \leq i \leq n$, n is equal to the number of FSMs, and e_i is an event of i-th FSM.

An element, e_i, of GE is null if i-th FSM is idle. Note that GE does not include the state of queues because events are external behavior abstracted away from internal behavior such as queue operations.

The GE consists of send and/or receive events of one or more subsystem because we allow true concurrency. A true concurrency can be expressed using a causality relation model. The causality relation model provides an answer to the question whether events occur independently (Visse, 1990). In order to verify the behavior of a concurrent system, techniques commonly employed are based upon interleaving of actions among subsystems in the concurrent system, thereby leading to state space explosion when the number of subsystems is large. In fact, we really do not need interleaving for those concurrent actions which do not affect each other (e.g., via datarace). It is possible to use a causality relation model to express true concurrency instead of using interleaving in order to test concurrent systems. We propose an approach to test concurrent behavior as it is.

4.2 Generation of Concurrent Paths

For simplicity, we shall consider the second visit to the initial state of an FSM as termination. A path is a sequence of events performed by the FSM starting and ending at the same initial state in a normal case. From the paths obtained from enumeration every possible path of an FSM, we select a set of paths among the FSMs communicating with one another. For the set of paths, we can make a vector of matching event sequences (in a path) ordered with respect to each FSM.

Definition 2: Path[i] is a set of event sequences, tours, that can be performed by i-th FSM, i.e. Path[i]= {Tour(i,1), ..., Tour(i,y), ...} where Tour(i,y) is y-th tour and is denoted by $e^1 \rightarrow e^2 \rightarrow \ldots \rightarrow e^m \ldots \rightarrow e^M$ for some natural number M.

Definition 3: A Concurrent Path (CP) is a tour (or event sequence) vector among FSMs. Namely, CP_k = <Tour(1,x), ... , Tour(i,y), ... , Tour(n,z)> where $1 \leq i \leq n$, n is equal to the number of FSMs, and k, x, y and z are some natural numbers.

CPs are obtained from $Path_1 \times \ldots \times Path_i \times \ldots \times Path_n$ where the operator '×' is the Cartesian product such that a send event is matched to its receive event in chronological order in every tour of Paths in FSMs.

An element, Tour(i,y), of CP is null if i-th FSM is idle. If there is no CP, k in Definition 3 becomes '0'. The meaning of CP is the same as *Multilogue* in (Araka, 1991, Kajiw, 1985).

The CP begins with the initial state and ends with a node which is not the initial node if there is a deadend such as deadlock, unspecified reception, and so forth. In this case, CP could contain some unpaired messages. If there is no deadend, the CP eventually returns to the initial state.

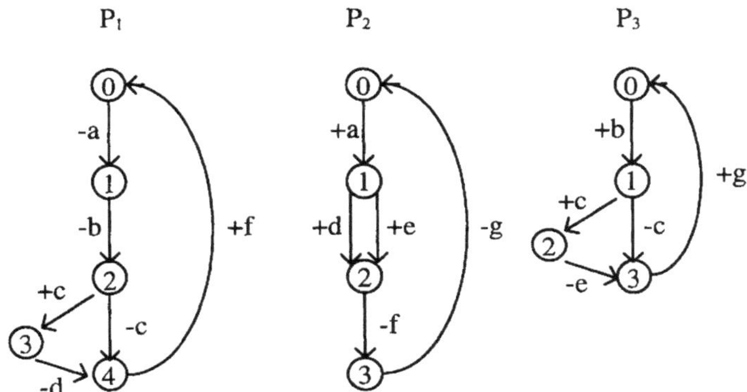

Figure 3 Example of a concurrent system.

For the example shown in Figure 3 which was adopted from (Kajiw, 1985), we can generate a set of individual Paths as follows:

Path[1] = {Tour(1,1), Tour(1,2)}
$= \{-a \rightarrow -b \rightarrow +c \rightarrow -d \rightarrow +f,\ -a \rightarrow -b \rightarrow -c \rightarrow +f\}$

Path[2] = {Tour(2,1), Tour(2,2)}
$= \{+a \rightarrow +d \rightarrow -f \rightarrow -g,\ +a \rightarrow +e \rightarrow -f \rightarrow -g\}$

Path[3] = {Tour(3,1), Tour(3,2)}
$= \{+b \rightarrow +c \rightarrow -e \rightarrow +g,\ +b \rightarrow -c \rightarrow +g\}$

The cross product of three paths and two tours per path gives eight concurrent paths. Let <x, y, z> represent a CP consisting of <Tour(1,x), Tour(2,y), Tour(3,z)> and 'Combi_*i*' mean the i-th combination of tours among paths. From this, we can obtain eight combinations:

Combi_1 = <1, 1, 1> = CP_3
Combi_2 = <1, 1, 2> = CP_1
Combi_3 = <1, 2, 1> = CP_3
Combi_4 = <1, 2, 2> = CP_5
Combi_5 = <2, 1, 1> = CP_6
Combi_6 = <2, 1, 2> = CP_4
Combi_7 = <2, 2, 1> = CP_2
Combi_8 = <2, 2, 2> = CP_4
CP_1 = <-a → -b → +c → -d → +f , +a → +d → -f → -g, +b → -c → +g>
CP_2 = <-a → -b → -c → +f, +a → +e → -f → -g, +b → +c → -e → +g>
CP_3 = <-a → -b → +c, +a → +e, +b → +c>
CP_4 = <-a → -b → -c → +f, +a → +d, +b → -c → +g>
CP_5 = <-a → -b → +c → -d → +f, +a → +e, +b → -c → +g>
CP_6 = <-a → -b → -c → +f, +a → +d, +b → +c → -e → +g >

Note that CP_3, CP_4, CP_5, and CP_6 reach deadends because these CPs do not end with the initial node.

Let us look at the traditional reachability analysis. It can be considered that the analysis consists of two levels of enumeration on:

1. paths among FSMs,
2. and events among the paths obtained.

Level 1 of enumeration in the traditional reachability analysis corresponds to the computation of CPs defined in Definition 3. In this level, we may encounter "state space explosion" which is not handled in this paper. The next step is trying to solve the "state space explosion" occurring in events enumeration (in level 2) among the paths obtained. Traditional reachability analysis enumerates one path out of a set of system (or FSM) interactions based on interleaving of events and each of which is tested individually. In this way, the global state space grows rapidly as the number and the complexity of the subsystems increase.

Furthermore, loops in a path could also lead to state space explosion since the potential number of distinct paths is infinite. Practically we can limit the number of times a loop is traversed.

4.3 Generation of Minimal Causality Path

Let's enumerate every possible event sequence by interleaving satisfying the following Concurrent Path:

CP_2 = <-a → -b → -c→ +f, +a → +e → -f → -g, +b → +c → -e → +g>.

In the CP, there are constraints imposed by causal relationships among the events. These consists of two kinds: ordering imposed by event sequences in CP and ordering imposed by matching send and receive events. So far in CP_2, we have the following constraints of the first kind:

- -a → -b → -c→ +f,
- +a → +e → -f → -g, and
- +b → +c → -e → +g.

The constraints of the second kind are: -a → +a, -b → +b, -c → +c, -e → +e, -f → +f, and -g → +g.

For CP_2, we can obtain event sequences satisfying the above constraints in a standard way using reachability tree. In Figure 4, notation (x, y, z) means a global state where x, y and z of P_1, P_2 and P_3, respectively, are integers representing sequential order which is the same value of 'm' in Definition 2. The initial global state is (0, 0, 0) and the final one is (4, 4, 4) in this example. In the left side of the tree, we have three event sequences ending with (4, 4, 4). Thus, there is a rectangle containing '3'. A leaf node with a circle denotes a transition to an intermediate node having the same global state as the leaf node. We have 36 event sequences in total as the result of summing up the values inside the rectangles.

Now let us construct a minimal length time-event sequence diagram to represent the above causal relationships.

Definition 4: A Minimal Causality Path (MCP) corresponding to a CP is a minimal length of Global Event sequences.

Minimal Causality Paths can be obtained in the same way that outlined in Section 3.2. The way is to calculate the linear logical time of events in a CP, and then obtain GEs for each logical time by collecting all events having the same logical time. Then MCP is a sequence of GEs aligned in the increasing order of logical time. For CP_2, we obtain an MCP corresponding to Figure 2. If we attach an identifier of FSM to an event as a suffix in case of causing ambiguity, then

MCP_2 = -a → (-b, +a) → (-c_1, +b) → +c_3 → -e → +e → -f → (+f, -g) →+g.

In the same way, we can obtain MCPs corresponding to CPs as follows:

MCP_1 = -a → (-b, +a) → +b → -c_3 → +c_1 → -d → +d → -f → (+f, -g) →+g.

MCP_3 = -a → (-b, +a) → +b.

MCP_4 = -a → (-b, +a) → (-c_1, +b) → -c_3.

MCP_5 = -a → (-b, +a) → +b → -c_3 → +c_1 → -d.

MCP_6 = -a → (-b, +a) → (-c_1, +b) → +c_3 → -e.

The notation (-b, +a) means that the two events occurred concurrently or mutually independently. Therefore, the order of occurrences of the two events does not matter. In addition, event '+a' could have occurred in the following circles spaced by synchronization delays in Figure 2 if a unit time assumption is relaxed to any natural numbers. In other words, event '+a' is concurrent with events such as '-b', '+b', '-c', '+c', and '-e' which could be aligned with event '+a' itself and the synchronization delays of event '+a' in Figure 2.

4.4 Generation of Combined Minimal Causality Path

Combined Minimal Causality Path (CMCP) is a global event-sequence diagram such that each of its paths is associated with a corresponding MCP. A CMCP is generated so that each global transition is associated with a set of events, which can be implemented

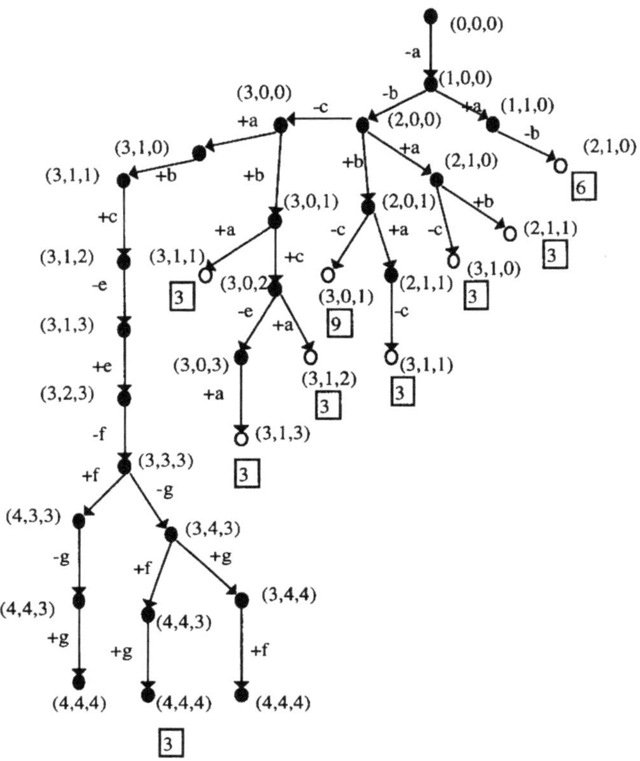

Figure 4 Generation of event sequences using reachability tree.

in parallel at the starting global state. Figure 5 shows the CMCP for the functional specification of Figure 3. The CMCP includes four MCP's described in Section 4.3. Four filled circles with the number *i* showing MCP_i in Figure 5 represent deadlocks.

The CMCP analysis is equivalent to the conventional reachability tree analysis while the computation times are enormously reduced. In order to measure the extent of reduction in our approach over the conventional way, let us show how many event sequences are represented by CPs. Using the same way that we obtained 36 event sequences for CP_2, we have 18, 3, 15, 6 and 12 event sequences for CP_1, CP_3, CP_4, CP_5, and CP_6, respectively. On the whole, the six CPs represent 90 event sequences. A CP is equivalent to an MCP under the assumption of unit logical time. If the number and the complexity of subsystems increase, the saving of testing effort becomes tremendous. For instance, 6 CPs represent over 10,000 event sequences in the example in (Kajiw, 1985) dealing with a concurrent system (modeled as CFSMs) consisting of two processors and two terminals of which numbers of states are 26, 14, 17 and 7 respectively.

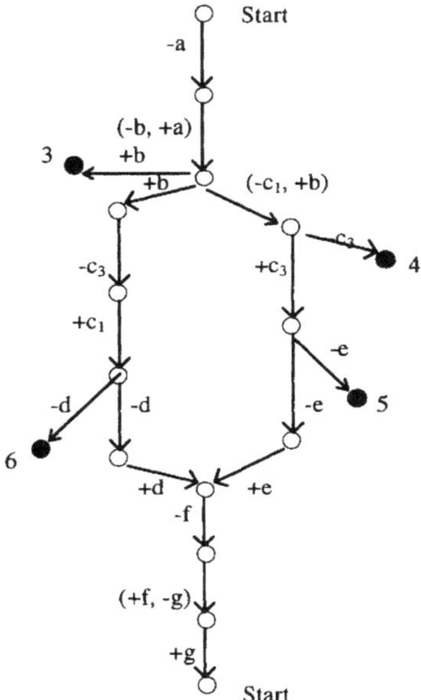

Figure 5 CMCP of the concurrent system in Figure 3.

4.5 Test cases generation and application

Test cases for a concurrent system are generated based on a specific test architecture. We view a concurrent system as a closed system which includes subsystems, the environment and operators if any. We assume that the external behavior of each subsystem in the concurrent system can be observed through one or more Point of Observations (POs) between subsystems. Subsystems under test are regarded as blackboxes. Subsystems not under test could include subsystems exposed as whiteboxes such as testers and operators.

In Figure 6 (a), we observe a trace and then decide whether the trace conforms to the system specifications. The CMCP given in Figure 5 is the minimal causality relation to which the trace should conform. Note the following characteristics of PO. Observing a message through PO implies that a message send event has occurred, but the corresponding receive event has not occurred yet. Therefore, in order to decide whether the receive event has actually happened, we need to wait for the subsequent observable events at the subsystem where the receive event is to occur. In reality, it is very hard to guess when the receive event has happened. For the time being, let us define Observability Rate as follows:
Definition 5: Observability Rate (OR) is the number of observable events divided by the total number of events in a computation.

Note that the computation in Definition 5 can be either the partial or total behaviors of a system. Because we can observe all message send events at the POs in the case of the

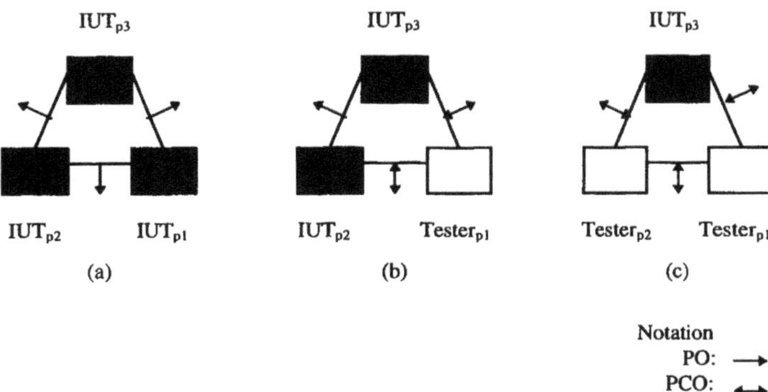

Figure 6 Test architecture.

computation shown in Figure 2 using the test architecture of Figure 6 (a), OR for Figure 6 (a), OR_a for CP_2, is 6/12.

Definition 6: A Stable State (SS) is a global state of a system in which the system can not proceed without inputs from the subsystems not under test.

Some of the subsystems may be testers as in (b) and (c). In the case of (b), we need to project the test cases obtained in the previous section with respect to P_1. Also, for the architecture of (c), projecting the test cases with respect to P_2 is needed as well. Because Tester$_{p1}$ in case (b) controls inputs such as '-a', '-b', '-c_1' and '-d' in addition to POs in (a), we can test P_2 and P_3 to see whether they conform to their specifications. Therefore, we have SSs in the event sequence: -a → (-b, +a) → -d and -a → (-b, +a) → (-c_1, +b). If you look at P_1 in Figure 3, you can understand the SSs easily. Let us work at the SS in the event sequence -a → (-b, +a) → (-c_1, +b). Tester$_{p1}$ controls inputs such as '-a', '-b' and '-c', and observes output such as '+f'. The other observation provided by the PO between IUT$_{p2}$ and IUT$_{p3}$ gives us some information described in the architecture (a). In case of the computation shown in Figure 2, OR_b for CP_2 is 7/12 because of '+f' which is not counted in OR_a for CP_2.

In case (c), Tester$_{p1}$ and Tester$_{p2}$, with more PCOs, provide us more SSs. The SS of IUT$_{p2}$ is the event sequence: -f → -g. OR_c of the computation in Figure 2 is 9/12, because of '+a' and '+e'. Therefore, the architecture (c) provides more confidence in testing than the architecture (b). If we count all CPs, the difference of OR of the three architectures becomes significant. Also, we have the advantage of being able to test wrong behavior of IUTs which was not possible in (a).

Definition 7: Controllability Rate (CR) of a system under test is defined as $\sum_{i=1}^{m} |Input|_i / \sum_{j=1}^{n} |Input|_j$ where m is the number of stable states of the system, n is the number of stable states of a white-box system corresponding to the system, and $|Input|_i$ and $|Input|_j$ are the number of allowed inputs at the i-th and j-th Stable States respectively.

According to Definition 7, we obtain $CR_a = 0 / (9 \times 1)$, $CR_b = (4 \times 1) / (9 \times 1)$, and $CR_c = (6 \times 1) / (9 \times 1)$. The higher CR is, the more controllable the system is. For example, in Figure 6, architecture (c) is more controllable than (b), and (b) is more controllable than (a).

Let us discuss briefly how to relax the assumptions about this model in order to make it more applicable to the real world situation. Note the value of 'd' in logical clocks for representing the duration of an operation in Section 3.1. So far, we have assumed that the value of 'd' is one unit time. If the value 'd' can be relaxed to any natural number or any real number, then this model has wider applicability.

5 CONCLUSION

In order to test and verify concurrent behavior of protocols running in a concurrent system, traditional approaches are to map the multiple modules into a single module and analyze the sequential behavior of the resultant single module. Other approaches are to use interleaving methods for the events in a concurrent system. Recently some attempts have been made to adopt a causality method or a true concurrency model in protocol verifications and testing for concurrent systems (Probs, 1991, Ulric, 1995).

By adopting a causality relation model, we proposed a non-heuristic approach adopting true concurrency model for testing asynchronous communicating systems as treating them blackboxes. The approach is succinct and formal compared with the previous work (Araka, 1991, Kajiw, 1985), deals with the unintentional nondeterminism either arising from the concurrent environment or due to the blackbox approach. Also, the approach avoids the state space explosion problem by representing a number of event sequences as a class, which we call Minimal Causality Path using logical clocks. The relationship between test case generation and application in a concurrent system was also discussed. Also new definitions such as Observability Rate, Stable State, and Controllability Rate were introduced to quantify some aspects of the systems. Our model is adequate to be a base model to investigate further work in this area.

For further study, we are working on enhancement of the model as follows:

- relaxing of the logical unit time to any natural number,
- resolving of the intentional nondeterminism in the specification of concurrent systems, and
- minimization of "state space explosion" in generating Concurrent Paths.

6 ACKNOWLEDGEMENT

The authors would like to thank anonymous reviewers for their helpful comments in revising this paper.

7 REFERENCES

Arakawa, N. and Soncoka, T. (1991) "A Test Case Generation Method for Concurrent Programs," *Proceedings of the Int'l Workshop on Protocol Test Systems.*

Cavalli, A., Kim, S. U., and Maigron, P. (1993) "Improving Conformance Testing for LOTOS," pp. 371 - 386, *Sixth Int'l Conference on Formal Description Techniques.*

Charron-Bost, B. (1989) "Combinatories and Geometry of Consistent Cuts: Application to Concurrency Theory," *The 3rd Int'l Workshop on Distributed Algorithms,* LNCS 392, Springer-Verlag, pp. 45 - 56.

Fidge, C. (1991) "Logical Time in Distributed Computing Systems," *IEEE Computer*, pp. 28 - 33.

ISO IS 9646 (1994) *OSI Conformance Testing Methodology and Framework.*

Kajiwara, M., Ichikawa, H., Itoh, M., and Yoshida, Y. (1985) "Specification and Verification of Switching Software," *IEEE Tr. on Communications*, pp. 193 - 198.

Kim, M. C., Chanson, S. T., and Vuong, S. T. (1993) "Concurrency Model and its Application to Formal Protocol Specifications," *IEEE INFOCOM*, pp. 766 - 773.

Kim, M. C., Chanson, S. T., Yoo, S. J. (1995) "Design for Testability for Protocols based on Formal Specifications," *Proceedings of the 8th Int'l Workshop on Protocol Test Systems.*

Lamport, L. (1978) "Time, Clocks, and the Ordering of Events in a Distributed System," *CACM*, pp. 558 - 565.

Lee, D., Sabnani, K. K., Kristol, D. M. and Paul, S. (1993) "Conformance Testing of Protocols Specified as Communicating FSMs," *IEEE INFOCOM*, pp. 115 - 127.

Luo, G., Bochmann, G. V., and Petrenko, A. (1994) "Test Selection Based on Communicating Nondeterministic Finite-State Machines Using a Generalized Wp-Method," *IEEE Tr. on Software Engineering*, pp. 149 - 161.

Mattern, F. (1988) "Virtual Time and Global States of Distributed Systems," *Parallel and Distributed Algorithms*, pp. 215 - 226, North-Holland.

Probst, D. K. and Li, H. F. (1991) "Using Partial-order Semantics to Avoid the State Explosion Problem in Asynchronous Systems," *2nd Int'l Conference on Computer-aided Verification.*

Raynal, M. (1992) "About Logical Clocks for Distributed Systems," *ACM Operating Systems Review*, pp. 41 - 48.

Raynal, M., Mizuno, M., and Neilsen, M. L. (1992) "Synchronization and Concurrency Measures for Distributed Communications," *12th Int'l Conference on Distributed Computing Systems*, pp. 700 - 707.

Taylor, R. N., Levine, D. L., and et al. (1992) "Structural Testing of Concurrent Programs," *IEEE Tr. on Software Engineering*, pp. 206 - 215.

Tai, K. C. and Carver, R. H. (1995) "Testing of Distributed Programs," *Handbook of Parallel and Distributed Computing*, ed. A. Zomaya, McGraw Hill.

Ulrich, A. and Chanson, S. T. (1995) "An Approach to Testing Distributed Software Systems," *Int'l Symposium on Protocol Specifications, Testing, and Verification.*

Verhaad, L., Tremans, J., Kars, P. and Brinksma, E. (1992) "On Asynchronous Testing," *Int'l Workshop on Protocol Test Systems*, pp. 14 - 27.

Vissers, C. (1990) "FDTs for Open Distributed Systems, A Retrospective and A Prospective View," *Int'l Symposium in Protocol Specifications, Testing and Verification.*

Young, M., and Taylor, R. N. (1988) "Combining Static Concurrency Analysis with Symbolic Execution," *IEEE Tr. on Software Engineering*, pp. 1499 - 1511.

8 BIOGRAPHY

Myungchul Kim received B.A. in electronics engineering from Ajou Univ. in 1982, M.S. in computer science from the Korea Advanced Institute of Science and Technology in 1984, and Ph.D. in computer science from the Univ. of British Columbia in 1992. Currently, he is with the Korea Telecom Research and Development Group as a managing director, chairman of Profile Test Specification - Special Interest Group of Asia-Oceania Workshop, and will be chair of the 10th Int'l Workshop on Testing of Communicating Systems. His interests include protocol engineering on telecommunications and multimedia.

Samuel T. Chanson received his Ph.D. degree in Electrical Engineering and Computer Sciences from the University of California, Berkeley in 1975. He was a faculty member at the School of Electrical Engineering, Purdue University for two years before joining the Department of Computer Science at the University of British Columbia where he was a professor and director of its Distributed Systems Research Group. In 1993 he joined the Hong Kong University of Science & Technology as professor and Associate Head of the Computer Science Department. He has consulted widely for industry and government on communication technologies. He has published about 100 technical papers in journals and conference proceedings, and has served on the program committee of many international conferences on distributed systems and computer communications. He was the conference co-chair of the first IFIP International Workshop on Protocol Test Systems (now called IWTCS), a co-chair of the 1994 International Conference on Protocol Specification, Testing and Verification, and the program co-chair of the IEEE International Conference on Distributed Computing Systems in May 1996. Dr. Chanson's research interests include protocol testing, ATM network management, multimedia applications, and Internet technologies. He is the director of Cyberspace Centre for Hong Kong software industry.

Sungwon Kang received a B.A. degree from Seoul National University in Korea in 1982 and received his M.S. and Ph.D. in computer science from the University of Iowa in U.S.A in 1989 and 1992. Since 1993, he has been a senior researcher at Korea Telecom Research and Development Group. Currently he is a guest researcher at National Institute of Standards and Technology of U.S.A. His research interests include communication protocol testing, program optimization and programming languages.

Jaehwi Shin received the B.S. and M.E. degrees in electronic engineering from Hanyang University and Seoul National University in 1991 and 1993, respectively. Currently he is with the Korea Telecom Research and Development Group as a member of Protocol Engineering Team. His research interests include intelligent networks, communication protocol testing, digital signal processing and circuit theory.

12

Checking unstable properties in distributed testing of communication protocols

*Antonio A.F. Loureiro** *Samuel T. Chanson*[†]
Son T. Vuong[‡] *Osvaldo S.F. de Carvalho**

Abstract

Protocol testing is a run-time activity and we can only hope to detect valid or invalid behaviors in an actual execution of a protocol implementation embedded in a testing environment. In this paper we focus our attention in the validation of dynamic unstable properties during the testing process and afterwards, during normal execution. Dynamic unstable properties define desirable or undesirable temporal evolutions of the behavior of a communication protocol. We shall present a novel on-line distributed algorithm and the corresponding design principles that will improve the testing process and the reliability of a protocol implementation by checking unstable properties in a distributed environment.

Keywords

Unstable properties, distributed protocol testing, distributed algorithms.

1 INTRODUCTION

The validation[1] of global predicates is a fundamental problem in distributed computing that has been used in different contexts such as design, simulation, testing, debugging, and monitoring of distributed programs [1, 7, 13, 12, 15]. A global predicate may either be stable or unstable. Informally, a stable predicate means that once it becomes true during a computation it will remain true after that, such as a system deadlock. An unstable predicate does not have this

---

*Departamento de Ciência da Computação, Universidade Federal de Minas Gerais, CP 702, 30161–970 Belo Horizonte, MG, Brazil. E-mail: {loureiro,vado}@dcc.ufmg.br. Osvaldo Carvalho is on leave at the Institut National des Télécommunications, Evry, France.

[†]Department of Computer Science, Hong Kong University of Science and Technology, Clear Water Bay, Hong Kong. E-mail: chanson@cs.ust.hk.

[‡]Department of Computer Science, University of British Columbia, Vancouver, BC, Canada V6T 1Z4. E-mail: vuong@cs.ubc.ca.

[1]The terms checking, detection, and validation will be used interchangeably in this paper when referring to the validation of desirable behaviors (i.e., predicates or properties) in a protocol.

characteristic. An example is a predicate that relates the length of two queues, each one in a different process. In fact, an unstable predicate may switch from valid to invalid and vice-versa during the execution.

The initial work in the detection of global predicates has concentrated on the validation of stable properties such as distributed termination [5, 6] and deadlock detection [2]. Chandy and Lamport [3] proposed an algorithm to take snapshots in a distributed system that became the basis of other algorithms which check stable properties. The word *snapshot* in this algorithm means the local state of a process P_i in the system. Therefore, when P_i receives a message to take a snapshot it records its local state and "relays the 'take snapshot' message along all of its outgoing channels" [3]. These 'snapshot' messages are used by a global monitor to construct only consistent global states (see Definition 5).

Note that a stable property defined in terms of global predicates can be checked by a global monitor which takes snapshots and constructs the consistent global states. If the stable predicate is found to be true in at least one consistent global state constructed from the snapshots taken in the individual processes then it can be inferred that it will remain in that state at the end of the algorithm. If the predicate is false in the global states constructed, then it was also false at the beginning of the algorithm [3].

Unfortunately this approach does not work for unstable predicates which may be true during an execution but not checked, or found to be true in some states but it may have never happened because the global monitor constructs all possible consistent global states (see [1] for an example of this situation).

The testing process is a run-time activity and we can only hope to detect valid or invalid behaviors in an actual execution of a protocol implementation embedded in a testing environment. In this paper we focus our attention in the validation of dynamic properties during the testing process and afterwards, during normal execution. Dynamic properties define desirable or undesirable temporal evolutions of the behavior of a communication protocol. We shall present a set of novel design principles that will improve the testing process and the reliability of a protocol implementation by checking desirable behaviors. The goals to be accomplished are summarized as follows:

▷ Protocol testing:
  - Mechanism to check global predicates based on local predicates.
  - Identification of consistent global states where any type of predicate can be checked.

▷ Protocol execution:
  - Obtain information to avoid the problem of state build-up.
  - Obtain information for use in exception handling.

The rest of this paper is organized as follows. Section 2 describes the formal model used in this paper. Section 3 discusses the tasks involved in the detection of dynamic properties. Section 4 describes the design principles related to the testing process, including the algorithm to detect the properties. Section 5 describes the design principles related to the execution of the protocol implementation. Section 6 discusses the related work. Finally, Section 7 presents the conclusions for this paper.

2 FORMAL MODEL

In this section we present some definitions that will be used in the algorithm described in Section 4. Our algorithm is based on the communicating extended finite state machine (CEFSM) model. First we present the nomenclature for identifiers that will be used in this paper.

Nomenclature 1 (Identifiers) Identifiers can have a subscript and/or a superscript both of which are used to indicate an element in a set. Let $\Box_i^x$ represent an identifier. The subscript i refers to the i-th element of $\Box$. For example, process P_i. The superscript refers to the x-th element of $\Box_i$ in the case $\Box_i$ is also a set.

Definition 2 (Communicating extended finite state machines) A communicating extended finite state machine is a labeled directed graph where vertices represent states and edges represent transitions. A designated vertex represents the initial state of the machine. Transitions are labeled with the pair $\frac{event}{action}$. An event is the reception of a message, or an internal event not related to a channel (e.g., a timeout or a service request from a service user). An action is the sending of a message, the starting of a timer, or the assignment of a value to a local variable. The communication between machines is asynchronous. The communicating channels between each pair of machines are not perfect so that messages can be reordered, corrupted or lost.

Formally, a communicating extended finite state machine P_i $(i = 1 \ldots r)$ is a five-tuple $P_i = (S_i, \Sigma_i, \delta_i, \lambda_i, s_i^0)$, where

- S_i is the set of local states of machine P_i.
- Σ_i is the set of events and actions. This is represented by the sets $\{\Sigma_{\mathcal{E}i}\}$ and $\{\Sigma_{\mathcal{A}i}\}$, respectively. There is also a restriction that machine P_i cannot send or receive messages from itself.
- δ_i is the transition function and is defined as δ_i: $S_i \times \Sigma_{\mathcal{E}i} \rightarrow S_i$.
- λ_i is the output function and is defined as λ_i: $S_i \times \Sigma_{\mathcal{E}i} \rightarrow \Sigma_{\mathcal{A}i}$.
- s_i^0 is the initial state of machine P_i.

In the labeled directed graph that represents a CEFSM, a message preceded by a + sign means that it was received, and a message preceded by a − sign indicates that it was sent. ■

Let P be a protocol specification comprised of processes $P_1, P_2, \ldots, P_r$. Each process is modeled as a CEFSM and they are interconnected by a set of communicating channels $C_1, C_2, \ldots, C_s$. Each process P_i $(i = 1 \ldots r)$ has a finite set of variables $V_i^1, V_i^2, \ldots, V_i^{k_i}$. In this model, the concept of a global state plays a fundamental role in the correctness of a communication protocol.

Definition 3 (Local state) The local state γ_i of a process P_i is defined as the contents of each local variable $(V_i^1, \ldots, V_i^{k_i})$ in P_i. ■

Definition 4 (Global state) A global state is an n-tuple of local states, one for each process P_i, and is represented as $\Gamma = (\gamma_1, \ldots, \gamma_n)$, where γ_i represents the local state of process P_i.

In this definition we do not include communication channels since we can encode them as part of each local state. ■

Definition 5 (Consistent global state) Informally, a global state is consistent if it could occur during an execution and a global clock in the system could be used to label precisely the total order of events. Formally, a global state $\Gamma = (\gamma_1, \ldots, \gamma_n)$ is consistent iff for any pair $(\gamma_i, \gamma_j) \in \Gamma$, then either $\gamma_i \prec \gamma_j$ or $\gamma_j \prec \gamma_i$ for $1 \leq i, j \leq n$ and $i \neq j$, according to the transitive closure of relation $\prec$ between local states. ■

The set of all consistent global states define exactly the states that could have happened in any computation with respect to the events that occurred in each process. Therefore, predicate values are meaningful only if evaluated in a consistent global state.

The set of all consistent global states Γ define a lattice structure $\mathcal{L}$ and its minimal element is the initial global state $\Gamma^0 = (\gamma_1^0, \ldots, \gamma_n^0)$. In the lattice $\mathcal{L}$ there is an edge from a node representing a global state $\Gamma^\alpha = (\gamma_1, \ldots, \gamma_i^a, \ldots, \gamma_n)$ to a node representing $\Gamma^{succ(\alpha)} = (\gamma_1, \ldots, \gamma_i^{a+1}, \ldots, \gamma_n)$ iff there exists an event e that P_i can execute in its state γ_i^a.

Now we present the definitions of local and global predicates that will be used in Section 4.

Definition 6 (Local predicate) A local predicate of a process P_i is a formula in propositional logic (i.e., a boolean expression) where each term of the formula is a local variable of P_i. The set of local predicates ϕ valid for protocol P can be expressed as:

$$\phi = \{\phi_1^1, \ldots, \phi_1^{p_1}, \phi_2^1, \ldots, \phi_2^{p_2}, \ldots, \phi_n^1, \ldots, \phi_n^{p_n}\} = \bigcup_{i=1..n} \phi_i$$ ■

Definition 7 (Global predicate) A global predicate of protocol P is a formula in propositional logic (i.e., a boolean expression) where each term of the formula is a local predicate of process P_i. A set of global predicates Φ valid for protocol P can be expressed as:

$$\Phi = \{\Phi_1, \Phi_2, \ldots, \Phi_t\}$$ ■

3 TASKS INVOLVED IN THE DETECTION OF UNSTABLE DYNAMIC PROPERTIES

The detection of unstable dynamic properties involves two tasks. The first one defines the ways a property (global predicate) can be expressed, and the second is the design of algorithms to detect such properties. Clearly the rules used for expressing the properties will guide the design of algorithms to detect such properties. For instance, a property can be defined in a general way [4] involving relations among variables in different processes. In this case we need to identify all the consistent global states, which can be represented by a lattice, to determine whether the property definitely or possibly occurred [1].

In Section 6 we compare the different solutions proposed in the literature to represent predicates and their algorithms and the representation used in this thesis with the algorithm proposed in Section 4.3.

4 DESIGN PRINCIPLES RELATED TO TESTING

This section has four parts. The first part describes how dynamic properties are represented (Section 4.1). The second part explains the basic principles used to detect dynamic properties (Section 4.2). The third part describes in details the algorithm to check dynamic properties based on local predicates (Section 4.3). Finally, the fourth part shows that for a specific type of protocol, namely 3-way handshake protocols, we can compute general properties using this algorithm (Section 4.4).

4.1 Representation of dynamic properties

Recall from the discussion in Section 1 that we are interested in checking dynamic properties during the testing phase as well as during normal execution of the protocol after the testing process. In this case, the representation of dynamic properties must consider two important requirements. First we need to check properties continuously, and second, a dynamic property should describe a valid trace for the protocol specification.

It is well known from automata theory that the languages accepted by finite automata are the languages defined by regular expressions. Furthermore, regular expressions can define infinite languages which, in our case, represent the valid traces for the protocol specification. From Definition 7 we can see that a global predicate p is a regular expression which defines a finite trace. If we want to extend the global predicate to represent an infinite language, we must include the operation for concatenation of expressions represented by the symbol $+$ as in p^+.

Definition 8 (Dynamic property) A dynamic property is a global predicate expressed in the form given in Definition 7 which may contain the operation for concatenation of expressions represented by the symbol $+$. ■

The way a global predicate is expressed is different from equivalence of a regular expression L and a finite automaton A_L in two aspects. First, in the case of language recognition there is only one symbol to be processed at each state of the automaton A_L. In the case of property detection we shall see that we may have a set of valid local predicates in each state of the automaton A_Φ that corresponds to the property Φ. Second, a dynamic property can only use the symbol $+$ for concatenation of expressions and not the symbol $*$. (See the discussion following Theorem 10 for not including the symbol $*$.)

From Definition 8 it follows that a dynamic property can be expressed by a deterministic finite automaton (DFA). In fact an automaton is a convenient representation for properties since:

- it represents the nature of communication protocols, that is, of reactive systems;
- it is compatible with the communicating extended finite state machines model used in this thesis to represent protocols;
- it allows the protocol behavior to be partitioned, that is paths and phases, where paths represent partial traces.

Figure 1–(a) shows a binary relation between local predicates and global predicates.

Let the binary relation μ_i $(i = 1 \ldots n)$ from ϕ_i to Σ be defined by

$$R_{\mu_i} = \{(\phi_i^x, \Sigma_j), \phi_i^x \in \phi_i, \Sigma_j \in \Sigma \mid \phi_i^x \text{ is a term of } \Phi_j\}.$$

The equivalent matrix representation M_{μ_i} for R_{μ_i} is given in Figure 1–(b). The entry $M_{\mu_i}[\phi_i^x, \Sigma_j] = 1$ iff ϕ_i^x is a term in the definition of the global predicate Φ_j. Otherwise it is zero.

Note that if column j of M_{μ_i} has all entries equal to zero it means that predicate Φ_j does not contain any local predicate of ϕ_i and this column can be removed. If line x of M_{μ_i} has all entries equal to zero it means that predicate ϕ_i^x does not appear in any global predicate. The designer should analyze this problem which may be an indication of error.

4.2 Detection of dynamic properties

To detect a property we will build the automaton A_j that represents the global predicate Φ_j. The idea of the detection algorithm is to superimpose onto each process P_i $(i = 1 \ldots n)$ that implements protocol P a local checking procedure that repeatedly performs the following steps:

(i) check and determine the valid local predicates in the set ϕ_i when process P_i goes to a new state, and move the automaton A_j to a new state if there are transitions associated with the valid local predicates;

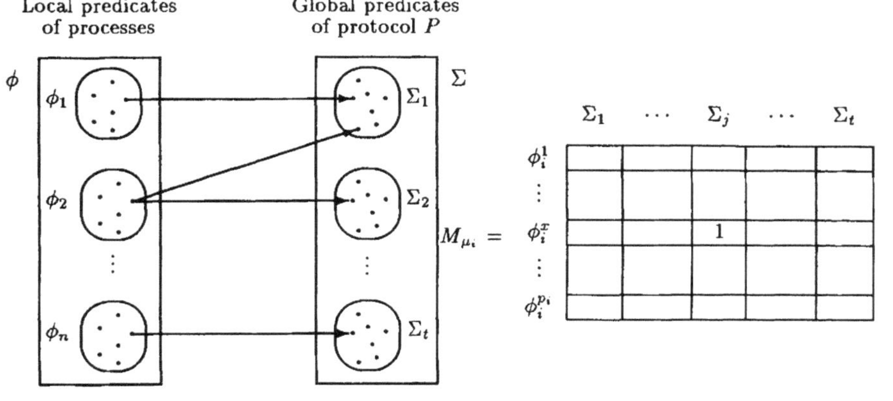

(a) Binary relation R_{μ_i} between local and global predicates.

(b) Matrix representation of binary relation R_{μ_i}.

Figure 1 Binary relation R_{μ_i}.

(ii) append the current state of the automaton A_j^i when P_i sends a message to another process; and

(iii) check the local predicates as in (i) when P_i receives a message from P_k and move the automaton A_j to a new state based on the current state and the state received from P_k.

The basic idea of this procedure is to use the automaton A_j to keep track of the protocol behavior. If the behavior exhibited by the protocol is valid, the automaton A_j, which represents property Φ, will eventually reach an accepting state. Of course if the protocol has, for instance, an unspecified reception, the procedure will not be able to check the property since it is superimposed onto each process that implements P. Therefore, the protocol should be checked for the common desired properties before applying this algorithm.

Note that this high level algorithm can be applied to any protocol or distributed program in general. In Section 4.3 we shall describe this algorithm in detail. The properties to be checked depend on each protocol and we shall assume that they are part of the input to the algorithm.

In the following we apply this high level algorithm to an example. Suppose we have two processes P_1 and P_2 that implement a protocol P such as the ISO Transport Protocol. Furthermore, each process implements a different class of the transport protocol. Given this scenario we would like to make sure, for instance, that the parameters negotiated during the connection establishment remain valid for the entire connection. Let $\phi_1 = \{\phi_1^1, \phi_1^2, \phi_1^3\}$ and $\phi_2 = \{\phi_2^1, \phi_2^2, \phi_2^3, \phi_2^4\}$ represent the set of local requirements for processes P_1 and P_2 respectively. Let us assume that the property that says that "the parameters selected must remain valid for the entire connection" can be expressed as $\Phi_j = (\phi_2^1 \vee \phi_2^2)^+ \wedge \phi_1^1 \wedge \phi_1^2 \wedge (\phi_2^4)^+$. This predicate can be represented by the automaton in Figure 2-(a). Figure 2-(b) depicts a possible computation for this protocol with the relevant events and messages shown. Next to each event is the set of local predicates that hold at that moment. The current state of the automaton is given inside a circle. Without loss of generality, assume that the protocol P executed the following computation according to the lattice of consistent global states as shown in Figure 2-(c): $(0,0), (0,1), (1,1), (1,2), (2,2), (2,3), (3,3), (4,3), (4,4)$

The initial state of the automaton for both processes is q_j^0. When the event e_2^1 occurs the state of the automaton A_j in P_2 goes to q_j^1 after checking the local predicates. At this point, a message $\langle m, q_j^1 \rangle$ is sent to P_1. Process P_1 receives the message and checks its local predicates. At this moment P_1 is in state q_j^0 and P_2 in state q_j^1. Note that the only possible transition in the automaton of P_1 is (q_j^1, ϕ_1^1, q_j^2) and therefore this transition is executed. If the local predicate ϕ_1^1 is not valid when event e_1^1 happens then we must take the automaton to its initial state since we cannot have a valid computation that started at a valid state with an invalid prefix. Then computation proceeds as before. When the event e_2^4 occurs, the state of the automaton A_j in P_2 goes to q_j^4 after checking the local predicates. At this point, a message $\langle m, q_j^3 \rangle$ was received from P_1. Again, the only possible transition in the automaton of P_2 is (q_j^3, ϕ_2^4, q_j^4) and therefore this transition is executed. When the state q_j^4 is reached, the property Φ_j is true in process P_2. Process P_1 will eventually detect this property if the local predicates associated with Φ_j remain true (i.e., ϕ_1^1 and ϕ_1^2) and P_2 sends a message to P_1. If the former condition does not hold it just reflects the temporal behavior of the protocol. If the latter condition does not happen it is because the protocol was not designed to send another message to P_1.

There are two important points that should be noted in the algorithm proposed. First, the properties do not specify interleaved behaviors of the protocol but conditions that should hold with time. This can be seen from the lattice of consistent global states of the distributed computation as shown in Figure 2-(c). All possible interleavings can be obtained from the lattice. Independent of which path of computation occurs, the property will hold iff along the path the local predicates of each process are valid when the checking is performed. Second, the checking procedure does not modify the behavior given by the protocol specification, it simply appends the current state of the automaton to each message sent.

4.3 Algorithm to check dynamic properties based on local predicates

The algorithm to check dynamic properties is given in Figure 3 and is divided into several parts as explained below.

Data structures and algorithm

▷ **Data structures**

The data structures used in the algorithm are described in the following:

- *StateOfAutomaton*$[1 \ldots t]$: each entry of this array represents the current state of the Automaton A_j $(j \ldots t)$ in the Process P_i.
- $LP[1 \ldots p_i]$: the x-th entry $(x = 1 \ldots p_i)$ of this array indicates whether the local predicate ϕ_i^x is true at the current state of Process P_i.

  The variable p_i represents the cardinality of the set ϕ_i as given in the Definition 6.
- *ValidLP*: set of valid local predicates when process P_i moves to a new state.
- *EndOfTransition*: boolean variable that indicates if there is more transition to be executed.

▷ **Initialization: Lines 1–3**

Initializes each automaton to its initial state that is by definition the state q_j^0.

▷ **Main part: Lines 4–9**

This is the main part of the program that repeatedly checks the properties when the process P_i goes to a new state or appends the array *StateOfAutomaton* to $\langle m \rangle$ when a message is sent.

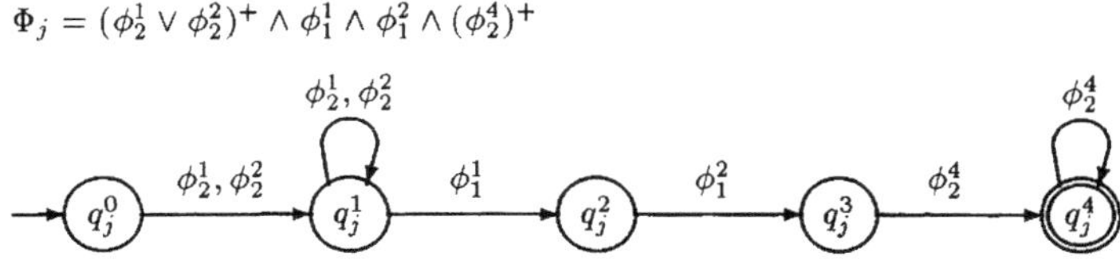

(a) Global property and the corresponding DFA.

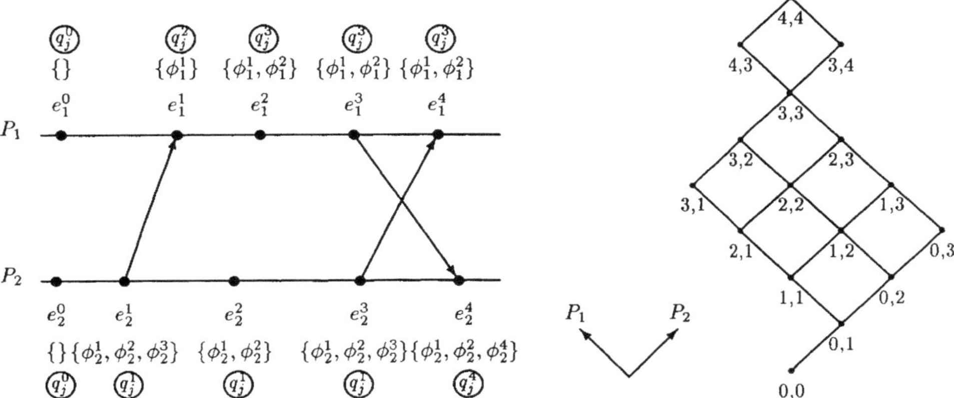

(b) Time-space diagram of the distributed computation with the valid local predicates at each state.

(c) Lattice of consistent global states of the distributed computation.

Figure 2 Checking of properties in a distributed computation.

The part that checks the properties can be seen as a function and is given in lines 10 to 46.

▷ Check each local predicate defined in P_i: Lines 10–12
When process P_i goes to a new state, we need to check each local predicate in order to evaluate the global properties. The truth-value of each local predicate depends on its definition. We give a generic function called *CheckLP* that checks a local predicate.

▷ Validation of global properties: Lines 13–44
In this part we check each property in the set Φ.

▷ Determine set of valid local predicates in Φ at the current moment: Lines 14–23
An automaton A_j^i can only execute a transition for a local predicate ϕ_i^x iff this local predicate is a term in Φ_j and is valid at the current moment in the process P_i. If the local predicate is a term of Φ_j but is invalid and there is a transition associated with ϕ_i^x at the current state of A_j^i then the automaton A_j^i must be reset. This guarantees the validity of Theorem 9.

Theorem 9 An execution (trace) that is validated according to the dynamic property Φ does not contain an invalid prefix (sequence of invalid local predicates) after its initial state q_j^0.

Proof. By definition, the initial state q_j^0 is a valid state for the automaton A_j. If A_j changes its state from q_j^0 to q_j^1, it is because there is a local predicate ϕ_i^x valid at state q_j^0 and a transition (q_j^0, ϕ_i^x, q_j^1) exists. If the local predicate ϕ_i^x is invalid then it is not possible to execute the transition. Suppose the current state of automaton A_j is q_j' and it is possible to associate a sequence of valid transitions with the path that led the automaton to this state. Now suppose that (q_j', ϕ_i^x, q_j'') is a possible transition at state q_j' but the local predicate ϕ_i^x is invalid so that it is not possible to execute this transition at this time. If we leave the automaton at this state it is possible this predicate may become true later on and we will execute this transition. That means that if the property Φ eventually holds after the state q_j' later on, Φ will be true but during the computation there was a false condition. To avoid this inconsistency, we reset the automaton to state q_j^0. □

▷ Determine next state of the automaton when it receives a message: Lines 24–26 At this point the current state of the automaton A_j is given by *StateOfAutomaton*[j] and a new state q_j^x is received. Therefore, one of these states has to be chosen so the process of checking the property can be carried on. Before showing how this state is determined, let us examine the situations in which an automaton A_j can and cannot move.

The automaton in P_i can move if there is a transition $\mathcal{T} = (q_j', \phi_i^x, q_j'')$ where q_j' is the current state of A_j^i and $\phi_i^x \in$ *ValidLP* (case i). The automaton cannot move when there is a transition $\mathcal{T}$ but $\phi_i^x \notin$ *ValidLP* (case ii), and when there is no transition $\mathcal{T}$ where $\phi_i^x \in \phi_i$ (case iii). If case (ii) occurs it means that the automaton must be reset as shown in Theorem 9. In case (iii), the automaton can only move from this state if there is a predicate ϕ_k^x in process P_k $(k = 1 \ldots n)$ that satisfies condition (i). The automaton must move until one of the conditions (ii) or (iii) happens.

Recall from Definition 8 that a global property may contain concatenation of expressions represented by the symbol +. Therefore state q_j^0 has the following characteristics: (a) it cannot be an accepting state for the automaton; (b) there is no transition that ends at q_j^0; and (c) it has just one transition that starts at q_j^0. Condition (a) follows from (b) but it was intentionally given to emphasize the characteristic of this state.

To determine the state of the automaton we need to enumerate the possible combinations in which processes P_1 and P_2 stopped moving (i.e., conditions (ii) or (iii) above). The combinations are:

Combination	1	2	3	4
P_1	(ii)	(ii)	(iii)	(iii)
P_2	(ii)	(iii)	(ii)	(iii)

Condition (ii) means a reset and condition (iii) means that the move depends on the other process.

Suppose that P_1 is the process that receives a message. (For P_2 is equivalent.) The first combination is trivial and the state of the automaton A_j^1 remains unchanged. The fourth combination shows that the next state of A_j^1 must be the state of the automaton A_j^2 received from the process P_2. The second and third combinations are symmetric. Let us analyze the second combination. The automaton A_j^1 is in the state q_j^0 and the automaton A_j^2 is in the state q_j^x. If the state q_j^x in A_j^2 can be reached directly from state q_j^0 (i.e., with transitions involving only local predicates in

P_2) then this situation is similar to the fourth combination and the next state of the automaton must be q_j^x. However, if the state q_j^x cannot be reached directly from q_j^0, it means that process P_1 has already contributed to this path, i.e., there is at least one transition with a local predicate of P_1 before reaching state q_j^x. Since there is no transition that ends at q_j^0 (condition (b) above) it means that the automaton A_j^1 was reset and therefore the next state of A_j^1 must be q_j^0 according to Theorem 9. This analysis also applies to the third combination.

Theorem 10 When a process P_i receives a message with the state of the automaton A_j in P_k the next state of automaton A_j in P_i can be uniquely determined.

Proof. Given above. □

Note that if we allow the concatenation of expressions represented by the symbol $*$ in the definition of global properties, then the conditions (a), (b), and (c) above do not hold anymore and we cannot apply Theorem 10 directly as stated. This does not seem to be an important restriction since if a local property can happen zero or more times, it is probably meaningful if it has to happen at least once.

▷ Determine whether the automaton can move: Lines 27–40
At this point we know the set of valid local predicates and the current state of the automaton A_j and would like to determine if the automaton can move to a new state. This must be done following the conditions (i), (ii), and (iii) described above.

▷ Check whether the current state of the automaton satisfies the global property: Lines 41–43
If the current state of the automaton A_j is an accepting state then the property Φ is valid at the current state of process P_i.

Complexity of the algorithm

We provide an analysis of the complexity of the algorithm to detect one property Φ_j in lines 13 to 44. Let p_i be the cardinality of the set of local predicates ϕ_i and e the cardinality of the set of transition functions δ_j. The algorithm has five sequential parts. The first part (lines 10–12) validates each local predicate ϕ_i^x. Suppose that each predicate can be checked in constant time. Then the validation of the local predicates can be carried out in time $O(p_i)$. The second part (lines 14–23) determines the set of valid local predicates in Φ_j which is also executed in time $O(p_i)$. The third part (lines 24–26) determines the next state of the automaton when it receives a message. There are four cases to be analyzed and this part is executed in constant time. The fourth part (lines 27–40) moves the automaton A_j^i, if possible. Since this depends on the number of transitions in A_j this is bounded by $O(e)$. The fifth part checks if the current state of the automaton satisfies the global property and this can done in constant time. Therefore, the five parts together are bounded by $O(p_i + e)$.

4.4 Special global states in the computation

In [1] Babaoğlu and Marzullo show that a property Φ involving relations among variables in different autonomous[2] processes (i.e., general predicates) can only be detected by building the lattice of consistent global states and then traversing it to determine whether Φ definitely or

[2] A non-autonomous process can only initiate a communication in response to a message received.

possibly is true. This problem arises because we are considering asynchronous distributed systems where the processes are autonomous and may execute at different speeds.

Begin of algorithm to check dynamic properties

Input:
- Set $A = \{A_1, A_2, \ldots\}$ of automata that represents the set of properties $\Phi = \{\Phi_1, \Phi_2, \ldots\}$ to be detected by process P_i.
- Set $\phi_i = \{\phi_i^1, \ldots, \phi_i^{p1}\}$ that represents the local predicates valid for process P_i.
- Matrix M_{μ_i} that represents the binary relation R_{μ_i}.

Output:
- Validation of each local predicate in ϕ_i.
- Validation of each dynamic property represented by an automaton in the set A.

/* Initialization */
foreach automaton $A_j \in A$ **do**
    StateOfAutomaton$[j] \leftarrow q_j^0$;
od;

/* Main part */
do forever
    (**when process** P_i **goes to a new state**) $\longmapsto$
        Check properties;
    (**when process** P_i **sends a message** $\langle m \rangle$) $\longmapsto$
        Appends the array *StateOfAutomaton* to $\langle m \rangle$;
od;

Figure 3 Algorithm to check dynamic properties (Part 1 of 3).

In a synchronous distributed system the coordination among processes happens in global synchronization points. Intuitively this means that the computations of the cooperating processes participating in a synchronization converge to a single global state since all computations have to reach that specific point (state). At that synchronization point, general properties can be evaluated. In asynchronous distributed systems there is a similar situation if there is only one autonomous process P_A participating in a synchronization and P_A is the last process to enter the synchronization state. This is exemplified in Figure 4 with two processes.

Suppose process P_2 is autonomous and P_1 is not. To execute a service in this protocol, P_2 sends a request to P_1 which may reply with a positive or negative response. If the response is positive, P_2 can go to a state $\mathcal{S}_2$ that represents "service accepted." Note that process P_1 is already in a state $\mathcal{S}_1$ that represents "service accepted" by the time P_2 receives the response message. Therefore P_2 was the last process to enter state $\mathcal{S}$ that is our synchronization state. In the executions of Figures 4-(a) and 4-(b) a positive response is represented by events $\boxed{e_2^2}$, and $\boxed{e_2^4}$ and $\boxed{e_2^5}$ respectively. The global states corresponding to these events in the time-space diagrams are in the lattices. This type of protocol is often called 3-way handshake.

Independent of which computation sequence occurred that contains state $\mathcal{S}_2$ (e.g., $[\ldots,(3,3),(4,3),(4,4),\ldots]$ or $[\ldots,(3,3),(3,4),(4,4),\ldots]$ in Figure 4-(b)), we can check general properties at this state since P_1 is in state $\mathcal{S}$ as well. Furthermore, P_1 is not autonomous and therefore will remain at $\mathcal{S}_1$. The same argument can be extended to systems consisting of two or more processes that are non-autonomous and only one autonomous process. This leads to the following theorem.

```
/*** Check properties ***/
/* Validation of each local predicate defined in $P_i$ */
foreach local predicate $\phi_i^x \in \phi_i$ do
    $LP[\phi_i^x] \leftarrow CheckLP(\phi_i^x)$;
od;
/* (P1) */

/* Validation of global properties */
foreach automaton $A_j \in A$ do

    /* Determine the set of valid local predicates in $\Phi$ at the current moment. */
    $ValidLP \leftarrow \{\}$;
    foreach local predicate $\phi_i^x \in \phi_i$ do
        if $M_{\mu_i}[\phi_i^x, \Sigma_j] = 1$ then
            if $LP[\phi_i^x] =$ true then
                $ValidLP \leftarrow ValidLP \cup \{\phi_i^x\}$;
            else /* go to the next iteration */
            fi;
        else /* go to the next iteration */
        fi;
    od;
    /* At this point set "ValidLP" contains the local predicates valid at the current
    moment in $P_i$. */
```

Figure 3 Algorithm to check dynamic properties (Part 2 of 3).

Theorem 11 3-way handshake protocols with two or more non-autonomous processes and only one autonomous process have a global state where general properties can be checked.

Proof. Given above. □

5 DESIGN PRINCIPLES RELATED TO THE EXECUTION

The algorithm to detect dynamic properties in Figure 3 provides two types of information when process P_i goes to a new state:

1. if each local predicate $\phi_i^x \in \phi_i$ is valid or not, and
2. if each global property $\Phi_j \in \Phi$ is valid or not.

The first information is provided in point (P1) and the second in point (P2) of the algorithm.

The fault/failure model [10] relates program inputs, faults, data state errors, and failures. A failure is a manifestation of a fault. Although it may or may not occur when there is a fault in the system.

If the protocol implementation runs in "detection mode" (i.e., with the detection algorithm present) we can use these two pieces of information to tackle the problems of state build-up[3] and exception handling when a fault is detected. Here there is no general solution since

[3] A reactive system such as a communication protocol interacts continuously with its environment. In

/* Determine the next state of the automaton if a message $\langle m, q_j^x \rangle$ was received from P_k ($k = 1 \ldots n \wedge k \neq i$). */

if received message $\langle m, q_j^x \rangle$ **then**
   StateOfAutomaton[j] $\leftarrow$ state according to Theorem 10;
fi;

/* Determine whether the automaton can move */

EndOfTransition $\leftarrow$ **false**; $\delta_j' \leftarrow \delta_j$;
while $\neg$*EndOfTransition* **do**
   if ($\exists$ (*StateOfAutomaton*[j], ϕ_i^x, q_j') $\in \delta_j'$)
      if $\phi_i^x \in$ *ValidLP* **then**
         if *StateOfAutomaton*[j] $= q_j'$ **then**
            $\delta_j' \leftarrow \delta_j' - (q_j', \phi_i^x, q_j')$;
         fi;
         StateOfAutomaton[j] $\leftarrow q_j'$;
      else *StateOfAutomaton*[j] $\leftarrow q_j^0$;
         EndOfTransition $\leftarrow$ **true**;
      fi;
   else *EndOfTransition* $\leftarrow$ **true**;
   fi;
od;

/* Check whether the current state of the automaton satisfies the global property. */

if *StateOfAutomaton*[j] $\in QF_j$ **then**
   /* *Global predicate* Φ_j *is valid at this state.* */
   /* (P2) */
fi;

od;

End of algorithm to check dynamic properties

Figure 3 Algorithm to check dynamic properties (Part 3 of 3).

these mechanisms depend on the semantics of each protocol. A possible strategy in the case of exception handling is to abort or reinitialize the implementation when "critical" local predicates and properties are not satisfied. This solution is better than allowing an erroneous behavior of the protocol implementation if no action is taken [11]. From the point of view of the other

this case valid input events may lead to an erroneous state in the protocol implementation. When the implementation reaches an erroneous state it may either continue to run but producing erroneous output or may crash and stop. An erroneous state was reached in the implementation because the values associated to its local variables and the contents of its communication channel are not correct according to constraints or requirements given in the protocol specification. Note that this behavior can only be identified if test sequences are as long as the length of the faulty path.

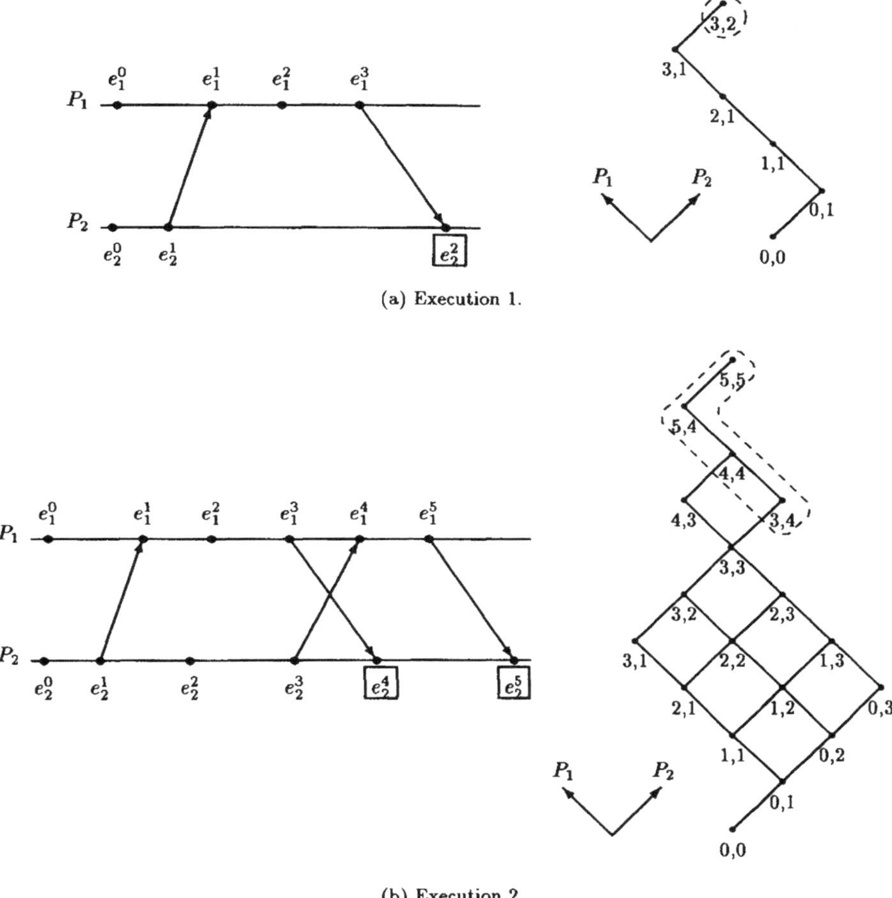

Figure 4 Special global states in asynchronous distributed systems.

cooperating processes, the action of aborting or reinitializing an implementation can be seen as a physical failure and therefore the protocol should handle it.

6 RELATED WORK

In the following we present the related work in chronological order.

Miller and Choi [9] define linked predicates "that can be ordered by the happened-before relation and are specified by expressions using the $\rightarrow$ operator." They use these predicates in breakpoints in a distributed debugger for halting the system. The halting algorithm is based on the Chandy-Lamport snapshot algorithm [3]. Spezialetti and Kearns [14] consider monotonic events that are similar to stable properties and discuss "conditions which must be met in order for specific assertions to be made about an event or the system state." Cooper and Marzullo [4] consider general properties which are intractable in practice since they involve building a lattice of consistent global states that can be exponential in the number of events in the system. Furthermore, all possible paths have to be checked. Hurfin, Plouzeau and Raynal [8] consider unstable nonmonotonic global predicates, called atomic sequences of predicates. These predicates describe global properties by causal composition of local predicates augmented with atomicity constraints that specify forbidden properties. Garg and Waldecker [7, 16] define properties in terms of boolean expressions using logic operators. The algorithm to detect these properties is implemented using a global monitor that collects information from all processes and evaluates the predicates. Venkatesan and Dathan [15] also define properties in terms of boolean expressions and give a distributed algorithm to detect these properties. However, the evaluation of properties is performed off-line and they assume that executions of the system are reproducible. Furthermore, they only consider FIFO channels.

The algorithm proposed in this paper considers properties expressed as boolean expressions with concatenation of expressions represented by the symbol +. We give a fully distributed detection algorithm that works on-line and does not modify the protocol specification.

In the case of a centralized monitor P_Φ, each process P_i $(i = 1 \ldots n)$ has to send a message to P_Φ so the property can be checked. Furthermore, if we want to use the information provided by the algorithm during normal execution (see Section 5) then the execution in process P_i has to be delayed. The algorithm proposed in this paper does not have the cost of sending extra messages since it is distributed and does not delay the execution of process P_i.

For an overview of the concepts related to this paper and some of the approaches proposed in the literature, the interested reader is referred to [13].

7 CONCLUSIONS

Verification and testing are two complementary techniques and often used during the protocol development cycle.

In the context of protocol testing, most of the properties that characterize valid or invalid behaviors cannot be expressed in terms of stable properties. In fact, these properties are more properly expressed in terms of desirable or undesirable temporal evolutions of the communication protocol behavior. These properties are inherently unstable since there is no guarantee that they will remain either valid or invalid in a protocol computation.

In this paper we have presented a new algorithm to detect dynamic unstable properties that can be used in the testing of distributed processes (modules) of a communication protocol. This algorithm provides two types of information that can be used for tackling two problems during program execution: state build-up and exception handling.

The algorithm is based on the observation that given a protocol specification there are multiple valid computations (traces) each of which can be defined by a causal precedence order. Dynamic properties are specified by stating conditions (using local predicates) that should hold on all possible computations.

We have also presented a new theorem that can be used to check general properties for a specific type of communication protocol, namely 3-way handshake protocols.

From the point of view of testing communication protocols or concurrent reactive systems the algorithm presented in this paper is an improvement over the ones presented in [7, 16] and [15].

REFERENCES

[1] O. Babaoğlu and K. Marzullo. Consistent global states of distributed systems: fundamental concepts and mechanisms. In Sape Mullender, editor, *Distributed Systems*, chapter 4, pages 55–96, ACM Press and Addison-Wesley, 1993.

[2] K. Mani Chandy, L.M. Haas, and J. Misra. Distributed deadlock detection. *ACM Transactions on Computer Systems*, 1(2):144–156, May 1983.

[3] K. Mani Chandy and L. Lamport. Distributed snapshots: Determining global states of distributed systems. *ACM Transactions on Computer Systems*, 3(1):63–75, February 1985.

[4] R. Cooper and K. Marzullo. Consistent detection of global predicates. In *Proc. of the ACM/ONR Workshop on Par. and Dist'd Debugging*, pages 167–174, Santa Cruz, CA, USA, 20–21 May 1991. Published as ACM SIGPLAN Notices, 26(12), December 1991.

[5] E.W. Dijkstra and C.S. Scholten. Termination detection for diffusing computation. *Information Processing Letters*, 11:217–219, August 1980.

[6] N. Francez. Distributed termination. *ACM Transactions on Programming Languages and Systems*, 2(1):42–55, January 1980.

[7] V.K. Garg and B. Waldecker. Detection of weak unstable predicates in distributed programs. *IEEE Trans. on Par. and Dist'd Sys.*, 5(3):299–307, March 1994.

[8] M. Hurfin, N. Plouzeau, and M. Raynal. Detecting atomic sequences of predicates in distributed computations. In *Proc. of the ACM/ONR Workshop on Par. and Dist'd Debugging*, pages 32–42, San Diego, CA, USA, 17–18 June 1993. Published as SIGPLAN Notices, 28(12), December 1993.

[9] B.P. Miller and J.-D. Choi. Breakpoints and halting in distributed programs. In *Proc. of the 8th Int'l Conf. on Dist'd Computing Sys.*, pages 316–323, San Jose, CA, USA, 13–17 June 1988.

[10] D. Richardson and M. Thomas. The RELAß model of error detection and its application. In *Proc. of the ACM SIGSOFT/IEEE 2nd Workshop on Software Testing, Analysis, and Verification*, Banff, Alberta, Canada, July 1988.

[11] J. Rushby. *Formal Methods and the Certification of Critical Systems*. Technical Report CSL–93–7, SRI International, Computer Science Laboratory, Menlo Park, CA, USA, 1993.

[12] B.A. Schroeder. On-line monitoring: A tutorial. *Computer*, 28(6):72–78, June 1995.

[13] R. Schwarz and F. Mattern. Detecting causal relationships in distributed computations: In search of the holy grail. *Distributed Computing*, 7(3):149–174, 1994.

[14] M. Spezialetti and J. P. Kearns. A general approach to recognizing event occurrences in distributed computations. In *Proc. of the 8th Int'l Conf. on Dist'd Comput. Sys.*, pages 300–307, San Jose, CA, USA, 13–17 June 1988.

[15] S. Venkatesan and Brahma Dathan. Testing and debugging distributed programs using global predicates. *IEEE Transactions on Software Engineering*, 21(2):163–177, February 1995.

[16] B. Waldecker and V.K. Garg. Detection of strong predicates in distributed programs. In *Proc. of the Third IEEE Symp. on Par. and Dist'd* , pages 692–699, Dallas, USA, 2–5 December 1991.

PART SIX

Test Generation 2

13

A rigorous and practical approach to service testing

L. Feijs.
Philips Research Laboratories,
Eindhoven University of Technology.
Address: Philips Research, Prof. Holstlaan 4, 5656AA Eindhoven,
Tel. +31 402742953, Email: feijs@natlab.research.philips.com

M. Jumelet.
University of Amsterdam.
Present affiliation: Data Sciences B.V. Amsterdam.

Abstract

We present an approach for the systematic development of abstract test cases for service testing. The approach is rigorous in the sense that the abstract test cases are derived from a rigorous specification. The approach is practical in the sense that it contains guidelines for choosing amongst the many possible execution paths. The method has been developed in practice. Tools are used for interactive simulation and for certain translation steps.

Keywords

Protocols, testing, TTCN, PSF, MSC, SDL.

1 INTRODUCTION

Distributed systems are most often organised along two more or less orthogonal decomposition principles: distribution and layering. Testing the components of distributed systems is an important issue, but because there are two decomposition principles, there are several kinds of testing (such as conformance testing and service testing). Much work has been done on protocol conformance testing, which is about testing protocol entities, that is, the implementations which are at the intersection of these two principles (a protocol entity is rougly speaking the local implementation dealing with one layer). Conformance testing is of key relevance for the correct operation of open systems, in particular if there are many different implementors.

On the other hand, there are many situations in which there are no multiple vendors (or not yet). Moreover the service to be provided by some layer may be completely new; in this case it is likely that priority is given to get this service as soon as possible properly implemented and tested and that the conformance issues are left for a later phase.

Therefore we believe it is appropriate to address the issue of 'service testing'. Service testing presents certain technical difficulties which make it different from conformance testing. In particular the state space of the system is the product of the state spaces of the components. The specification of a service is more difficult than the specification of a single protocol entity. Certain concepts and techniques originally developed for conformance testing can be fruitfully applied to service testing too (with adaptations).

Moreover it is well known (Knightson 1993) that in practice conformance does not imply interoperability. Amongst several reasons, this is because in practice, testing can not be exhaustive. The term 'interoperability testing' is often used for the trials to see if the system as a whole really works. This is considered supplementary to conformance testing.

We use the term 'service testing' to stress our point of view that it is more than a supplement to fill the holes left by conformance testing, and that service testing has a value in its own right. Service testing deals with the compatibility between the artifacts created by the decomposition principle of layering (the layers), without assumptions about the protocols that implement a layer. The service is tested against a service specification.

We use terminology from the world of conformance testing (Knightson 1993). Some additional terms will be proposed for the purposes of service testing.

2 SURVEY OF THE TEST DEVELOPMENT METHOD

The approach to develop abstract test cases is rigorous (the abstract test cases are derived from a rigorous specification) and practical (it contains guidelines for choosing execution paths). The main steps of the method are divided into two groups: (1) preparation steps, relevant for the whole test suite, and (2) steps to be done for each test goal. The following steps must be done once:

- rigorous specification of the service, using a formalism which allows simulation. This serves as a model for the external behaviour of the service (not of the real implementation). We used the process specification formalism PSF (Mauw, Veltink 1993).
- making the test architecture explicit.
- extraction of a finite state diagram, which we call the service state transition diagram (S-STD). It serves as a model of the service as perceived by one local user.
- performing a further reduction of the S-STD, keeping only particular states, which we call 'stable states' and distinguishing the user rôles of Initiator and Follower. The resulting diagram is called rôle state transition diagram (R-STD).
- establishing test goals, in particular by choosing the ability to sucessfully perform certain transitions (one goal for each transition from the Follower's R-STD).

The following steps must be done for each test goal:

- draw a rôle transition path (RTP), which is an outline of two or more simultaneous walks through the R-STDs. This is needed as a guidance when interactively performing the next step (simulation).
- simulate a 'straight' execution, by which we mean a sequence of events in which the service behaves as expected.

- construct a sequence chart (SC) from the simulation trace. The SC serves for documenting the test (useful when analysing log files later and for debugging). Moreover the SC is used as an intermediate format before making the next step.
- construct an initial TTCN description. This is essentially the same as the 'straight' execution, with alternatives '`?OTHERWISE FAIL`' added for each step.
- simulate alternatives: in general the service can have internal non-determinism, most of which is because the service model also describes the behaviour under 'bad conditions' (i.e. the underlying medium or network fails).
- complete the TTCN description by assigning verdicts `PASS` or `INCONC` to the alternatives and editing the initial TTCN description.

The SC formalism used is a simplified (synchronous) version of MSC (CCITT 1992); within Philips such SC used to be indicated by the somewhat confusing term 'interworkings'. Test execution is outside the scope of this paper; for a few remarks on it see Section 15.

3 EXAMPLE OF A SERVICE

In this section we will sketch a service offered by a small hypothetical system. It is a true toy example, devised in order to explain our test approach; it is kept deliberately simple in order to fit in the scope of the present report. While working on this paper we discovered that this kind of service is more realistic than we initially thought, because the Wehkamp vending service (Wehkamp 1996) has some similar characteristics.

The service is a distributed service, as shown in Figure 1. There is a service interface which runs as a horizontal line through the architecture. If it is our goal to test this service, the implementation under test (IUT) is the shaded area in Figure 1.

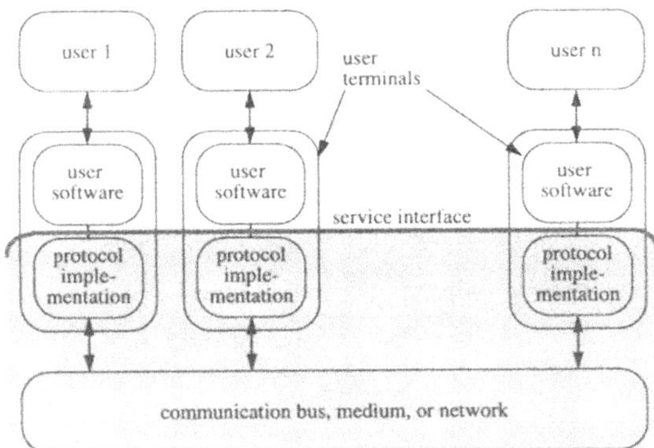

Figure 1: Typical System Architecture.

In our toy example, the service is meant to support processes of negotiating between two or more users about the price of certain objects. One user can offer some object (for example a painting or some other piece of art) and associate a price with it. Later, as time passes, he may decide to decrease the current price. Other users can bid for the object and later increase there current bid. If the system finds a match, the object is considered sold and the two users involved are informed. The abstract service primitives (ASPs) are:

1. offer(object,price): offer an object and require that at least 'price' (in some agreed unit currency like thousands of Guilders) be paid for it.
2. decrease(N): decrease the required price by N (only for a user who offered an object).
3. flush(): remove all bidders in order to start the negotiation process over again (only for a user who has offered an object at present)
4. bid(object,price): try to acquire 'object' for 'price'. The mechanisms to find out whether indeed someone is offering 'object' are not included in the service.
5. increase(N): increase the required price by N (only for a user who is bidding).
6. stop(): go to the idle state (withdraw offer or bidding).
7. ok(): indication from the service to the user to confirm 'offer' and 'bid'.
8. notok(): indication from the service to inform the user that a previous command offer, or 'bid' is not accepted. Also used to inform a user that a previous command like 'decrease', 'flush', or 'increase' was issued in a state not suitable for execution of that command.
9. sold(price): indication from the service to inform both an offering and a bidding user that the current object has been sold for 'price' (this example does not allow for simultaneous offering of objects, which of course could be modeled).

When specifying a service like this, it is often fruitful to consider it as a single process (even if it is distributed in reality). This view is used in Figure 2, where we show one typical run of the system. There are two users, called Jelle and Sjeng (two typical Dutch names) who are negotiating for a Degas (a typical impressionist French painting). In the run shown, the answer 'sold(4)' is sent to Jelle first, but the other order, where Jeng gets his 'sold(4)' answer first is possible too (this does not follow from Figure 2, but such facts could follow from a rigorous specification as in Section 4).

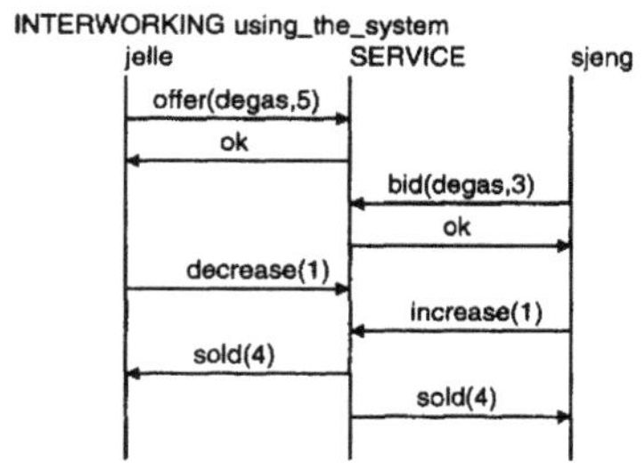

Figure 2: Example run of the system.

4 STEP: RIGOROUS SPECIFICATION OF THE SERVICE

This section is about the rigorous service specification, using PSF (Mauw, Veltink 1993). Type checking and simulation tools for PSF are publicly available. It is based on algebraic principles: signatures and equations, both for data types and for processes. It builds upon the theory of initial algebras for data types and upon the algebra of communicating processes (ACP) for the processes.

The advantage of having a rigorous specification is that it is much more precise than an informal description and that the simulator can be used to generate event-traces which are automatically correct. There is also a price to be paid for this: writing and reading rigorous specifications, requires certain skills and significant investments. We use a number of additional diagrams to make the model as readable and intuitive as possible: (S-STD, R-STD, SC, cf. Figures 5, 6 and 2 respectively)

The service interface can be viewed as a kind of application programmer's interface (API) which hides a communication mechanism and which can be used as a basis for building a system with user interface for object-offering and bidding. The atomic actions `c` and `a` model the commands and the answers which are sent to the service and received from the service, respectively. The commands and the answers are parameterised: their first parameter tells which user performs the request or gets the answer. The second parameter carries the real information of the abstract service primitive. The entire service interface layer is modeled as a single process `S` which accepts service requests (commands) from the users and gives answers to the users. It has an additional parameter of a type called `USERLIST`. This parameter is carried along for every recursive invocation of `S`, and in a certain sense, it contains the state of the service layer as a whole.

The PSF specification imports the data types used (`Nat`, `Obj`, `User`, `Bool`, `Info`, `State`, `UserList`). The internal state of the service is modeled as a list of four-tuples, one for each user (called Peter, Karel, Sjeng and Jelle). For each user the state contains, apart from the user-identity, (1) its main state (`idle`, `opening`, `offering` etc.), (2) its current price proposal (a natural number), (3) its object of interest (either for selling or buying). The specification is fixed for a number of four users. It is obvious how to extend this to any arbitrary number of users. The initial process is specified as:

```
S( cons(peter,idle,0,NONE
 , cons(karel,idle,0,NONE
 , cons(sjeng,idle,0,NONE
 , cons(jelle,idle,0,NONE,empty))))
 )
```

The rest of the specification is one big equation for `S(L)`. The right-hand side of this equation extends over several pages (of which we show a small fragment), since it is a sum of many terms: four terms about the events which can happen in state `idle`, three terms about the events which can happen in state `opening`, and so on.

We show the term which describes what happens if user U offers object (painting) P for a price of N, provided this user is in state idle. The equation must be valid for all alternative values of U,P and N and therefore the term is described using three `sums`. The `sum` construct generalises the `+` operator of alternative composition (choice). Please read `sum( U in USERs, xyz(U))` as $\sum_{U \in \text{USERS}} \texttt{xyz}(U)$ which is the same as `xyz(peter) + xyz(karel) + xyz(sjeng) + xyz(jelle)`.

There are a number of auxiliary functions which work on data types (specified algebraically but not included here). For a user U and state-list L, `state`(U,L) can be used to look up the individual state of this user. There are also three 'assignment' functions: `setobject`, `setstate`, and `setcurrent`. For example `setobject`(U,P,L) changes the object of interest for user U into P in the state-list L, returning a modified state-list.

After commanding an `offer`(P,N), a user is transferred to state `opening` (where it will be sorted out if he will get an `ok` or a `notok`). After commanding a `bid`(P,N), a user is transferred to state `trying`. After commanding a `stop` in state `idle`, nothing happens.

```
S(L) = sum( U in USERs,
       sum( P in OBJs,
       sum( N in NATs,
            [state(U,L)=idle] ->
            [eq(P,NONE)=false]->
            c(U,offer(P,N)) . S( setobject(U,P
                               , setstate(U,opening
                               , setcurrent(U,N,L)))
          )))                  ) +

       sum( U in USERs,
       sum( P in OBJs,
       sum( N in NATs,
            [state(U,L)=idle] ->
            [eq(P,NONE)=false]->
            c(U,bid(P,N)) . S( setobject(U,P
                             , setstate(U,trying
                             , setcurrent(U,N,L)))
          )))                ) +

       sum( U in USERs,
            [state(U,L)=idle] ->
            c(U,stop) . S(L)
          ) +

       sum( U in USERs, sum( P in OBJs, sum( N in NATs,
            [state(U,L)=idle] ->
            ( c(U,decrease(N)) . a(U,notok) . S(L)
            + c(U,flush)       . a(U,notok) . S(L)
            + c(U,increase(N)) . a(U,notok) . S(L)
          ) )              )                ) +
```

The terms for `opening`, `offering`, `trying`, `bidding` and `finalising` are not included.

5 STEP: MAKING THE TEST ARCHITECTURE EXPLICIT

The test architecture can be different from the system architecture (like in Figure 1) which describes the system in normal use. A typical test architecture is shown in Figure 3. There is one special point of control and observation (PCO) which corresponds to the possibility of provoking 'bad conditions' (making the underlying communication bus, medium or network fail). Sometimes this must be done by a kind of manual intervention, e.g. by unplugging certain wires. In the present paper we assume the comfortable situation of a centralised test system. When testing the system, it is often fruitful to organise the tester

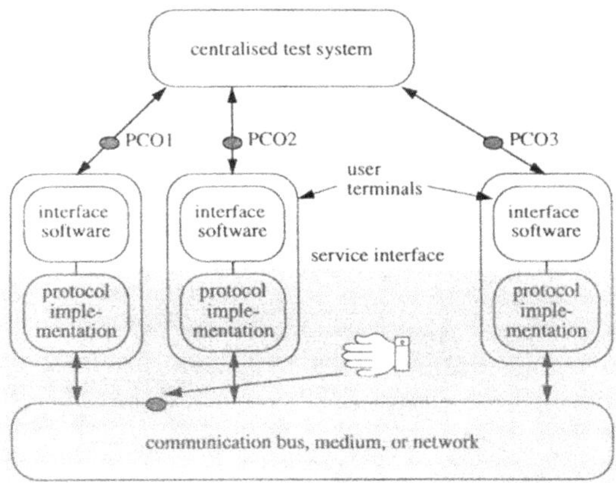

Figure 3: Typical Test Architecture.

as a single process. Because there are distinct PCOs for the devices, it is convenient now to consider each device as a single process. This view is used for example in Figure 4 below. The service is split into two processes, one for each user's PCO (so now 'jelle' and 'sjeng' do not refer to the real users, but to their respective terminals).

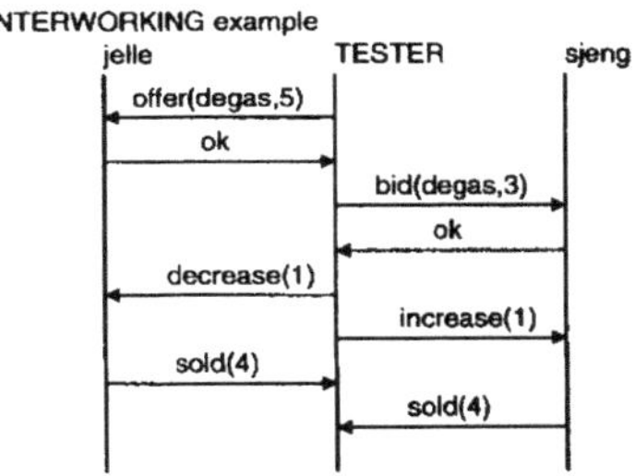

Figure 4: Test Run of the System.

6 STEP: EXTRACTION OF THE S-STD

The diagram of Figure 5 gives a survey of the entire service in a single diagram. We call this a service-STD (S-STD). The transitions are labeled with information about conditions, communications steps and internal actions. If possible this information is organised in two parts, separated by a horizontal line: the text above the line describes a command or a condition and the text below the line is the result (an answer or an internal action). In many projects this kind of diagram has been drawn already long before the test-phase of the system, since it is useful for other purposes as well. The diagram is compact and attractive from an intuitive point of view, and thus it is nice to have such a diagram next to the rigorous model.

The diagram is not rigorous: the descriptions of the conditions and the internal actions are informal and they leave room for all kinds of wrong interpretations.

In order not to complicate the diagram too much, certain actions which have no effect and which immediately return `notok` are not shown fully; by way of reminder there is just an unlabeled transition for the states in which such transitions are possible.

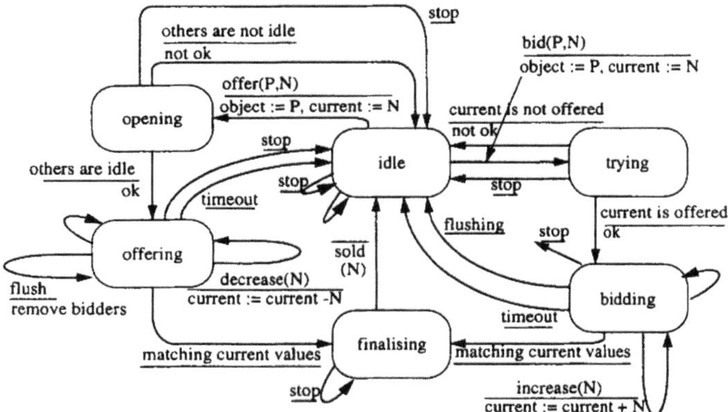

Figure 5: Finite State Diagram of the Service.

7 STEP: FROM S-STD TO R-STD

Next we perform a further reduction of the S-STD, keeping only particular states, which we call 'stable states' and distinguish the user rôles of Initiator and Follower. The resulting diagram is called rôle state transition diagram (R-STD). This is done twice: for the initiator and for the follower (Figure 6). We say that a state s is *stable* if whenever the system under test (SUT) is in this state, it will not move to another state s', unless by an external event. For example state `trying` is not stable because the service itself will find out that either an `ok` or a `notok` is to be issued. This definition of 'stable' is not completely rigorous; for example a `stop` command may force the transition from `trying` to idle too. But as a heuristic, the concept seems useful (tests will still be derived from simulation runs of the full PSF model and the loss of information in these diagram is therefore not harmful).

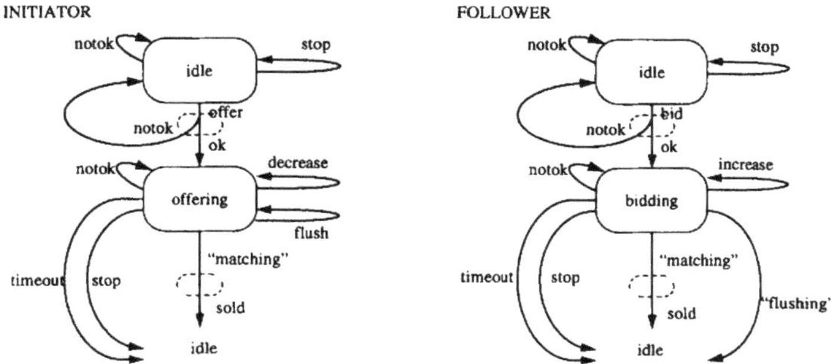

Figure 6: Role-State Transition Diagrams.

Note that idle is shown twice in each diagram; of course one occurrence is enough, but the present layout is convenient later, since test cases are drawn as walks trough the diagram. The present layout leads to walks which proceeds, roughly speaking, from top to bottom in the diagram.

8 STEP: ESTABLISH TEST GOALS

The R-STDs provide an excellent starting point for establishing the test goals. Of course one needs to have some idea about the degree of coverage required. Full test coverage is out of the question anyhow and the choice usually is a trade-off between the costs of testing on one hand and the costs of undetected defects on the other hand. Here we propose one particular choice which in some situations gives a reasonable compromise and which moreover provides some handles for managing the test development process. But of course other choices could be made here as well.

The criterion proposed here to make one abstract test case (ATC) for each transition of the Follower's R-STD. The test goals are numbered in Figure 7 (starting with the most important transitions). We add some explanation for the ATCs numbered 9, 10 and 11. These correspond to the commands **decrease**, **flush** and **increase**, all of which must yield **notok** as an answer. Similarly for 12, 13, 14 and 15 (**offer**, **decrease**, **flush** and **bid**). This has a number of advantages.

- the approach gives rise to a finite number of tests, 15 in this example (this number will increase according to the complexity of the service);
- all transitions will be executed at least once, so for certain classes of defects a complete coverage is achieved for the follower (if the service is not too complex of course); in particular these include the defect class where the transition is completely forgotten and bad code (opcode traps etc.) which causes a crash in an implementation;

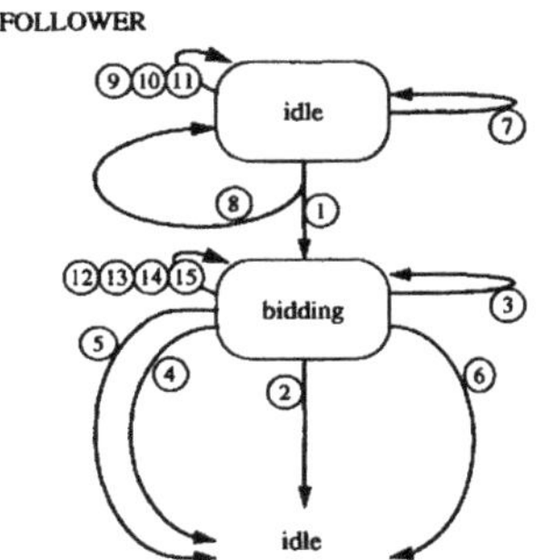

Figure 7: Survey of Test Goals.

- many services can be viewed upon as if their main purpose is that the initiator can bring the follower from one state to another; so, it is likely that a test campaign which leads the follower through all its transitions has the effect that at the end effectively the initiator has gone trough most of its transitions too (which can be checked, by marking each transition in the Initiators R-STP which is used in any of the ATCs of these 15 goals).

The followings steps are to be performed for each of the test goals.

9 STEP: DRAW A RÔLE TRANSITION PATH (RTP)

Let us demonstrate the development of ATC_2, which is the transition with condition "matching" and answer `sold`. In Figure 8 the RTP for the follower is shown (right-hand side). This path must be devised first. Based on this, one needs a certain path for the initiator in order to make the planned walk through the follower's R-STD happen. In practice it turned out convenient to draw these diagrams by hand, using a sufficient supply of paper copies of R-STD pairs (typically one initiator-follower pair on each sheet). For the coloring scheme we refer already to Figure 9. In order to have some heuristics for designing the follower's RTP we borrowed some concepts from the world of automated test generation for purposes of conformance testing, see e.g. (Sidhu, Leung 1989). We adapted them slightly; moreover we sometimes did not apply them completely rigorously according to the original definitions. As a heuristic and also for documentation purposes we found them very useful. Next to the concept of 'test-goal', which in our case coincides with a transition in the R-STD, we use three more concepts:

- transferring sequence, or preamble (a sequence of transitions in the R-STD leading from `idle` to the begin-state of the test goal).
- test goal (one of the transitions chosen in Section 8).
- unique input/output sequence (a sequence of commands and answers which uniquely distinguishes the end-state of the test goal from all other states - if possible).

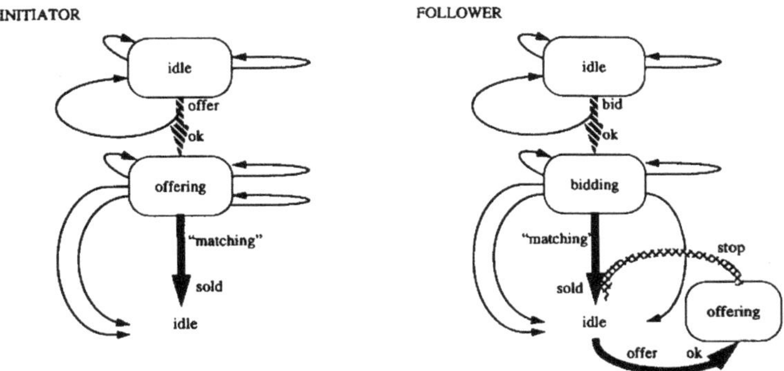

Figure 8: RTP for the Follower (and Initiator too) for ATC_2.

- synchronising sequence, or postamble (to return the system again in the `idle` state).

We used a kind of coloring scheme for indicating the four sub-segments of the RTPs. This is summarised in Figure 9. In practice it turned out convenient to use colored pens for this (green for the transferring sequence, red for the test goal, yellow for the unique input/output sequence, and blue for the synchronising sequence. For the RTPs of Figure 8

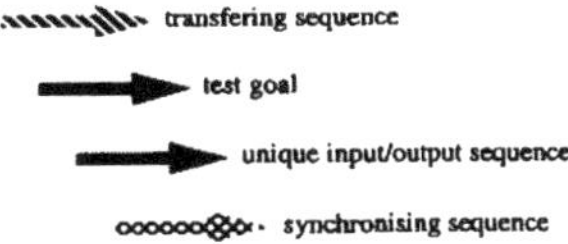

Figure 9: Coloring Scheme used in RTP.

the motivation is as follows. Since the goal is to see if the move from `bidding` to `idle` based on "matching" and with answer `sold` can be made properly, the follower needs to go to `bidding` first. Therefore the transferring sequence consists of one transition (`idle` $\stackrel{\text{bid, ok}}{\longrightarrow}$ `bidding`) in the follower's R-STD (corresponding to two transitions in the S-STD). The test goal is one transition (`bidding` $\stackrel{\text{matching, sold}}{\longrightarrow}$ `idle`) in the follower's R-STD (again corresponding to two transitions in the S-STD).

It is not sufficient that the proper answer (`sold`) has been obtained. It has to be checked whether the new state is `idle` indeed. This is done using a sequence of commands and answers which is able to uniquely identify `idle` (or at least excludes most other states). In this case it is not hard: `idle` can be characterised as the only state in which the command `offer` must be performed successfully (provided certain conditions concerning the other users hold). So the unique input/output sequence is (`idle` $\stackrel{\text{offer, ok}}{\longrightarrow}$ `offering`). And in order to leave the system in an orderly state, ready for a next test, it is brought back to state `idle` by means of `stop`. So the synchronising sequence is (`offering` $\stackrel{\text{stop}}{\longrightarrow}$

idle). Note that for the purpose of the unique input/output sequence, the follower has to switch to an initiator's role temporarily.

10 STEP: SIMULATE A 'STRAIGHT' EXECUTION

Next we use the PSF simulator to get a trace of the intended behaviour of the service. This can be done interactively; for each step, the simulator gives a clickable list of possible alternatives in a special menu window. The number of choices for each step is quite large: typically about 60 alternatives for this example service. The trace appears in another window, from which it is easily copied to a file for further processing.

This is demonstrated for ATC_2 again, using the RTP of Figure 8 as a road-map. Letting `jelle` play the initiator's rôle and `sjeng` the follower's rôle, and choosing to negotiate for a `manet`, the following trace was obtained:

```
atom c(jelle, offer(manet, s(0)))
atom a(jelle, ok)
atom c(sjeng, bid(manet, s(s(s(0)))))
atom a(sjeng, ok)
atom INT(sjeng, jelle)
atom a(jelle, sold(s(s(0))))
atom a(sjeng, sold(s(s(0))))
atom c(sjeng, offer(manet, s(s(s(s(s(0)))))))
atom a(sjeng, ok)
atom c(sjeng, stop)
```

This trace is correct by definition (if the PSF model is right); using the simulator one may make mistakes in the sense of not reading the road-map properly, but the trace found is still a valid trace for the given service (even it is not of help for the test goal).

11 STEP: FROM SIMULATION TRACE TO SC

By means of simple conversion tools, the trace can be transferred to a sequence chart in textual format. This is edited in order to make the begin-state and the end-state of the transition of the test goal explicit. Then there is another tool which makes a postcript file. The latter file only serves for documentation purposes. It is a useful help when trying to understand the TTCN-GR resulting in subsequent phases. As before, the service is split into two processes, one for each user's PCO, so 'jelle' and 'sjeng' do not refer to the real users, but to their respective terminals.

12 STEP: CONSTRUCT AN INITIAL TTCN DESCRIPTION

Since TTCN is the de facto standard for ATC description, we aim at producing graphical TTCN (TTCN-GR) next. It is not hard to construct an initial TTCN-GR description from the sequence chart automatically. We built a simple 'translator' from SC to TTCN-GR, programmed in Gofer. In general, the result is not a correct test yet, but it serves

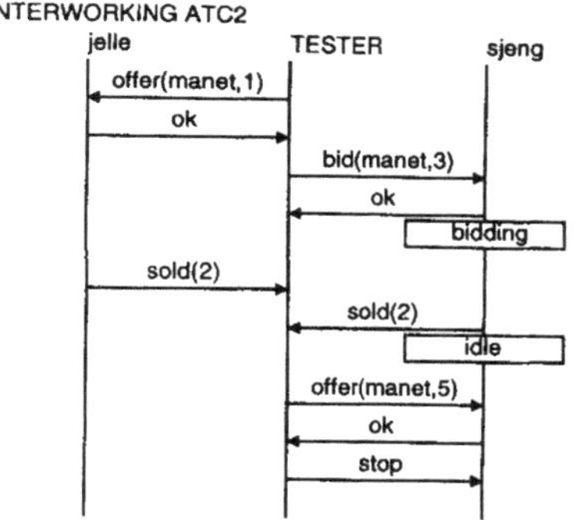

Figure 10: Sequence Chart of ATC_2.

as a draft for further development. E.g. we could remove some of the preliminary (PASS) verdicts and turn the last of them into PASS (and alternatives must be taken into account).

```
+----------------------------------------------------------------------------+
|Test Case ATC_2                                                             |
+----------------------------------------------------------------------------+
|Test Case Name : ATC_2                                                      |
|Group          : 1/                                                         |
|Purpose        : Transition from bidding to idle for follower (draft)       |
|Default        :                                                            |
|Comments       :                                                            |
+----------------------------------------------------------------------------+
|Nr | Label | Behaviour Description   | Constraints Ref | Verdict | Comments|
+----------------------------------------------------------------------------+
|0             jelle!offer(manet,1)                                          |
|1              jelle?ok                                   (PASS)            |
|2               sjeng!bid(manet,3)                                          |
|3                sjeng?ok                                 (PASS)            |
|4                 jelle?sold(2)                           (PASS)            |
|5                  sjeng?sold(2)                          (PASS)            |
|6                   sjeng!offer(manet,5)                                    |
|7                    sjeng?ok                             (PASS)            |
|8                     sjeng!stop                                            |
|9                    sjeng?OTHERWISE                      (FAIL)            |
|10                 sjeng?OTHERWISE                        (FAIL)            |
|11                jelle?OTHERWISE                         (FAIL)            |
|12               sjeng?OTHERWISE                          (FAIL)            |
|13             jelle?OTHERWISE                            (FAIL)            |
+----------------------------------------------------------------------------+
```

13 STEP: SIMULATE ALTERNATIVES

The service may have internal non-determinism, or depend on timing properties or network states which cannot be fully controlled. So there are alternative behaviours, not captured yet. The simulator can be used to find such alternatives (they appear in the simulator's menu). Some intuition and extra information about the physical test circumstances is needed as input. So we run the simulator again, looking for alternatives. After line 3 (in the above TTCN-GR), `sjeng` could make the internal step 'timeout':

```
atom c(jelle, offer(manet, s(0)))
atom a(jelle, ok)
atom c(sjeng, bid(manet, s(s(s(0)))))
atom a(sjeng, ok)
atom int(sjeng)
```

This is acceptable behaviour, but it does not help in judging the outcome of the test goal, so it is judged `INCONC`. There is another branch: the service may give `sold` to `jelle` first and to `sjeng` next, or it may do so in the reverse order:

```
atom c(jelle, offer(manet, s(0)))
atom a(jelle, ok)
atom c(sjeng, bid(manet, s(s(s(0)))))
atom a(sjeng, ok)
atom INT(sjeng, jelle)
atom a(sjeng, sold(s(s(0))))
atom a(jelle, sold(s(s(0))))
atom c(sjeng, offer(manet, s(s(s(s(0))))))
atom a(sjeng, ok)
atom c(sjeng, stop)
```

The generated TTCN will already accept this trace too, since the TTCN semantics assumes that the `sold` answer for `sjeng` is buffered until `jelle`'s `sold` answer is consumed.

14 STEP: COMPLETE THE TTCN DESCRIPTION

Several lines have to be added because of the alternatives discovered. A timer is added to find out if the `sold` answers do not happen. The timeouts are `INCONC`. Some of the preliminary verdicts generated first are removed; others are turned into final verdicts.

```
+---------------------------------------------------------------------+
|Test Case ATC_2                                                      |
+---------------------------------------------------------------------+
|Test Case Name : ATC_2                                               |
|Group          : 1/                                                  |
|Purpose        : Transition from bidding to idle for follower        |
|Default        :                                                     |
|Comments       :                                                     |
+---------------------------------------------------------------------+
```

```
|Nr | Label | Behaviour Description    | Constraints Ref | Verdict | Comments|
+------------------------------------------------------------------------+
|0           jelle!offer(manet,1)                                         |
|1            jelle?ok                                                    |
|2             START T1 (10 sec)                                          |
|3              sjeng!bid(manet,3)                                        |
|4               sjeng?ok                                                 |
|5                jelle?sold(2)                                           |
|6                 sjeng?sold(2)                                          |
|7                  sjeng!offer(manet,5)                                  |
|8                   sjeng?ok                           PASS              |
|9                    sjeng!stop                                          |
|10                  sjeng?OTHERWISE                    FAIL              |
|11                ?TIMEOUT T1                          INCONC            |
|12                sjeng?OTHERWISE                      FAIL              |
|13               ?TIMEOUT T1                           INCONC            |
|14               jelle?OTHERWISE                       FAIL              |
|15              sjeng?OTHERWISE                        FAIL              |
|16          jelle?OTHERWISE                            FAIL              |
+------------------------------------------------------------------------+
```

15 TOOL ASPECTS AND TEST EXECUTION

The PSF simulator plays a central role in our approach. Probably one could use LOTOS or SDL simulators as well. The Philips 'Interworking Tool' (Mauw, Winter 1993) was used to make the diagrams of the type shown in 2, 4 and 10. STDs, R-STDs were made using standard drawing tools (Framemaker) and the RTPs were drawn by hand with color pencils (Framemaker for the paper). We built some simple preprocessing tools with `cpp` (macro preprocessor) and `lex` (scanner generator) to convert traces to sequence charts and a simple 'translator' from sequence charts to TTCN, programmed in Gofer. In our project the ATCs were manually translated to C and in this step several practical issues were resolved too, like accessing the PCOs and mapping the packet structure to the TTCN events. There are options for further automation; e.g. translating TTCN to C can be automated, maybe using ASN-1 for the message-layout.

16 OPEN ISSUES

In general, there may be more users and there may be more sub-services. In that case various extensions of the approach are possible. Due to space limitations we will not discuss them here.

There are a number of subtle technical problems which arise in practice because the synchronous communication model of PSF does not formally match the asynchronous model of TTCN, which need not perfectly match the typical event-based mechanisms used in a real service. We used intuitively right ad-hoc solutions for these issues and for the real-time aspects, but there are clearly options for further research and improvement.

17 CONCLUDING REMARKS

The proposed method has been developed in two phases when we had to generate tests for new services developed for a Philips product. In an earlier phase of this project, PSF had been used already (most of the PSF was not developed by the authors of the present paper).

The test-development activities proceeded in two phases. In the first phase we devised tests for one layer of service, which we did almost entirely by hand, checking only a few tests with the simulator (we had a PSF model for that service too). The tests were translated to C and applied to the implemented service. In the second phase (in the context of the same project), we addressed another more complex layer of service and we decided to better exploit the available specifications and tools. For this purpose we developed the Gofer translator from sequence charts to TTCN, and used simulation traces as the starting point for all tests. In this way we developed about 30 ATCs. Most of these tests were never put into execution because of shifting priorities and other boundary conditions of the project.

We found the method of test development workable in practice. Moreover the approach seemed very manageable, in the sense that we could tell in advance how many test cases we were going to develop and after some time we were able to predict how long a typical test case would take.

The strong point of the method is that there is a precise model from which easily correct traces are derived, leading to a good starting point for editing the final TTCN text of the ATC. The method does not demand that everything is formal or automated, but it is a mix of automated and manuals steps. Therefore it is always possible to insert specific solutions for specific problems not built-in to the models or the tools.

Acknowledgements: the authors want to thank J. Bergstra, P. Boonen, A. Engel, J.F. Groote, R. de Jong, R. Koymans, A. Mikes, J. van de Pol, M. Reniers, O. Sies, J. Springintveld, A. Timmermans, and E. Zondag for the cooperation, help and discussions.

18 REFERENCES

CCITT (1992) Message Sequence Chart (MSC), Rec. Z.120, Study Group X (1992).

ISO (1991) International Organisation for Standardisation, OSI conformance testing methodology and framework, International standard IS 9646.

Knightson K.G. (1993) OSI protocol conformance testing, McGraw-Hill, Inc., 1993.

Mauw S., Veltink G. (1993) Algebraic specification of communication protocols, Cambridge university press, 1993.

Mauw S., Winter T. (1993) A Prototype toolset for interworkings, Philips Telecommunication Review, 51(3), (Dec. 1993), pp. 41–45.

Sidhu D.P., Leung T.K. (1989) Formal methods in protocol testing, a detailed study, IEEE trans. on Softw. Eng., Vol. 15, No. 4, pp. 413–426.

Wehkamp (1996) Wehkamp Homeshopping, interactive world wide web application, at URL `http://www.wehkamp.nl/`.

14

Selective Generation of Symmetrical Test Cases

F. Michel and P. Azéma and K. Drira
LAAS-CNRS
7, av. du Colonel Roche, 31077 Toulouse Cedex, France.
Phone: (+33) 61-33-62-00. Fax: (+33) 61-33-62-65.
email: {fmichel,azema,drira}@laas.fr

Abstract
In this paper, system symmetries are shown to efficiently reduce the size of the complete test suite required for checking the implementation conformance. The reduction is obtained by eliminating all the symmetrical test cases during the generation of the test suite. The method is applied over the theoretical model of conformance testing developed by [Bri88] using the implementation relation **conf**. In this model, symmetries are defined as bijective renamings of actions which preserve conformance properties, and allow test folding. The symmetries may concern either architectural properties, or execution sequences, or data structures, respectively. By exploiting data or state symmetries, infinite complete test suites may be reduced into finite one.

Keywords
Generation of test cases, conformance testing, symmetrical systems

1 INTRODUCTION

This study merges two topics : conformance testing and validation of symmetrical systems.

Conformance testing concerns the process of checking whether a system implementation conforms to its specification when this implementation is only accessed by its interaction points (black box hypothesis).

On the other hand, the validation of symmetrical system [PHJ91] has been studied in the field of verification where behaviour descriptions are freely available. Several dedicated formalisms [Lam91, ID93] have been designed to verify structural properties [Lam91, Sch94] or temporal logic formulae by means of model checking [ECJ93, EE93, EE95].

The main motivation of these works lies in the reduction, modulo system symmetries, of the state space made untractable by combinational explosion. In testing framework, this motivation can be reformulated as : how to reduce the size of the test suite required for testing the implementation conformance.

The reduction process is rather simple. Consider, for instance, the transport protocol between two entities E_1 and E_2. If the implementation of both entities are exactly the same, then testing the connection establishment by E_2 is worthless if the connection establishment by E_1 has already been tested : an architectural symmetry exists between E_1 and E_2. In the same way, if the implementation of a buffer behaves independently of conveyed values then one needs only to test the buffer for one particular value : a data symmetry exists among all the values. More generally, if one knows system symmetries, then one can eliminate all test cases symmetrical to an already generated test case.

Though this approach is closed to test selection [BTV91] regarding the expected goal, it differs from the latter in three points : first, it is performed during the generation of the test suite, then, it does not use external knowledges but extra informations : system symmetries, finally, it does not weaken the conformance relation since only redundancies caused by symmetries will be eliminated. That is why, our approach could be identified as a selective generation of test cases by means of system symmetries.

Formally, the chosen theoretical framework is the model of synchronous testing developed by [Bri88, Tre89] over LOTOS. In fact, we will use the version of [Tre92] where systems are described by labelled transition systems (LTS). The implementation relation to be tested is the relation **conf**. **conf** is defined from testable properties S **after** σ **must A**. S **after** σ **must A** means that every states reached after the sequence of actions σ in the system S must accept an action in the set **A**.

A complete test suite for **conf** can be generated automatically as an unique LTS, the so-called *canonical tester*. This derivation is based on the construction of an intermediate object : the `failure tree` or its complementary version the `success tree` (remark : this is not exactly the acceptance tree defined in [Hen85]).

Finally, the canonical tester is applied in a synchronous way to the implementation and a verdict is returned by observing deadlocks over the composition.

This paper defines a symmetrically selective generation of test cases within this model.

First, in order to study what are the possible consequences of a selective generation, we will parameterize the whole method by sets of properties to be tested; these sets will be called domains. Therefore, the implementation relation will be denoted $\leq_D$ where D is an arbitrary domain of testable properties, (**conf** corresponds to a particular domain). $I \leq_D S$ means that if S verifies a property in D then I verifies the same property. The main result is that if D is a domain of properties verified by the specification and if D can be represented by a `success tree` then the whole generation method is still valid. Besides, we notice that if $D' \subseteq D$ then the test suite required for $\leq_{D'}$ will be smaller than the test suite required for $\leq_D$. So, in this context, a selective generation of test cases consists in finding a subset D' of the initial domain D such that :

- $I \leq_D S \Leftrightarrow I \leq_{D'} S$ (the reduction does not weaken the implementation relation)
- D' can be represented by a `success tree` if D can be represented by a `success tree` (if the generation of test cases can be applied over D then it can be applied over D')

The reduction will be achieved by means of symmetries.

Symmetries are defined over a LTS by means of bijective renamings of actions : two states in the system are symmetrical iff they verify the same testable properties up to a particular renaming of actions, two sequences of actions are symmetrical iff they reach symmetrical states. Then the notion of group of symmetries allow us to build the equivalence relation "is equal up to a symmetry".

In the case of conformance testing, two LTS are studied : the implementation and the specification. That is why, the hypothesis (H1) is added : " *(H1) : the considered symmetries are common to the specification and the implementation*".

The reduced domain of properties, D', is defined as a set of representatives of D for the equivalence relation "is equal up to a common symmetry". D' fulfills both requirements : under hypothesis (H1), $I \leq_D S \Leftrightarrow I \leq_{D'} S$ and a quotient `success tree` representing D' can be build from a `success tree` attached to D by cutting symmetrical branches. Consequently, the canonical tester derived from the quotient `success tree` represents a reduced complete test suite for $\leq_D$.

We may note that our method relies strongly on (H1), but how to identify symmetries over a partially-known implementation ? In fact, (H1) can be established by exploiting three kind of informations about the system : architectural, data and state symmetries.

Data and state symmetries are particularly interesting since they allow us to test systems lying beyond the operational scope of the initial method. Indeed, a finite-branching `success tree` reduced by data symmetries may be computed even if the initial `success tree` is infinite-branching. In the same way, a finite-depth `success tree` reduced by state symmetries may be processed even if the initial `success tree` has an infinite depth.

In this paper, symmetry handling is assumed to be supported by the designer. That is, the designer is able to provide group of symmetries and to implement a decision procedure for the relation "is equal up to a symmetry".

The remaining of the document is organized as follows: section 2 recalls basic notions about LTS, section 3 explains the parameterized test model, section 4 introduces symmetries into this model, section 5 details the symmetrically selective generation, section 6 instantiates our results in the case of architectural, data and state symmetries.

2 LABELLED TRANSITION SYSTEMS

Let us consider L a set of (visible) actions and τ a particular symbol representing the internal computing. L^* denotes the set of action sequences, ϵ the empty sequence. $\mathcal{P}(L)$ denotes the power set of L. $\mid \sigma \mid$ denotes the length of a sequence σ.

A labelled transition system (LTS), B, is a triple $(\mathcal{Q}, \rightarrow, s_0)$ where $\mathcal{Q}$ is the set of states, s_0 is the initial state, and $\rightarrow \subseteq \mathcal{Q} \times L \cup \{\tau\} \times \mathcal{Q}$ is the transition relation. $\mathcal{Q}$ and L need not be finite.

We note :$\forall a \in L \ : \ s \stackrel{a}{\Longrightarrow} s' \stackrel{def}{\Leftrightarrow} s \stackrel{\tau^* a \tau^*}{\longrightarrow} s'$ (abstraction of the internal computing)

$\forall \sigma = a_1 \ldots . a_n \in L^* \ : \ s \stackrel{\sigma}{\Longrightarrow} s' \stackrel{def}{\Leftrightarrow} s \stackrel{a_1}{\Longrightarrow} \ldots \stackrel{a_n}{\Longrightarrow} s'$ and $s \stackrel{\epsilon}{\Longrightarrow} s' \stackrel{def}{\Leftrightarrow} s = s' \text{ or } s \stackrel{\tau^*}{\longrightarrow} s'$

$s \stackrel{\sigma}{\Longrightarrow} \stackrel{def}{\Leftrightarrow} \exists s' \ : \ s \stackrel{\sigma}{\Longrightarrow} s'$ $\qquad B \stackrel{\sigma}{\Longrightarrow} s \stackrel{def}{\Leftrightarrow} s_0 \stackrel{\sigma}{\Longrightarrow} s$

$B \stackrel{\sigma}{\Longrightarrow} \stackrel{def}{\Leftrightarrow} s_0 \stackrel{\sigma}{\Longrightarrow}$ $\qquad Trace(B) =_{def} \{\sigma \in L^* / B \stackrel{\sigma}{\Longrightarrow}\}$ $\qquad out_B(s) =_{def} \{a / s \stackrel{a}{\Longrightarrow}\}$.

A deterministic tree B is a particular LTS such that the relation $\{ \ (s, \sigma) \ / \ B \stackrel{\sigma}{\Longrightarrow} s\}$

is one-to-one. In this case, a state is called a node, $\mid \sigma \mid$ is the depth of s, and the set of states whose depth is k, is the kth level of the tree.
Remark : transition relations $\rightarrow$ and $\Rightarrow$ are equal over a deterministic tree.
In the sequel, S (resp. I) is the specification (resp. implementation) LTS.

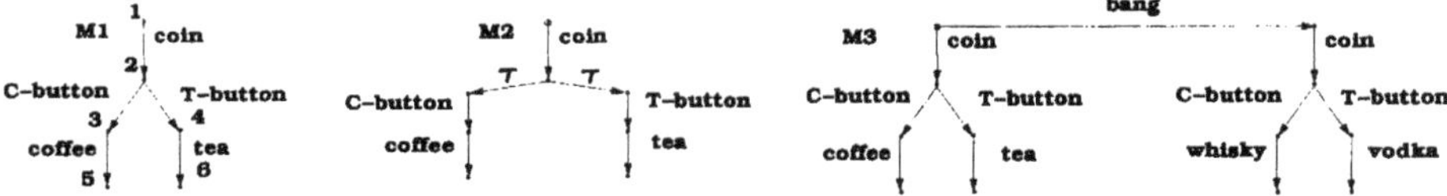

Figure 1 LTS of vending machines

Example : LTS `M1`, `M2`, `M3` in figure 1 describes three vending machines. States in `M1` are numbered. The deterministic tree, `M1`, is a vending machine delivering either coffee or tea for one coin. The selection is performed by pushing the right button. The machine `M2` decides internally whether you can have coffee or tea. `M3` adds a new behaviour to `M1` : when you kick the machine (`bang`), it starts delivering whisky or vodka instead of respectively coffee or tea.

3 SYNCHRONOUS TESTING

Conformance testing consists in determining whether the implementation conforms to its specification by only using implementation interaction points with the environment. In synchronous testing, test cases are run in a synchronous way over the implementation. Here, the interactions are represented by the visible actions of the LTS.

In this section, implementation relations together with test generation defined in [Tre92] will be parameterized.

3.1 Parameterized implementation relations

The testable properties are **after** σ **must A** where $\sigma \in L^*$ and $\mathbf{A} \in \mathcal{P}(L)$. A LTS, B, satisfies **after** σ **must A** , we note B **after** σ **must A** , iff :

$$\forall s \,:\, B \stackrel{\sigma}{\Longrightarrow} s \Rightarrow \exists a \in \mathbf{A} \,:\, s \stackrel{a}{\Longrightarrow}$$

Intuitively, B **after** σ **must A** means that B must accept an action in **A** after σ.
Example : The following properties are satisfied by the vending machines :
`M1` **after** `coin` **must** {`T-button`} `M3` **after** `coin` **must** {`bang`,`C-button`}
But `M2` does not satisfy **after** `coin` **must** {`T-button`}.

The relations specifying the implementation conformance are called implementation relations. They are all built according to the same pattern. First, choose a domain of pairs $(\sigma, \mathbf{A})$ (the properties to be tested) then claim that I conforms to S iff all properties in D satisfied by S are also satisfied by I. Formally :

Definition 31 (Implementation Relations) *Let $D \subseteq L^* \times \mathcal{P}(L)$ be a domain*
$I \leq_D S \stackrel{def}{\Leftrightarrow} \forall(\sigma, \mathbf{A}) \in D \,:\, S$ **after** σ **must A** $\Rightarrow I$ **after** σ **must A**

Remark: the **conf** relation is expressed by : I **conf** $S \stackrel{def}{\Leftrightarrow} I \leq_{Trace(S) \times \mathcal{P}(L)} S$

Our aim is to reduce D to the smaller domain D' such that $I \leq_D S \Leftrightarrow I \leq_{D'} S$. Symmetries will be exploited for this purpose, nevertheless structural reductions are already possible. Let D and D' be two domains, we note :
$S[D] =_{def} \{(\sigma, \mathbf{A}) \ / \ (\sigma, \mathbf{A}) \in D$ and S **after** σ **must** $\mathbf{A}$ $\}$ (prop. in D satisfied by S)
$D' \sqsubseteq D \overset{def}{\Leftrightarrow} (\ D' \subseteq D$ and $\forall(\sigma, \mathbf{A}) \in D, \ \exists(\sigma, \mathbf{A}') \in D' \ : \ \mathbf{A}' \subseteq \mathbf{A} \ )$
Then, it holds that:

Property 31 (Structural Reductions)
(1) $I \leq_D S \Leftrightarrow I \leq_{S[D]} S$ *and* *(2)* $\forall D' \sqsubseteq S[D] \ : \ I \leq_D S \Leftrightarrow I \leq_{D'} S$

Sketch of proof : (1) As $False \Rightarrow Any$, only properties satisfied by S are discriminating.
(2) From : $\forall$ LTS B : $\mathbf{A}' \subseteq \mathbf{A}$ and B **after** σ **must** $\mathbf{A}' \Rightarrow B$ **after** σ **must** $\mathbf{A}$ □

The test generation takes benefit of these reductions.

3.2 Parameterized test generation

The derivation takes as input a particular tree, the `success tree`. The `success tree` is a compressed form of the domain D. Part 1 deals with the notion of `success tree`. Part 2 recalls briefly the process of synchronous testing (derivation, application, verdict). Part 3 presents the fundamental result about the validity of the parameterized test generation. Part 4 shows how the parameterized test generation is instantiated when $\leq_D$ is **conf**.

Success trees

A `success tree` F is a deterministic tree labelled by a mapping $must_F()$ such that :

$$\forall s \ node \ : \ must_F(s) \subseteq \mathcal{P}(L) \qquad \text{and} \qquad out_F(s) \subseteq \cup_{\mathbf{A} \in must_F(s)} \mathbf{A}$$

For instance, figure 2 depicts three `success trees`.

`Success trees` are compressed forms of domains. Indeed, a domain D can be attached to each `success tree` F in this way :

$$(\sigma, \mathbf{A}) \in D \ \Leftrightarrow \ (\ \exists s \ : \ F \overset{\sigma}{\Longrightarrow} s \text{ et } \mathbf{A} \in must_F(s) \ )$$

On the other hand, some domains may not be represented by a `success tree`. For instance `{(a,{b}),(ac ,{b})}` cannot be associated with any `success tree`, since the condition $out_F(s) \subseteq \cup_{\mathbf{A} \in must_F(s)} \mathbf{A}$ will never be fulfilled.

Test Derivation and application

As this process will not be modified by the symmetrical method, we will only recall its principles. However, a more detailed presentation can be found in appendix 1.

It is assumed that a `success tree`, F, attached to the domain D is available. First, a complete test suite can be systematically derived from F. This test suite is computed as a particular LTS called the canonical tester. Then, the canonical tester is synchronized with the implementation over their common actions. Finally, a verdict is returned by observing the resulting deadlocks. The verdict is a success iff all the observed deadlocks are caused by the canonical tester.

In the sequel, the whole process is denoted by a function **Test(F,I)**. The function outputs "success" iff the obtained verdict is a success.

Fundamental result

The validity of the method relies on the following theorem :

Theorem 31 *If D is a domain of properties satisfied by S and if F is a* `success tree` *representing D then :* **Test(F,I)** *outputs "success" iff* $I \leq_D S$

The proof relies on the same arguments that in [Tre92].

Testing conf

For **conf**, a `success tree` can be computed automatically from the specification S.

At first, the underlying deterministic tree associated with the sequences in Trace(S) is built : for all σ in Trace(S), a node, $\hat{\sigma}$, is created in F, and the transition relation in F is defined by : $\forall \sigma, \sigma.a \in Trace(S) : \hat{\sigma} \xrightarrow{a} \widehat{\sigma.a}$.

Then, the mapping $must_F$ is processed by : $must_F(\hat{\sigma}) =_{def} \Psi(\ OUT_S(\sigma)\ )$ where $OUT_S(\sigma) =_{def} \{out_S(s)\ /\ S \stackrel{\sigma}{\Longrightarrow} s\}$, and Ψ is a representative selection over set families :

$$\Psi((A_i)_{i\in I}) =_{def} \{\cup_{i\in I}\ \{a_i\}\ /\ a_i \in A_i\}$$

In the example of the vending machines, it holds that :

$$\begin{aligned}\Psi(\ OUT_{M2}(\texttt{coin})\ ) &= \Psi(\ \{\{\texttt{T-button}\}, \{\texttt{C-button}\}\}\ ) \\ &= \{\{\texttt{T-button}, \texttt{C-button}\}\}\end{aligned}$$

Intuitively, $must_F(\hat{\sigma})$ is composed of sets **A** verifying for all state s in S reachable by σ there exists an action a in **A**, such that $s \stackrel{a}{\Longrightarrow}$.

Consequently, it can be easily shown that the domain D' attached to the resulting `success tree` verifies : $D' \sqsubseteq S[Trace(S) \times \mathcal{P}(L)]$

Particularly, D' is a set of properties satisfied by S, hence :

Test(F,I) outputs "success" $\Leftrightarrow I \leq_{D'} S$ (theo. 31) $\Leftrightarrow I \leq_{Trace(S)\times\mathcal{P}(L)} S$ (pro. 31-2)
$\Leftrightarrow I$ **conf** S (by def,)

Example : `success trees` `F1`, `F2` and `F3` from `M1`, `M2` and `M3` are depicted in Figure 2.

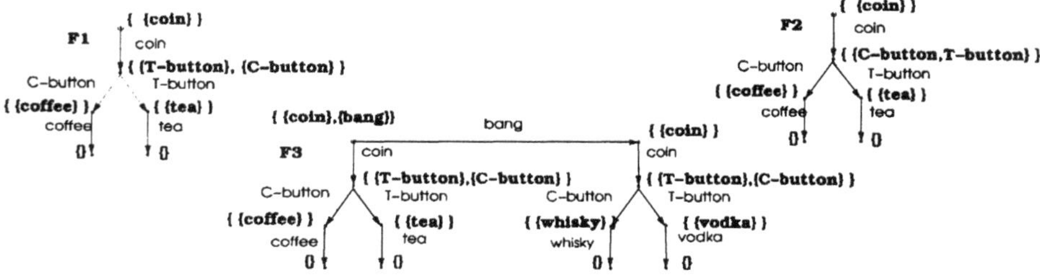

Figure 2 `success trees`

4 TEST SYMMETRIES

This section explains how to introduce the notion of symmetries in the field of testing.

4.1 Definition

Informally, symmetries express assertions such as : `M1` behaves in the same manner for coffee or tea, or else, after having performed `bang`, `M3` delivers whisky instead of coffee and vodka instead of tea.

This action changing is expressed by a bijective renaming φ over L. φ is extended to action sequences and sets by :

$$\varphi(\epsilon) = \epsilon \qquad \varphi(a_1...a_n) = \varphi(a_1)....\varphi(a_n) \qquad \varphi(\mathbf{A}) = \{\varphi(a)/a \in \mathbf{A}\}$$

Two states in the LTS are symmetrical iff they behave in the same way up to φ. As the behaviour is defined in terms of testable properties, a binary relation T_φ over states is a test symmetry up to φ iff :

$$sT_\varphi s' \Rightarrow \forall \sigma \in L^*,\ \forall \mathbf{A} \subseteq L\ :\ s\ \textbf{after}\ \sigma\ \textbf{must}\ \mathbf{A}\ \Leftrightarrow s'\ \textbf{after}\ \varphi(\sigma)\ \textbf{must}\ \varphi(\mathbf{A})$$

Example : In the example of the vending machine, M1, let f be the renaming :

a	coffee	tea	C-button	T-button	*other action* x
$f(a)$	tea	coffee	T-button	C-button	x

$T_f = \{(1,1),(2,2),(3,4),(5,6),(4,3),(6,5)\}$ is a test symmetry up to f among states in M1.

If the LTS is deterministic, a test symmetry can be easily extended to sequences : two action sequences are related by T_φ iff they reached two states related by T_φ. Formally :

$$\sigma T_\varphi \sigma' \Rightarrow \forall \sigma_1 \in L^*,\ \forall \mathbf{A} \subseteq L\ :\ B\ \textbf{after}\ \sigma.\sigma_1\ \textbf{must}\ \mathbf{A}\ \Leftrightarrow B\ \textbf{after}\ \sigma'.\varphi(\sigma_1)\ \textbf{must}\ \varphi(\mathbf{A})$$

Finally, the definition of test symmetries over any LTS is given by :

Definition 41 *$T_\varphi \subseteq L^* \times L^*$ is a test symmetry up to φ upon B iff :*

$$\sigma T_\varphi \sigma' \Rightarrow \quad \forall \mathbf{A} \subseteq L\ :\ B\ \textbf{after}\ \sigma\ \textbf{must}\ \mathbf{A}\ \Leftrightarrow B\ \textbf{after}\ \sigma'\ \textbf{must}\ \varphi(\mathbf{A})$$
$$\textit{and}\ \forall a \in L\ :\ \sigma.a\ T_\varphi\ \sigma'.\varphi(a)$$

Example : $T_f =_{def} \{\ (\sigma, f(\sigma))\ /\ \sigma \in Trace(M1)\ \} \cup \{\ (\sigma,\sigma')\ /\ \sigma,\sigma' \notin Trace(M1)\}$ is a test symmetry up to f over M1 and M2. Furthermore, if g is defined by :

a	whisky	coffee	tea	vodka	*other action* x
$g(a)$	coffee	whisky	vodka	tea	x

then $T_g =_{def} \{\ (\sigma, bang.g(\sigma))\ or\ (bang.g(\sigma),\sigma)\ /\ \sigma, bang.g(\sigma) \in Trace(M3)\} \cup$ $\{\ (\sigma,\sigma')\ /\ \sigma,\sigma' \notin Trace(M3)\}$ is a test symmetry up to g over M3.

4.2 Group of symmetries and equivalence relation

Let T_φ be a test symmetry up to φ over B, T_ψ be a test symmetry up to ψ over B. An homomorphism maps permutations to test symmetries :

- $ID = \{(\sigma,\sigma)/\sigma \in L^*\}$ is a test symmetry up to the identity renaming, over B.
- $T_\varphi \circ T_\psi$ is a test symmetry up to $\psi \circ \varphi$ over B.
- T_φ^{-1} is a test symmetry up to φ^{-1} over B.

So, group structure can be brought to test symmetries. A group of test symmetries is a set of test symmetries containing ID and closed for the composition and inversion of relations. For each group $\mathcal{G}$ of test symmetries, a relation "is symmetrical to", denoted $\equiv$, can be defined as :

$$\forall \sigma, \sigma' \in L^* \;:\; \sigma \equiv \sigma' \stackrel{def}{\Leftrightarrow} \exists T_\varphi \in \mathcal{G},\ \sigma T_\varphi \sigma'$$

By using group laws, it holds that : $\equiv$ is an equivalence relation.
Example : $G_1 =_{def} \{ID, T_f\}$ et $G_3 =_{def} \{ID, T_g\}$ are groups of test symmetries, we note the attached equivalence relations respectively $\equiv_1$ and $\equiv_3$.

5 A SYMMETRICAL SELECTIVE GENERATION

First, let us assume the following hypothesis : I and S have common test symmetries, that is, there exists $\mathcal{G}$ such that :

$\mathcal{G}$ is a group of symmetries over I and over S (H1)

The induced relation "is symmetrical to" is denoted $\equiv$. The implementation relation to be tested is $\leq_D$ and F is a `success tree` attached to D. The goal of this section is to exploit symmetries of $\mathcal{G}$ in order to find a domain D' verifying :

D' $\subseteq$ D (C1)	$I \leq_D S \Leftrightarrow I \leq_{D'} S$ (C2)
D' can be represented by a `success tree` F' (C3)	

Indeed, (C1) implies F' smaller than F implies the canonical tester from F' is smaller than the canonical tester from F.

Furthermore, if **Test(F,I)** is valid, that is, D is a domain of properties verified by S, then as a subset, D' is also a domain of properties verified by S, hence :
Test(F',I) outputs "success" $\Leftrightarrow I \leq_{D'} S$ (theorem 31) $\Leftrightarrow I \leq_D E$ (condition (C2))

In part one, the candidate domains are defined as sets of representatives w.r.t. $\equiv$, i.e., domains whose intersections with each equivalence classes in $D/\equiv$ is a singleton. By definition, these domains satisfy (C1). Moreover, they are also shown to fulfill (C2).

Part two provides a construction of a quotient `success tree` from the `success tree` representing D. The domain attached to the quotient `success tree` is a candidate domain. Hence, this domain fulfill conditions (C1) to (C3).

5.1 Symmetrical implementation relations

$\equiv$ is extended to pairs $(\sigma, \mathbf{A})$ by : $(\sigma, \mathbf{A}) \equiv (\sigma', \mathbf{A}') \stackrel{def}{\Leftrightarrow} \exists T_\varphi \in \mathcal{G} \;:\; \sigma T_\varphi \sigma'$ et $\mathbf{A}' = \varphi(\mathbf{A})$

The symmetrical implementation relation consists in testing only a single property by classes of symmetrical properties. So, let us consider a domain D' of representatives in D w.r.t. $\equiv$ (building a set of representative is selecting a member in each classes). The symmetrical implementation relation, denoted $(\leq_D / \equiv)$, is defined by :

Definition 51 (Symmetrical implementation relation) $I\ (\leq_D / \equiv)\ S \stackrel{def}{\Leftrightarrow} I \leq_{D'} S$

This definition does not depend on the chosen representative domain. Indeed, if E is another set of representatives, it holds that : $I \leq_{D'} S \Leftrightarrow I \leq_E S$.

By definition, if a testable property P is satisfied by S (resp. I), then all properties symmetrical to P are also satisfied by S (resp. I). Hence :

Theorem 51 *If (H1) is verified then :* $I \leq_D S \Leftrightarrow I\ (\leq_D / \equiv)\ S$

5.2 Quotient success trees

By definition, a `success tree` F' is a quotient `success tree` of F w.r.t $\equiv$ iff the domain D' attached to F' is a representative domain w.r.t. $\equiv$ of the domain D.

This relationship is illustrated by the following diagram :

$$\begin{array}{ccccc} & D & \xrightarrow{\text{symmetrical reduction}} & D' & \\ \text{attached to} & \downarrow & & \downarrow & \text{attached to} \\ & F & \xrightarrow{\text{symmetrical reduction}} & F' & \end{array}$$

As some domains can not be represented by `success trees`, we have to show that such a quotient `success tree` exists. The proof is achieved by construction :

The quotient `success tree`, F', is built level by level : level k is constructed from lowers levels. Formally, a level series $(F'_k)_{k \geq 0}$ is defined inductively by :

Initialization k=0 : let us consider $D_0 =_{def} \{(\epsilon, \mathbf{A})\ /\ \mathbf{A} \in must_F(\hat{\epsilon})\ \}$, and D'_0 a set of representatives in D_0 w.r.t. $\equiv$. F'_0 is the `success tree` reduced to the root $\hat{\epsilon}$, and such that $must_{F'_0}(\hat{\epsilon}) =_{def} \{\mathbf{A}\ /\ (\epsilon, \mathbf{A}) \in D'_0\}$.

Induction step : F'_{k+1} is built from F'_k according to the following operations :

1. $\texttt{Next}_k$ is the domain of pairs $(\sigma.a, \mathbf{A})$ such that

   - $\hat{\sigma}$ is a leaf of F'_k,
   - $\mathbf{A} \in must_F(\widehat{\sigma.a})$ and $a \in \cup_{A \in must_{F'_k}} \mathbf{A}$,
   - $\forall \sigma' \in Trace(F'_k),\ \forall \mathbf{A}' \in must_{F'_k}(\hat{\sigma'})\ :\ (\sigma.a, \mathbf{A}) \not\equiv (\sigma', \mathbf{A}')$

2. D'_{k+1} is a set of representatives in $\texttt{Next}_k$ w.r.t. $\equiv$,
3. F'_{k+1} is the `success tree` F'_k plus the level attached to D'_{k+1}, that is : $\forall\ (\sigma.a, \mathbf{A}) \in D'_{k+1}$: state $\widehat{\sigma.a}$ and transition $(\hat{\sigma}, a, \widehat{\sigma.a})$ are added in F'_{k+1}, and $\mathbf{A}$ is added in $must_{F'_{k+1}}(\widehat{\sigma.a})$

The limit of this series is the quotient `success tree`, F'. The demonstration is performed inductively by showing that for all k, F'_k is a quotient `success tree` of F_k (The subtree of F whose all node have a depth $\leq k$). □

We note that this construction does not require that F is finite. Thus, quotient (and sometimes finite) `success tree` of infinite `success trees` can be considered. However, if F is finite, then the previous construction becomes an algorithm building automatically the quotient `success tree` from the initial `success tree` F.

Remark : the algorithm complexity relies strongly on determining the relation $\equiv$. Since the size of groups of symmetries may be important, computing $\equiv$ may be very hard. However, some particular kind of groups can be handled simply, as shown by some formalisms dedicated to symmetrical systems [Lam91].

Example : Quotient `success trees` F1' and F2' of F1 and F2 w.r.t. $\equiv_1$ and F3' of F3 w.r.t. $\equiv_3$ are depicted in Figure 3.

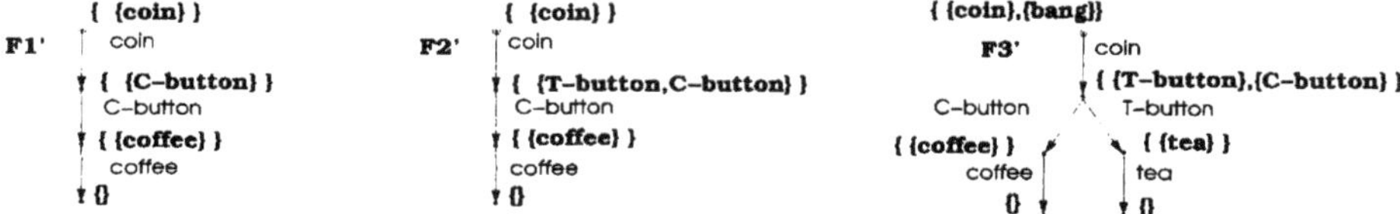

Figure 3 Quotient `success trees`

6 THREE SORTS OF SELECTIVE GENERATION

The operational application of the method requires to clarify two points :

- how to detect test symmetries over a partially-known implementation ? The exploitation of structural information will allow us to determine test symmetries. There are three kind of informations : architectural, data and state symmetries.
- how to compute a quotient `success tree` when the initial one is infinite ? We will explain how to compute a finite-branching quotient `success tree` from an infinite-branching `success tree` in case of data symmetries and how to compute finite-depth `success tree` from an infinite-depth `success tree`.

To study both problems, a formal description P in the Process Algebra CCS [Mil80] is assumed to be given. By the operational semantics of CCS, a L.T.S., B, can be derived. Both problems can be reformulated as :

the detection problem : how to detect test symmetries over B, if B is not available, (P plays the role of the implementation).

The computation problem : how to compute the quotient `success tree` from P when the initial `success tree` cannot be used (P plays the role of the specification).

6.1 Architectural symmetries

Detection problem

An architecture of n parallel processes can be expressed by the following CCS term :

$$(E_1[I_1] \mid E_2[I_2] \mid \ldots \mid E_i[I_i] \mid \ldots \mid E_n[I_n]) \setminus L_{int} \qquad \text{where}$$

E_i is CCS expression describing the behaviour of the ith isolated process, I_i is a renaming function describing the interface of the ith process, and L_{int} is the set of action only used for the internal synchronization among processes. L_{int} expresses the encapsulation of actions.

In this context, an architectural symmetry is a pair (φ_I, φ_A) such that : φ_I is a permutation among process indexes, $\{1, \ldots, n\}$, φ_A is a bijective renaming among actions, and (φ_I, φ_A) leave the architecture unchanged :

E_i and $E_{\varphi_I(i)}$ are syntactically equal and $I_{\varphi_I(i)} = \varphi_A \circ I_i$ and $\varphi_A(L_{int}) = L_{int}$

The set of architectural symmetries is a group, and each architectural symmetries can be turned into a test symmetry :

If (φ_I, φ_A) is an architectural symmetries then $T_{\varphi_A} = \{(\sigma, \varphi_A(\sigma)) \ / \ \sigma \in L^*\}$ is a test symmetries up to φ_A.

Computation problem

The number of architectural symmetries is always finite, and so is the number of induced test symmetries. Consequently, if the initial `success tree` is infinite then the quotient `success tree` is also infinite. However, architectural symmetries allows us to decrease the size of a finite `success tree` by applying the construction of section 5.2.

6.2 Data Symmetries

In order to represent data, the value-passing version of CCS described in [Mil89] will be used in this section. [Mil89] defines a semantics of this calculus by a translation into CCS without data. Besides, the obtained expression may be infinite-branching. Data symmetries will be introduced in the value-passing CCS. Then, a symmetrical translation of the value-passing expression is defined. The interest of the symmetrical translation is twofold : on the one hand, the symmetrical translation preserves the testable properties up to test symmetries, on the other hand, the obtained expressions are generally finite-branching. Consequently, if F' is the `success tree` built from the result of the symmetrical translation, then F' is a quotient `success tree` and F' is likely to be finite-branching.

Value-passing CCS

Let `Val` be a set of values, and X a *finite* set of variables.

The value-passing is described by the following expressions : $a(x)$ is the input action which inputs a value from gate a and stores it in x, and $\overline{a}(e)$ is the output action which outputs the result of expression e to gate a.

Data processing is expressed by the operator : `if b then P` with `b` boolean expression. If the result of b is true then the process behaves like P else it performs nothing.

The system expressions must not contain free variables, that is, each variable x occuring free in a boolean or output expression is bounded by an input $a(x)$.

Example : Let $X = \{x, y\}$ and $Val = \mathbb{R}$, the agent Max receiving two value from gate `in` and returning their maximum to gate `out` is expressed by :

$$Max \stackrel{def}{=} in(x).in(y).(\ if\ x > y\ then\ \overline{out}(x).0\ +\ if\ x \leq y\ then\ \overline{out}(y).0\ )$$

[Mil89] proposes a semantics of this calculus via a translation into basic CCS (CCS without data). It is based on the study of all the possible assignments of variables. An assignment is a partial function from variables to values. Let f and g be two assignments :

- $Dom(f)$ is the definition domain of f,
- $f = g$ iff $Dom(f) = Dom(g)$ and for all x in $Dom(f)$: $f(x) = g(x)$,
- $\emptyset$ is the assignment such that $Dom(f) = \emptyset$,
- $f[v/x]$ is the assignment such that $Dom(f[x/v]) = Dom(f) \cup \{x\}$ and f is redefined at x by $f(x) = v$

f is said to be total iff $Dom(f)$ =X. If b is an expression, $b.f$ denotes the expression where each variable x is substituted by the value $f(x)$. Moreover, an evaluation function `eval` is assumed to be available. if f is a total assignment, b a boolean expression and e an output expression then `eval(b.f)` is either true or false and `eval(e.f)` is a value in `val`. The input of a value v is denoted by a_v and the output by $\overline{a_v}$ in basic CCS.

The translation, defined in this section, differs a little from the one proposed by [Mil89], since assignments are performed in a lazy way. However, the result is the same. $Trad(f, P)$ the translation of P, under assignment f, is defined by structural induction :

$$Trad(f, a(x).P) = \sum_{v \in Val} a_v.Trad(f[v/x], P)$$
$$Trad(f, \overline{a}(e).P) = \overline{a_v}.Trad(f, P) \text{ where } \texttt{v=eval(e.f)}$$
$$Trad(f, if\ b\ then\ P) = \begin{cases} Trad(f, P) & \text{if } \texttt{eval(b.f)=true} \\ 0 & \text{if } \texttt{eval(b.f)=false} \end{cases}$$
$$Trad(f, P + Q) = Trad(f, P) + Trad(f, Q) \qquad Trad(f, P \mid Q) = Trad(f, P) \mid Trad(f, Q)$$
$$Trad(f, P \setminus L) = Trad(f, P) \setminus L' \text{ with } L' = \{a_v \ / \ a \in L,\ v \in Val\ \}$$
$$Trad(f, P[R]) = Trad(f, P)[R'] \text{ with } R'(a_v) = R(a)_v \qquad Trad(f, 0) = 0$$

The translation over the expression describing the system is called by $Trad(\emptyset, P)$.

To simplify, two restrictions are added over the considered value-passing CCS :

- the operator $\stackrel{def}{=}$ is forbidden, (no recursion is allowed).
- an input variable can not be used twice, for instance, expression $a(x) \ldots .a(x).P$ must be rewritten in $a(x) \ldots .a(x').P[x'/x]$.

Example : $Trad(\emptyset, Max) = \sum_{v1 \in \mathbb{R}} in_{v1}.(\sum_{v2 < v1} in_{v2}.(\overline{out_{v1}}.0 + 0) \ + \sum_{v1 \leq v2} in_{v2}.(0 + \overline{out_{v2}}.0) \ )$

Detection problem

The system is represented by an expression P in value-passing CCS. $b_1, \ldots, b_n$ denote the boolean expressions and $e_1, \ldots, e_m$ the output expressions occuring in P.

A data symmetry is a permutation φ over `Val` such that :

1. $\forall\ f$ total assignment, $\forall 1 \leq i \leq n \ : \ eval(b_i.f) = eval(b_i.(\varphi \circ f))$
2. $\forall\ f$ total assignment, $\forall 1 \leq j \leq m \ : \ \varphi(eval(e_j.f)) = eval(e_j.(\varphi \circ f))$

The whole set of data symmetries is a group and can be turned into test symmetries :

- φ is extended to ground actions by :
  $\varphi(a_v) =_{def} a_{\varphi(v)}$ and $\varphi(\overline{a_v}) =_{def} \overline{a_{\varphi(v)}}$ and $\varphi(\tau) =_{def} \tau$
- $T_\varphi =_{def} \{(\sigma, \varphi(\sigma)) \ / \ \sigma \in L^*\}$ is a test symmetry up to φ over $Trad(\emptyset, P)$.

Example : Data symmetries in `Max` are strictly increasing bijections over $\mathbb{R}$.

Computation problem

The guideline is to translate only one case per symmetrical cases. The notion of symmetrical cases is specified by a symmetry relation, denoted $\equiv$, among assignments :

$$\forall f, g \text{ assignments } : f \equiv g \overset{def}{\Leftrightarrow} \exists\varphi \text{ data symmetry} : \varphi \circ f = f'$$

$\equiv$ is an equivalence relation and a set of assignment representatives can be built by the sequence $(AT_i)_{0 \leq i \leq |X|}$: $AT_0 =_{def} \{\emptyset\}$ and for all i in $\{0, \ldots, |X| - 1\}$
AT_{i+1} is a set of representatives w.r.t $\equiv$ for $\{f[v/x] \ / \ f \in AT_i \ , \ x \notin Dom(f) \ and \ v \in Val\}$.

$AT_{|X|}$ is a set of assignment representatives w.r.t. $\equiv$, denoted AT_{sym}.

The symmetrical translation, denoted $Trad_{sym}$, is a translation where only the assignments in AT_{sym} are authorized. Its definition is the same that $Trad$ except for the rule :

$$Trad_{sym}(f, a(x).P) = \sum_{v \ : \ f[v/x] \in AT_{sym}} a_v.Trad_{sym}(f[v/x], P)$$

The interest of the symmetrical translation lies in the property illustrated by the following diagram :

$$\begin{array}{ccc} Trad(\emptyset, P) & \xrightarrow{\text{symmetrical reduction}} & Trad_{sym}(\emptyset, P) \\ \texttt{success tree} \downarrow & & \downarrow \texttt{success tree} \\ \text{F} & \xrightarrow{\text{symmetrical reduction}} & F' \end{array}$$

Indeed, it can be shown that the symmetrical translation preserves all the testable properties modulo data symmetries :

1. $Trad(\emptyset, P)$ **after** σ **must A** $\Rightarrow \exists\varphi$ symmetry : $Trad_{sym}(\emptyset, P)$ **after** $\varphi(\sigma)$ **must** $\varphi(\mathbf{A})$
2. $Trad_{sym}(\emptyset, P)$ **after** σ **must A** $\Rightarrow Trad(\emptyset, P)$ **after** σ **must A**

Consequently, the `success tree` built from $Trad_{sym}(\emptyset, P)$ is a quotient `success tree` of the `success tree` attached to $Trad(\emptyset, P)$.

Example : In the case of `Max`, $AT_{sym} =_{def} \{f_1, f_2, f_3, f_4, f_5\}$ where $f_1 : x \to 0$, $f_2 : y \to 0$ $f_3 : x \to 0, \ y \to 1$, $f_4 \ : \ x \to 1, \ y \to 0$ and $f_5 : x \to 0, \ y \to 0$, hence :
$Trad_{sym}(\emptyset, Max) = in_1.in_0.(out_1.0 + 0) + in_0.(\ in_1.(0 + out_1.0) + in_0.(0 + out_0.0) \ )$

6.3 State Symmetries

Detection problem

State symmetries express assertions such as : after sequence σ the system is in the same state that after σ'. Formally, it corresponds to the test symmetries up to the identity defined by : $T_{id} = \{ \ (\sigma.\sigma_e, \sigma'.\sigma_e) \ / \ \sigma_e \in L^*\}$.

State symmetries are more difficult to detect than architectural or data symmetries. However, they can be stated, if some states in the system are observable.

Computation problem

State symmetries are useful to cut loops which make `success tree` F depth infinite.

Since the depth of F is infinite, it is impossible to first build F and then build the quotient `success tree` F'. On the other hand, the construction of section 5.2, needs only the kth level of F to build the kth level of F'. Hence, the construction of F can be overlaped incrementally with the construction of F' : At each step of the following procedure, level k of F is built from the CCS expression describing the system and then level k of the quotient `success tree` is derived.

If there exists a finite-depth quotient `success tree` then this procedure will terminate although F is infinite.

Example : Suppose that M1 is reinitialisable, we note `Mr` the new machine :

$$Mr \stackrel{def}{=} \texttt{coin.(C-button.coffee.}Mr + \texttt{T-button.tea.}Mr)$$

That is, after `coin.T-button.tea` or `coin.C-button.coffee`, the system is in the initial state. So, we have the following state symmetries T^c_{Id} and T^t_{Id} such that :

$\forall \sigma \in L^*$: $\epsilon.\sigma\ T^c_{Id}$ `coin.C-button.coffee.`σ
$\epsilon.\sigma\ T^t_{Id}$ `coin.T-button.tea.`σ

Figure 4 depicts the construction steps of the quotient `success tree` Mr'.

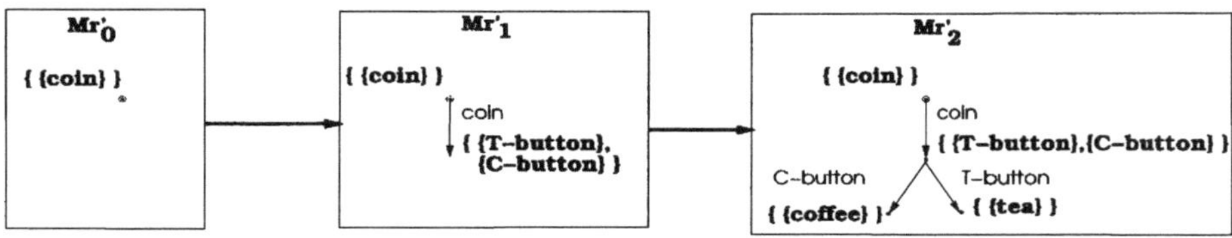

Figure 4 Construction steps

7 CONCLUSION

The paper have proposed a mean to drastically reduce the complete test suite required for a implementation relation : the symmetrically selective test generation. Within the selected theoretical framework developed by [Bri88, Tre92], test symmetries have been introduced. Their exploitation results in a reduction of the size of the `success tree` and so the canonical tester representing the complete test suite.

For the proposed method, system symmetries are needed as extra informations, and the test selection is performed over the `success tree`, at the beginning of the generation.

The latter point is very interesting since sometimes it allows us to avoid the generation of infinite test suites. The first point has an advantage : as extra and not external informations are used, the implementation relation is not weakened by the reduction. On the other hand, it has a drawback, the extra informations must be detected. Consequently, we have explained how to detect and exploit symmetries in three cases : architectural, data and state symmetries. Data symmetries are demonstrated to overcome problems caused by infinite branching in some cases and state symmetries appear to be a good solution to cut infinite-length test cases.

In this paper, symmetry processing is not detailed : we suppose that the designer is able to provide symmetries and to implement a decision procedure for the relation "is symmetrical to". However, some tools already manage symmetry processing [ID93,

GCH91]. Consequently, their integration into the symmetrically selective test generation is being considered. Furthermore, these tools deal with architectural symmetries [GCG94], and their scope extension to data and state symmetries would be very interesting.

REFERENCES

[Bri88] E. Brinksma. A theory for the derivation of tests. In S. Aggrawal and K.Sabani, editors, *Protocol Specification Testing and Verification*, volume VIII. Elsevier Science Publishers B.V., North-Holland, 1988.

[BTV91] E. Brinksma, J. Tretmans, and L. Verhaard. A framework for test selection. In *Proc. of the 11th Symposium on Protocol Specification, Testing Specification, Testing and Verification*, Stockholm, June 17-20 1991. international IFIP WG6.1.

[ECJ93] T. Filkorn E.M. Clarke and S. Jha. Exploiting symmetry in temporal logic model checking. In *Computer Aided Verification*, pages 451–462. Lecture Notes in Computer Science 697, June-July 1993.

[EE93] A. Sistla E. Emerson. Symmetry and model checking. In *Computer Aided Verification*, pages 463-478. Lecture Notes in Computer Science 697, June-July 1993.

[EE95] A. Sistla E. Emerson. Utilizing symmetry when model checking under fairness assumptions: an automata-thoeirc approach. In *Computer Aided Verification*, pages 279–292. Lecture Notes in Computer Science 939, June-July 1995.

[GCG94] G. Franceschinis G. Chiola and R. Gaeta. Modeling symmetric computer architectures by swns. In *Application and Theory of Petri Nets*, pages 139–158. Lecture Notes in Computer Science, 1994.

[GCH91] G. Franceschinis G. Chiola, C. Dutheillet and S. Haddad. On well-formed coloured nets and their symbolic reachability graph. In *High-Level Petri Nets*, pages 373–396. Springer-Verlag, 1991.

[Hen85] M. Hennessy. Acceptance trees. *Journal of the Association for Computing Machinery*, 32(4):896–928, October 1985.

[ID93] C. Ip and D. Dill. Better verification through symmetry. In *Int. Symp. on Computer Hardware Description language and their Application*, April 1993.

[Lam91] Claude Dutheillet Lamonthezie. *Symétrie dans les Réseau Colorés*. PhD thesis, Université Paris 6, 1991.

[Mil80] Robin Milner. *A Calculus of communicating Systems*, chapter Syntax and Semantic of CCS, pages 77–78. LNCS 92. Springer-Verlag, 1980.

[Mil89] Robin Milner. *Communication and Concurrency*. Prentice Hall, 1989.

[PHJ91] L.O. Jepsen P. Huber, A.M. Jensen and K. Jensen. Reachability trees for high-level petri nets. In *High-Level Petri Nets*, pages 319–350. Springer-Verlag, 1991.

[Sch94] K. Schmidt. Symmetries of petri nets. Informatik-Berichte 33, Humbolt-Universität zu Berlin, 1994.

[Tre89] J. Tretmans. Test case derivation from lotos specification. In *Proc. 2nd International Conference on Formal Description Techniques. FORTE'89*, Vancouver, CANADA, 1989.

[Tre92] J. Tretmans. *A Formal Approach to Conformance Testing*. Phd thesis, University of Twente, The Netherlands, 1992.

APPENDIX 1 TEST GENERATION

This appendix details the derivation and the application of a complete test suite, the canonical tester, w.r.t. $\leq_D$. F is assumed to be a **success tree** representing a domain D. The derivation consists of three steps :

The canonical tester: The canonical tester T attached to F is the L.T.S. verifying :

- States in T are states $\hat{\sigma}$ of F, the indexed states $\hat{\sigma}_A$ where $\mathbf{A} \in must_F(\hat{\sigma})$, and **dead**, a deadlock state **dead**.$\hat{\epsilon}$ is the initial state of T,
- The transition relation is ruled by ($\hat{\sigma}$ is a state in F) :

$$\frac{\mathbf{A} \in must_F(\hat{\sigma})}{(\hat{\sigma}, \tau, \hat{\sigma}_A) \text{ transition in T}}(Choice)$$

$$\frac{a \in \mathbf{A} \text{ and } \widehat{\sigma.a} \text{ state in } F}{(\hat{\sigma}_A, a, \widehat{\sigma.a}) \text{ transition in T}}(Action1) \quad \frac{a \in \mathbf{A} \text{ and } \widehat{\sigma.a} \text{ not a state of } F}{(\hat{\sigma}_A, a, dead) \text{ transition in T}}(Action2)$$

The (Choice) rule represents the internal choice of the tester among properties to be tested. Rules (Actioni) represent the expecting observation of action a.

The property $out_F(s) \subseteq \cup_{\mathbf{A} \in foncF(s)} \mathbf{A}$ of the **success tree** ensures that T is really a L.T.S., that is, all the states $\hat{\sigma}$ or $\hat{\sigma}_A$ are reachable. This means that no property in D is lost by T.

Example : Figure 5 depict the canonical tester T3' attached to the quotient **success tree** F3'.

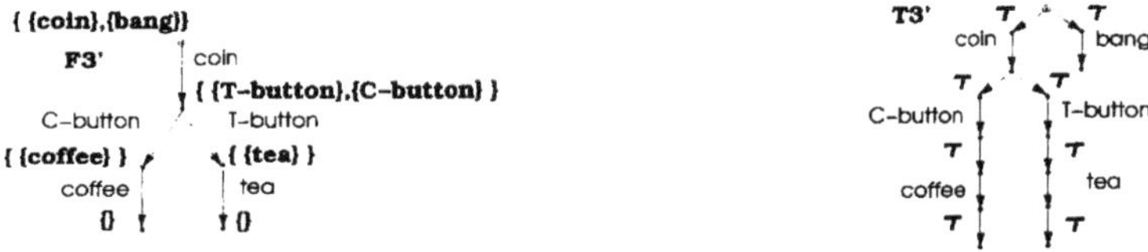

Figure 5 Canonical tester

Synchronous application to the implementation : Tests are run over the implementation by synchronization on common actions. The result of this synchronization is the L.T.S. $T \parallel I$ such that :

- states in $T \parallel I$ have the form $s \parallel s'$ where s is a state in T and s' is a state in I. The initial state in $T \parallel I$ is $s_0 \parallel s'_0$ where s_0 is the initial state in T (i.e. $\hat{\epsilon}$) and s'_0 the initial state in I.
- The transition relation is ruled by :

$$\frac{s \stackrel{a}{\Longrightarrow} p \quad \text{and} \quad s' \stackrel{a}{\Longrightarrow} p'}{s \parallel s' \stackrel{a}{\Longrightarrow} p \parallel p'}$$

Verdict : the verdict is obtained by observing deadlocks. A deadlock is observed over a S.T.E. B after a sequence σ iff :

$$\exists s : (B \stackrel{\sigma}{\Longrightarrow} s \text{ and } \forall a : s \stackrel{a}{\not\Longrightarrow})$$

In this case, we note $B \stackrel{\sigma}{\Longrightarrow} deadlock$.

The verdict is "success" iff all the deadlocks are caused by the tester :

$$\forall \sigma \in L^* : T \parallel I \stackrel{\sigma}{\Longrightarrow} deadlock \Rightarrow T \stackrel{\sigma}{\Longrightarrow} deadlock$$

15
Automation of test case derivation in respect to test purposes

A. Guerrouat[a] and H. König[b]

[a]University Otto von Guericke of Magdeburg
Dept. of Computer Science
P.O. Box 4120, 39016 Magdeburg, Germany
Email: guer@cs.uni-magdeburg.de

[b]Brandenburg Technical University of Cottbus
Dept. of Computer Science
P.O. Box 101344, 03013 Cottbus, Germany
Email: koenig@informatik.tu-cottbus.de

Abstract

Test case derivation and test verdicts specification in respect to given test purposes are an important issue in conformance testing. Since test purposes are commonly expressed in an informal way an automation of this process is not possible. In this paper, we show how this process can be automated for restricted classes of test purposes using a knowledge-based approach. Our approach is founded on a partial formalization of the test purposes that permits an automated derivation of test cases and the deduction of test verdicts. The adaptation of test cases to a given test method is also addressed. An example that illustrates the approach is given.

Keywords

Conformance testing, test purposes, test case derivation, knowledge-based systems

1 INTRODUCTION

A formal (or informal) specification of a protocol defines conformance requirements on the implementation of the protocol. The conformance between specification and implementation is defined by means of conformance relations that are deduced from the protocol specification. These relations state the requirements which have to be fulfilled by the implementation to be conform to the specification. In order to assess that an implementation is in compliance with its specification, conformance testing is needed. This is carried out by applying a *test suite*. It is well-known that the classic test suite generation methods for both finite state machines (FSMs) and labelled transition systems (LTSs) generate a huge number of *test cases* [Brin88,

Shen90]. They lack clearly identified test purposes and mostly require a complete specification. The execution of the test cases should improve the likelihood of interworking and interoperability between systems developed by different vendors. Economic and practical considerations, however, considerably restrict the number of test cases that can be carried out. Testers work within the constraints of budget and tight time schedules [Lai95]. Since it is impossible to apply every possible test case under a given time constraint, priority has to be given to test the most important and vulnerable parts of the protocol by defining appropriate test purposes.

In practice, an implementation under test (IUT) is often tested on a functional basis. Experienced testers are capable to assess which conformance requirements are more important to test or which functions of the IUT may be likely incorrectly implemented. Appropriate test cases that match with test purposes addressing conformance requirements or given functions have to be deduced. This implies that the conformance requirements referred to by the test purposes are identifiable with those ones described in the formal specification. For each test event a verdict has to be specified to indicate whether the expected outcome indicates a correct or incorrect behavior of the IUT. The derivation of test cases is manually done in practice because test purposes are informally described. Examining the use of test purposes in practice it seems possible to us to partially formalize certain classes of test purposes in order to structure and to classify them in such a way that their identification based on the conformance requirements described by a formal specification becomes feasible. A partially formal definition and identification of test purposes requires the assistance of a human expert. He can be assisted by knowledge-based systems to store his knowledge and to simplify further development steps.

In this paper, we present a knowledge-based approach for automatically deriving test case for a given test purpose. The approach is based on a partial formalization of test purposes that permits an automated identification of test cases and the specification of test verdicts. The adaptation of the test cases to a given test method is also addressed. We use the SELEXPERT tool [Guer95] for the knowledge-based analysis. It has been designed to select test cases based on the estimation of their fault coverage and cost. It has also been applied to select test cases taking into account data flow aspects. In this paper, we want to extend the use of SELEXPERT for the identification of test purposes to derive test case and test verdicts. In addition, the approach provides a more user-friendly utilization in the sense that it gives more freedom to a user to express his test purposes compared to other approaches (e.g. [Robl92]).

The rest of the paper is organized as follows. Section 2 introduces basic elements related to test purposes and describes the problem of the test case derivation w.r.t. test purposes. In Section 3, we give an overview of the principle of the approach and discuss some classes of test purposes. Section 4 presents the principle of test purpose formalization and of test case derivation and test verdicts specification. In Section 5, we describe the knowledge-based procedure to derive a test case for a given test purpose. An example that illustrates the approach is presented. The test case adaptation to a given test method is discussed in Section 6. Finally, some concluding remarks are given.

2 PRELIMINARIES

2.1 Basic concepts

In the ISO conformance test methodology ISO9646 [ISO91] test cases are organized in a hierarchical structure in which a test suite represents the top level of this hierarchy. Each test case belongs to a precise test purpose. Test purposes are defined informally. There exists no formally defined relationship between test purposes and test cases. Therefore, it is very difficult to exploit test purposes for an automatic derivation of test cases.

A *test purpose* is a prose description of a narrowly defined objective of testing focusing on a single conformance requirement as specified in the protocol specification. In the context of FMCT [FMCT95], the set of test purposes *TP* is defined as a finite subset of the set of conformance requirements *R*.

Conformance requirements are the requirements, a conforming implementation has to fulfill. They are divided in two groups: static conformance requirements and dynamic conformance requirements. The first ones represent requirements on possible combinations of values of the formal parameters of the specification whereas the second ones define requirements on the observable behavior of the implementation. In FMCT, the dynamic conformance requirements are defined as requirements implied by the formal specification and the implementation relations. A specification *S* is expressed as a set of conformance requirements $R_S=\{r_1,r_2,...,r_n\}$. An element $r_i \in R_S$ represents a single conformance requirement that may be addressed by a test purpose. An implementation *I* dynamically conforms to a specification *S* if *I* satisfies all conformance requirements in R_S. This is referred as *satisfaction* relation and denoted by *sat*: $(\forall r_i \in R_S\text{: } I \textit{ sat } r_i)$.

In this paper, the meaning of conformance is implicitly based on this type of relation since the associated verdicts are essentially based on the fulfillment of the conformance requirements by the IUT as indicated by the test purposes. In addition, we confine ourselves to dynamic conformance requirements, since these represent the main part of a protocol.

A *test case* refers to a given *test architecture* [ISO91] and consists of

- a test purpose
- a preamble
- a test body
- test verdicts.

A *preamble* is a sequence of events that transfers the IUT into a desired state from which the test body will start. A *test body* consists of sequences of test events according to the test purpose. A *test verdict* indicates the conformance of the IUT with respect to the test purpose. The verdict *pass* is assigned to the outcome of a test event if it matches the test purpose, otherwise the verdict *fail*. If no clear statement on conformance the verdict *inconclusive* is associated. The postamble is optional [ISO91], therefore, it is not considered here.

2.2 Description of the problem

Let be given a set of test purposes *TP*, and a test architecture *TA*. To derive a test case TC_i for a given test purpose the following 5-tuple $TC_i=\langle TP_i,TA_i,PA_i,TB_i,V_i\rangle$ has to be determined in which $TP_i \in TP$ denotes the test purpose corresponding to a conformance requirement $r_j \in R_S$, TA_i the test architecture, PA_i the preamble, TB_i the test body, and V_i the test verdicts. The main issue to automate the derivation procedure consists in mapping a given test purpose to the respective conformance requirements $r_j \in R_S$ in the specification. Unfortunately, this problem cannot be generally solved unless the set of test purposes TP is reasonably restricted and adequately formalized. This is due to the fact that test purposes are usually informal and may express different kinds of conformance requirements, e.g. requirements on the dynamic behavior, requirements referring to the test context and so on.

3 PRINCIPLE OF THE APPROACH

In this section, we first give a brief overview of our approach to automate the test case derivation w.r.t. test purposes. Then, we examine certain classes of test purposes to illustrate the principle of the approach. We discuss the possibility of their formalization and the feasibility of an automatic identification of test purposes when these are restricted to some 'typical' classes.

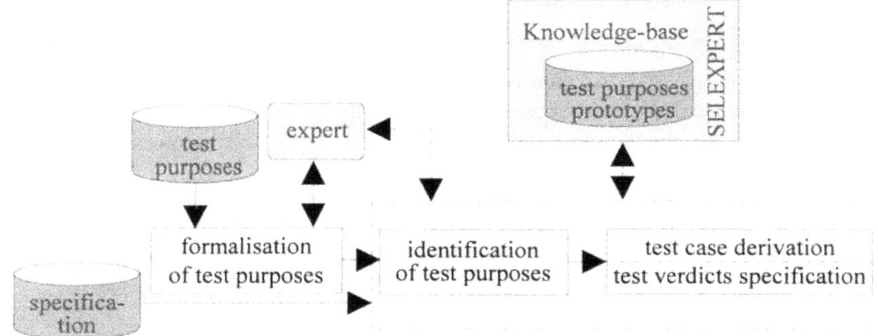

Figure 1 Overview of the approach.

The principle of our approach is depicted in Figure 1. The approach is based on the assumption that test purposes can be partially formalized and classified related to their syntax and objectives as will be discussed below. A prototype of each class is stored as a fact in the knowledge-base of SELEXPERT. It describes the syntactical basic structure of the test purposes of this class and the conformance requirements they are related to. With each class, an algorithm is associated for deriving the test case from the specification presented as behavior tree and for specifying the verdicts. The behavior tree is supposed to be deterministic. It can be obtained, for instance, by simulation [Guer95, Ulri93]. If such a knowledge-base is set up SELEXPERT can be used to automatically derive test cases and test verdicts from a given set of test purposes in connection with the respective formal description. The derivation procedure may be supported by a human expert. The assistance of the human expert is needed if the automatic identification of a test purpose fails, i.e. no test purpose prototype matches the given one. In this case, a new prototype may be added to the knowledge-base which is thus stepwise extended.

Our approach to formalize test purposes was inspired by considering representations of test purposes in practice and literature. We give some examples here.

First, we present some test purposes used in a conformance test center for testing *Profibus* protocols. *Profibus* (process field bus) [DIN91] is the German fieldbus system used for process-oriented applications in the automation technique area. The testing comprises among other the *fieldbus message specification* (FMS) using the *remote test method* [ISO91]. FMS represents a sublayer of the application layer. The conformance testing takes place by applying test cases such as those given in Table 1.

Table 1 Test purposes for FMS of *Profibus*

Test case identifier	*Test purpose*
satv0100	send *write-with-type-request* to IUT and await *write-with-type-response* from IUT
sciy1100	send *initiate-request* to IUT and await *abort-request* from IUT
sciv9100	send *initiate-request* to IUT and await *initiate-confirmation* from IUT

The words in italic of the test purposes in Table 1 denote the data concerned by the test purpose. The other words can be considered as keywords that characterize the test purpose. The keywords possess certain semantics in the sense that they, for instance, indicate an order in which the interactions, i.e. the data, must occur. Because of the simplicity of the representation in which these test purposes are expressed, it should be possible to define a formal structure for such a type of test purposes. This formal structure can be used as a prototype of the test purposes to automatically identify test purposes that follow the same structure. Thus, the identification of the conformance requirements addressed by the test purpose with corresponding ones described in the specification can be automated to derive the test case and test verdicts.

In [Brin91], a notation (*after-must*) based on the LTS formalism was introduced to express test purposes. This notation is used in the form: (*after* σ *must* A). σ denotes an event sequence, i.e. a trace, leading to a given state, and $A \subseteq L$ a subset of the *label set* L of the LTS. A represents a set of events that may be performed by the given process. If A is restricted to one element, i.e. an event, the test purpose addresses a single conformance requirement in this case. If test purposes are supposed to be expressed in this notation, they can be identified by the keywords *after* and *must*, and by exploiting the indicated path σ that refers to the given conformance requirement included in A. Furthermore, this notation may be enriched to become more familiar for an ordinary user, as in the example: "after σ the reaction of IUT must A" and so on.

In other works, especially those exploiting the structure information of formal specifications (e.g. [Ulri93, Velt92]), test purposes generally concern a process or a module of the specification as in the example: "check the correct execution of process *connection_request*". A process or a module describes a function of the specified system in this case. Since these representations are not very complex, it is possible to define a formal structure for such a class of test purposes. However, this supposes that the specification may be represented in a suitable form, i.e. as an hierarchy of processes.

4 FORMALIZATION OF TEST PURPOSES

In this section, we show how formal structures can be defined for different classes of test purposes. For lack of space, we discuss here only some examples to demonstrate the principle.

There exist two types of test purposes:

- test purposes concerning static conformance requirements (not considered here as mentioned previously);
- test purposes referring to dynamic conformance requirements focusing in general on a certain property of the specified system (e.g. a function), or on a single conformance requirement, as in the example: "after receiving *event*1 the IUT must respond with *event*2". A conformance requirement may also concern timing aspects as, for instance, "test whether *event* is returned within time limit of *n* (seconds)". The timing aspect depends the given test architecture and particularly refers to the propagation delays of messages and so on [Chan92].

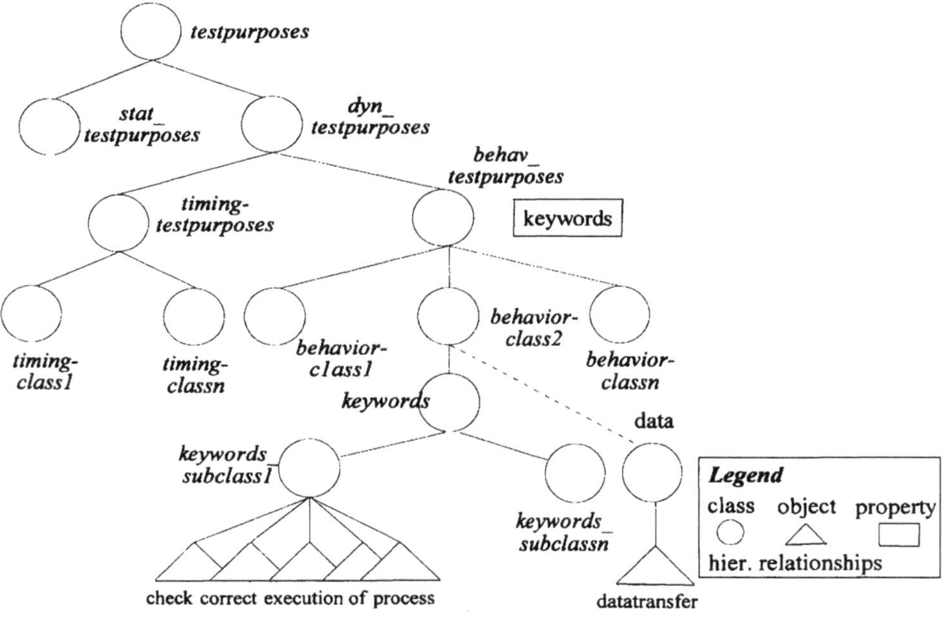

Figure 2 Knowledge representation of test purposes.

The knowledge representation formalism used in SELEXPERT is the object paradigm [Rumb91]. Figure 2 shows the representation of test purposes using this paradigm. The two types of test purposes indicated above constitute classes, noted as *stat_testpurposes* and *dyn_testpurposes*, respectively. They are subclasses of a superclass *testpurposes* that comprises all test purposes. *Dyn_testpurposes* is again organized into two other subclasses: *behav_testpurposes* and *timing_testpurposes*. These subclasses comprise test purposes concerning the behavior and the timing aspect, respectively. Each subclass is characterized by keywords that represent a *property* in respect to the object paradigm. The hierarchical structure of the test purposes facilitates the exploration of the different subclasses when identifying a given test purpose. Some of these subclasses will be described in the rest of this section. For convenience, we adopt a *Backus-Naur form* (BNF) like-notation for describing the formal structure of a class of test purposes, as in the following case:

⟨*dyn_testpurposes*⟩→ ⟨*behav_testpurposes*⟩ | ⟨*timing_testpurposes*⟩

To derive an appropriate test case in respect to a given test purpose, we assume that the specification is given as a finite behavior tree*, i.e. a finite set of traces. It can be obtained, for instance, by simulation [Ulri93, Guer95]. A test case consists of a set of traces.

4.1 Test purposes related to behavior aspects

To demonstrate the principle of test case derivation and test verdicts specification we consider here two example subclasses of test purposes, denoted by *behaviorclass*1 and *behaviorclass*2. They are principally distinguished with respect to their objective and syntax. They belong to class *behav_testpurposes* indicated above.

Behaviorclass1

This subclass of test purposes is supposed to have the following syntax:

⟨*behaviorclass*1⟩→⟨*keyword*1⟩ (⟨*event*1⟩) ⟨*keyword2*⟩ (⟨*event*2⟩)

⟨*keyword*1⟩ and ⟨*keyword*2⟩ are keywords, (⟨*event*1⟩) and (⟨*event*2⟩) the events that represent the conformance requirement addressed by the test purpose†. An example for this test purpose class would be "after receiving (ICONreq) the IUT must accept (ICONconf)".

A test purpose that can be expressed in this form does not explicitly indicate the localization of the conformance requirement referred to in the specification. An event sequence leading to the given conformance requirement has to be determined in the specification such that its length is minimal. It constitutes the preamble of the test case.

The fact that the event sequence to be adopted as preamble has to be minimal permits us to find the conformance requirement given in the test purpose with minimal effort, i.e. with relatively low cost. In fact, a conformance requirement usually may occur at different places, i.e. branches, of the behavior tree of the specification. This is due to the recursive character of protocols, e.g. through the process calls (see e.g. [Ulri93]). Thus, a given conformance re-

* We especially address LOTOS specifications here, but also specifications in other formal descriptions techniques whose semantics can be expressed as behavior trees.

† The parentheses aim at distinguishing the data concerned by the test purpose.

quirement can be found in different ways, i.e. event sequences. The optimal event sequence has to be selected in our case to be considered as preamble.

Now, we show how the different components of a test case TC_i corresponding to a test purpose TP_i in respect to the structure given above, can be derived.

The *preamble* $PA_i(TC_i)$ consists of only one event sequence that can be obtained as follows:

$$PA_i(TC_i)=\sigma_i \in S \text{ with } (\sigma_i=s_0 \text{—} e_{i1} \rightarrow s_{i1} \text{—} e_{i2} \rightarrow s_{i2} \ldots s_{ip-1} \text{—} e_{ip} \rightarrow s)$$

with (s_i—e→s_j) denotes a transition t where s_i represents its current state, e an event, and s_j the tail state of t

such that

- s_0 is the initial state of the specification S;
- $(\exists\ \sigma' \in S)$ such that $(\sigma'=s_0 \text{—} e_{i1} \rightarrow s_{i1} \text{—} e_{i2} \rightarrow s_{i2} \ldots s_{ip-1} \text{—} e_{ip} \rightarrow s \text{—} \langle event1 \rangle \rightarrow s')$[‡] ;
- $length(\sigma_i) \leq length(\sigma_j)$ for all $j \neq i$, with $\sigma_j \in S$ and $(\sigma_j=s_0 \text{—} e_{j1} \rightarrow s_{j1} \text{—} e_{j2} \rightarrow s_{j2} \ldots s_{jq-1} \text{—} e_{jq} \rightarrow s_{jq})$ such that
  $\sigma_j' \in S$ with $(\sigma_j'=s_0 \text{—} e_{j1} \rightarrow s_{j1} \text{—} e_{j2} \rightarrow s_{j2} \ldots s_{jq-1} \text{—} e_{jq} \rightarrow s_{jq} \text{—} \langle event1 \rangle \rightarrow s'')$.

Note that for convenience, in the event sequences above states are explicitly represented. Furthermore, $\langle event1 \rangle$ refers to the event indicated in the test purpose and *length*(...) denotes the number of events of the given event sequence. We have omitted, to explicitly represent the *gates* associated to the events in the test sequences given above for more abstraction at this level.

The *test body* represents a set of event sequences that can be obtained in relation to the preamble determined above and the test purpose as follows:

$$TB_i(TC_i)=\{\sigma_i \in S \mid \sigma_i=s_{i0} \text{—} e_{i1} \rightarrow s_{i1} \text{—} e_{i2} \rightarrow s_{i2} \ldots s_{im-1} \text{—} e_{im} \rightarrow s_{im}\}$$

such that

- $(\forall \sigma_i \in TB_i(TC_i))\ ((s_{i0} \equiv s) \text{ and } (e_{i1}=\langle event1 \rangle))$;
- $(\exists \sigma_k \in TB_i(TC_i))$ with $(\sigma_k=s_{k0} \text{—} e_{k1} \rightarrow s_{k1} \text{—} e_{k2} \rightarrow s_{k2} \ldots s_{km-1} \text{—} e_{km} \rightarrow s_{km})$ such that $(e_{k2}=\langle event2 \rangle)$.

Note $\langle event1 \rangle$ and $\langle event2 \rangle$ indicated above refer to the events addressed by the test purpose, and s refers to the state indicated in the preamble $PA_i(TC_i)$.

To specify the verdicts $V_i(TC_i)$ w.r.t. the test purpose $TP_i(TC_i)$ for the test body $TB_i(TC_i)$ an algorithm *VerdictBehavClass1* is applied [Guer96]. It has as input the test body $TB_i(TC_i)$, the test purpose $TP_i(TC_i)$, the behavior tree of the specification S, and the test architecture TA_i. The output of the algorithm are the test verdicts $V_i(TC_i)$ that are specified for the test body $TB_i(TC_i)$. The algorithm first determines the conformance requirement addressed by the test purpose. This is facilitated by the derived preamble $PA_i(TC_i)$. Then the corresponding verdicts are associated by considering the possible interactions of the IUT at the different PCOs (points of control and observation) of TA_i in relation to the test purpose and the specification.

[‡] For convenience, the parentheses delimiting the data (i.e. events) in the test purpose are omitted.

According to the object concept the different components of a test case are considered as *properties* associated with the object TC_i. For their determination, *methods* have been attached to these properties. The algorithm *VerdictBehavClass1*, for instance, is attached as a *method* to the *property* V_i of the object TC_i.

Behaviorclass2

Now, we consider the structure of example subclass *behaviorclass*2 that address a process of a specification.

⟨*behaviorclass*2⟩→⟨*keyword*1⟩ (⟨*function*1⟩) [⟨*keyword*2⟩ (⟨*function*2⟩)]

The test purpose "check the correct execution of process (wait)" is an example for this class.

(⟨*function*1⟩) indicates a process[§] name referred to by the test purpose. (⟨*function*2⟩) is facultative; it is ignored in the example given above. It represents another process name in which (⟨*function*1⟩) is defined. This is necessary if the peer entities of a protocol specification may include functions that can be ambiguously referenced.

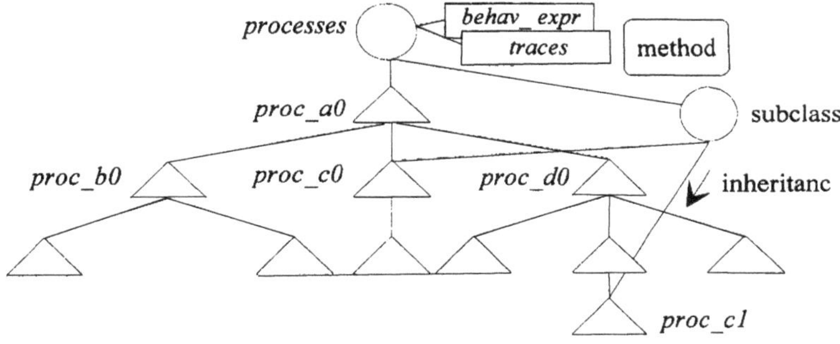

Figure 3 Representation of a LOTOS specification using the object paradigm.

To allow the identification of a process, the specification is represented as an hierarchy of processes (Figure 3). Considering the object paradigm, each (sub)process is viewed as object that is characterized by two main properties: *behav_exp* and *traces*. The first includes the definition of the process whereas the second indicates the resulting behavior of *behav_exp* as a set of traces. The latter is obtained by attaching a *method* to the property *traces*. The *method* calls a simulator to evaluate the given behavior expression *behav_exp* to generate the behavior tree. The latter represents the behavior described by the specification, if *proc_a*0 is subject of simulation, and a partial behavior if another subprocess is concerned by simulation. The value obtained for the property *traces* may be automatically *inherited* by means of the *inheritance mechanisms* [Rumb91]. Thus, the processes possessing the same definition can be attached to a same subclass. They *inherit* the same value to the property *trace*, e.g. *proc_c*0 and *proc_c*1. The representation of a specification as a hierarchy of objects facilitates the exploration of the

[§] This especially concerns a LOTOS specification.

processes of the specification and therefore, the identification of the one addressed by the test purpose.

The *preamble* $PA_i(TC_i)$ consists of exactly one event sequence, i.e. a trace, that is obtained as follows:

$$PA_i(TC_i)=\sigma\in S \text{ with } (\sigma=s_0 \xrightarrow{e_{k1}} s_{k1}\ldots s_{kr-1} \xrightarrow{e_{kr}} s_{kr})$$

such that

$(\forall\sigma'_i\in P)$ where $(\sigma'_i=s_{P0} \xrightarrow{e_{i1}} s_{i1}\ldots s_{iq-1} \xrightarrow{e_{iq}} s_{iq})$, implies

- $(s_0 \xrightarrow{e_{k1}} s_{k1}\ldots s_{kr-1} \xrightarrow{e_{kr}} s_{kr} \xrightarrow{e_{i1}} s_{i1}\ldots s_{iq-1} \xrightarrow{e_{iq}} s_{iq}\in B_S)$ with $(s_{kr}\equiv s_{P0})$;
- $(\forall\sigma_j\in Q)$ $(length(\sigma)\leq length(\sigma_j))$.

Note that $B_S=\{\sigma_1,\sigma_2,\ldots,\sigma_n\}$ represents the behavior tree of the specification S, where s_0 denotes its initial state, and $P=\{\sigma'_1,\sigma'_2,\ldots,\sigma'_m\}$ is the behavior described by the given process. σ_i denotes a trace, and σ'_i a subsequence of events w.r.t. a corresponding trace of B_S. Q is a finite set of traces that relates to P and B_S as follows:

- $P=\{\sigma'_i \mid (\sigma'_i=s_{P0} \xrightarrow{e_{i1}} s_{i1}\ldots s_{iq-1} \xrightarrow{e_{iq}} s_{iq})$ with $(i=1\ldots m)\}$;
- $Q=\{(\sigma_j=s_0 \xrightarrow{e_{j1}} s_{j1}\ldots s_{jl-1} \xrightarrow{e_{jl}} s_{jl}\in B_S) \mid (\forall\sigma'_i\in P)$
  $(s_0 \xrightarrow{e_{j1}} s_{j1}\ldots s_{jl-1} \xrightarrow{e_{jl}} s_{jl} \xrightarrow{e_{i1}} s_{i1}\ldots s_{iq-1} \xrightarrow{e_{iq}} s_{iq}\in B_S)$ with $(s_{jl}\equiv s_{P0})\}$.

The *test body* $TB_i(TC_i)$ consists of the behavior described by the process that is referred to as (⟨*function*1⟩) in the test purpose, i.e. $TB_i(TC_i)=\{\sigma'_1,\sigma'_2,\ldots,\sigma'_m\}$. The corresponding verdicts $V_i(TC_i)$ can be deduced by applying an appropriate algorithm *VerdictBehavClass*2 [Guer96] similar to that used for the test purposes subclass *behaviorclass*1. It considers the obtained test body $TB_i(TC_i)$ described by the process P, the behavior tree of the specification S, the test purpose $TP_i(TC_i)$, the determined preamble $PA_i(TC_i)$, and the test architecture TA_i. It first identifies the events that represent reactions of the IUT to given events of process P. After that, a verdict *pass* is specified for the reaction in question, if it is described in the process. Otherwise an *inconclusive* or *fail* verdict is specified depending on description in the specification of the reaction by considering the different PCOs of the test architecture TA_i.

4.2 Test purposes related to timing aspects

The timing aspect is an important factor in testing. It is especially introduced by the propagation delays and message loss. Propagation delays that are implied by the underlying service-medium may be the cause that a tester observes test events in an order different from the IUT side. Even though the underlying service is usually supposed to be sufficiently reliable, message delays and message loss may happen in practice which cause the expiration of the respective timer, thus leading to an inconclusive verdict [Chan92].

We again consider two example subclasses of test purposes: *timingclass*1 and *timingclass*2. They belong to the test purposes class *timing_testpurposes*. The reason of distinguishing these subclasses is the same as that indicated in Section 4.1.

Timingclass1

The test purposes belonging to this subclass are supposed to address conformance requirements without considering their emplacements in the specification as explained above. In other words, the path leading to the concerned conformance requirement is not explicitly given. We consider that the occurrence of the conformance requirement that can be reached with the lowest cost implicitly represents the objective of the test purpose.

$\langle timingclass1\rangle \rightarrow \langle keyword1\rangle\ (\langle event\rangle)\ \langle keyword2\rangle\ (\langle timevalue\rangle)$, e.g.

"check if (ICONconf) is returned within time limit (3)" where the time limit is expressed in a given time unit, e.g. in seconds.

$\langle keyword1\rangle$ and $\langle keyword2\rangle$ identify the test purpose. $(\langle event\rangle)$ is the event name to which the test purpose is related. $(\langle timevalue\rangle)$ represents the period in which the IUT must react to the event $(\langle event\rangle)$. After expiring $(\langle timevalue\rangle)$ a time-out will occur for the timer associated to $(\langle event\rangle)$. $(\langle event\rangle)$ and $(\langle timevalue\rangle)$ are referred to as the data. Note that such a structure gives no indication about the localization of the test event $(\langle event\rangle)$. The event $(\langle event\rangle)$ is automatically identified in the protocol specification in such a way that the event sequence leading to it is optimal.

The preamble $PA_i(TC_i)$ of the test case TC_i consists of exactly one event sequence that is determined as follows:

$$PA_i(TC_i)=\sigma_i \in S \text{ with } (\sigma_i = s_0 \text{—} e_{i1} \rightarrow s_{i1} \text{—} e_{i2} \rightarrow s_{i2} \ldots s_{ip-1} \text{—} e_{ip} \rightarrow s)$$

such that

- s_0 represents the initial state of the specification;
- there exists a transition t $(t=s_{ip} \text{—} e_{ik} \rightarrow s_{ik})$ with $(s_{ip}=s)$ such that

  $(\sigma'_i = s_0 \text{—} e_{i1} \rightarrow s_{i1} \text{—} e_{i2} \rightarrow s_{i2} \ldots s_{ip-1} \text{—} e_{ip} \rightarrow s_{ip} \text{—} e_{ik} \rightarrow s_{ik} \text{—} \langle event\rangle \rightarrow s' \in S)$;
- $length(\sigma_i) \leq length(\sigma_j)$ for all $j \neq i$, with $(\sigma_j = s_0 \text{—} e_{j1} \rightarrow s_{j1} \text{—} e_{j2} \rightarrow s_{j2} \ldots s_{jq-1} \text{—} e_{jq} \rightarrow s_{jq} \in S)$ such that
  - o $\exists$ a transition t' $(t'=s_{jq} \text{—} e_{jk} \rightarrow s_{jk'})$;
  - o $\sigma_j' = s_0 \text{—} e_{j1} \rightarrow s_{j1} \text{—} e_{j2} \rightarrow s_{j2} \ldots s_{jq-1} \text{—} e_{jq} \rightarrow s_{jq} \text{—} e_{jk} \rightarrow s_{jk} \text{—} \langle event\rangle \rightarrow s'' \in S$.

Note that $\langle event\rangle$ indicated above refers to the test event given in the test purpose.

The *test body* $TB_i(TC_i)$ of the test case TC_i consists of a set of event sequences that is obtained in connection with the preamble, the test purpose, and the specification as follows:

$$TB_i(TC_i)=\{\sigma_j \in S \mid (\sigma_j = s_{j0} \text{—} e_{j1} \rightarrow s_{j1} \text{—} e_{j2} \rightarrow s_{j2} \ldots s_{jm-1} \text{—} e_{jm} \rightarrow s_{jm})\} \text{ such that}$$
$$\forall \sigma_j \in TB_i(TC_i)\ (s_{j0} \equiv s)$$

Note that s above refers to the state indicated in the preamble $PA_i(TC_i)$ and m represents the length of the given event sequence.

As for the behavior classes we apply an appropriate algorithm *VerdictTimingClass1* [Guer96] to deduce the verdicts $V_i(TC_i)$ for the derived test body $TB_i(TC_i)$ by taking into consideration the preamble $PA_i(TC_i)$, the test purpose $TP_i(TC_i)$, the specification S and the testing architecture TA_i. The algorithm first determines an event in the specification for which (⟨*event*⟩) addressed by the test purpose represents a reaction. Then a timer whose value is indicated in the test purpose is associated to the given event. After that, the possible verdicts in respect to the time value indicated in the test purpose are specified. Particularly, verdicts depending on the occurrence of the time-out are assigned to (⟨*event*⟩). All possible interactions of the IUT at the different PCOs of the testing architecture are considered.

Timingclass2

Finally, we consider another test purposes subclass *timingclass*2. This subclass supposes in contrast to the subclass *timingclass*2 that the path leading to the conformance requirement addressed by the test purposes is explicitly given. This is used to deduce the corresponding preamble.

⟨*timingclass*2⟩→⟨*keyword*1⟩(⟨*eventsequence*⟩)⟨*keyword*2⟩(⟨*event*⟩)⟨*keyword*3⟩(⟨*timevalue*⟩),

An example of this subclass would be: "check if after (ICONreq; ICONind; ICONresp; ICONconf; IDISreq; IDISind; IDISreq; ICONreq) the event (ICONind) is returned within time limit (5)".

(⟨*eventsequence*⟩) represents a sequence of events leading to the event (⟨*event*⟩) to which the test purpose is related. It will be exploited to deduce the preamble. (⟨*timevalue*⟩) is the period within which the IUT has to react to the event (⟨*event*⟩) as explained for *timingclass*1. This subclass is characterized by the fact that a trace (⟨*eventsequence*⟩) is explicitly given which leads to the test event (⟨*event*⟩) concerned by the test purpose. This is relevant when a particular attention is accorded to a precise occurrence of the given event.

In the manner of subclass *timingclass*1, we can deduce the various components of a test case TC_i in respect to test purposes following the structure of *timingclass*2.

5 TEST CASE DERIVATION PROCEDURE

In this section, we present the test case derivation procedure for a given test purpose. It is represented in Figure 4.

The procedure starts with the identification of the test purpose. It supposes that the test purpose belongs to a certain prototype class which defines its formal structure as described in Section 4. The variety of instances of a class of test purposes corresponds to the different ways to express a given type of test purposes. This can be done by inserting different syntax units, i.e. words, supposed to be less relevant than the keywords on which the definition of the test purpose is based.

After parsing the test purpose by means of a specific parser routine, the keywords as well as the data concerned by the test purpose are identified. To recognize the objective of the test purpose the keywords are compared with the test purpose prototypes stored as classes in the knowledge base. This operation can be assisted by a human expert if the automatic recognition of the test purpose fails. In this case, the test purpose may be possibly reformulated with the help of an expert and then added to the knowledge base as a new prototype. The identification of a test purpose allows to deduce an appropriate test case and test verdicts. This consists of the determination of the properties *preamble*, *test body* and *test verdicts.* Methods that are attached to them are executed each time the identification of a test purpose is carried out.

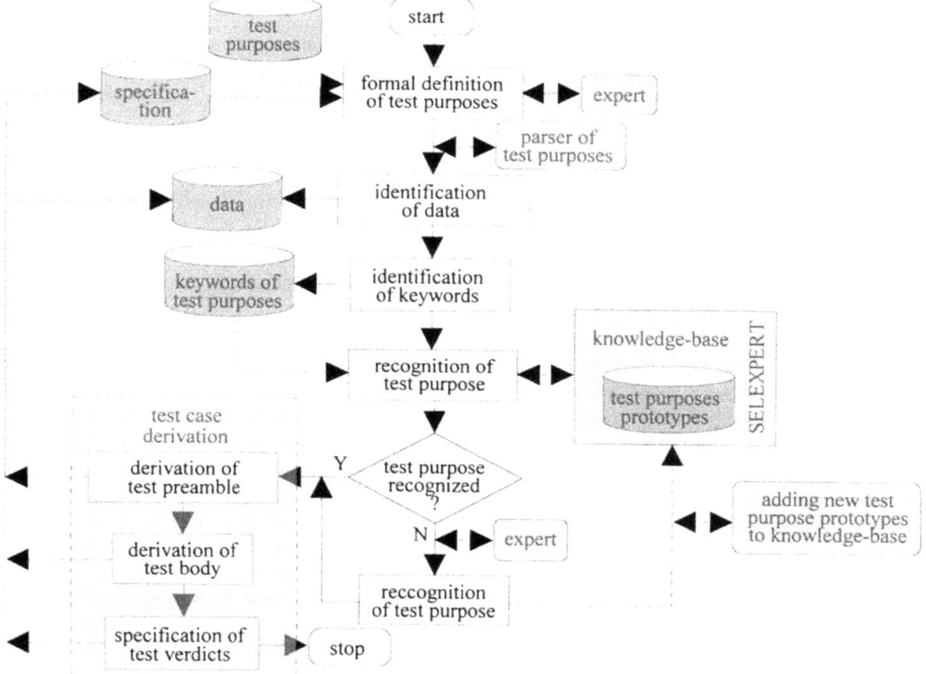

Figure 4 Procedure of the test purpose identification and test case derivation.

Note that the role of an expert consists in assisting the definition of the test purposes and supporting their identification in certain cases. The cooperation of the knowledge engineer and the expert will help to formalize these test purposes, i.e., to assign the respective formal structure that can be used as test purpose prototypes when identifying new test purposes.

Example

We illustrate the principle of the approach for the INRES protocol [Hogr92]. The test purpose *TP* is defined as follows: "check the correct execution of process (*datatransfer*)". *datatransfer* is supposed to represent the *data transfer phase* that is specified as a process. The formal structure of this test purpose corresponds to that associated to the subclass *behaviorclass*2 (see Section 4). The keywords of an instance of this subclass are supposed to be identifiable with the ones characterizing the given test purpose class. To identify the latter the hierarchy of classes consisting of the different subclasses (*behaviorclass*1, *behaviorclass*2, etc.) has to be explored (see Figure 2). A formal structure as indicated in Section 4 corresponds to each one of them. To identify the test purpose, the parser has to distinguish the two parts of the test purpose: keywords and data (here *datatransfer*). If the test purpose, i.e. its keywords, coincides with one of the instances (e.g. *keywords_subclass*1), the next step can start that determines the different components of the test case. Otherwise, the assistance of a human expert becomes indispensable.

We suppose the test purpose *TP*(*TC*) concerning the process *datatransfer* as indicated above. A test case that corresponds with *TP*(*TC*) is deduced as follows:

```
MSAP!CR
   ISAP_R?ICONind
      ISAP_R!ICONresp
         MSAP?CC
MSAP!DT
   ISAP_R?IDATind        pass
   ISAP_R?IDISreq        inconclusive
   ISAP_R?otherwise      fail
   MSAP?otherwise        fail
```

The test event sequence "MSAP!CR...MSAP?CC" represents the preamble *PA*(*TC*) of the test case *TC*. "MSAP!DT...MSAP?otherwise" is the test body. The test verdicts *V*(*TC*) has been deduced for the test body *TB*(*TC*) as explained above. The preamble is optimal in the sense that its length in test events is minimal as indicated in Section 4. The preamble has no verdicts since it is not addressed by the test purpose. Only the behavior addressed by the test purpose was deduced. Note that we have considered here a test architecture whose PCOs are called *ISAP_R* and *MSAP* situated at the upper and lower tester of the test method, respectively. In addition, *otherwise* is used to indicate unexpected reactions of the IUT at the PCOs.

6 TEST CASE ADAPTATION IN RESPECT TO TEST METHODS

In this section, we describe the procedure to adapt a derived test case to a given test method. This is needed because test cases are usually derived from a formal specification without taking into consideration a given test method.

A specification such as that of an OSI protocol describes the behavior at the service access points. The points where the tester can control and observe the IUT are the PCOs that may not coincide with the service access points belonging to the IUT. The upper tester controls and observes the PCOs connected to the upper access point whereas the lower tester those connected to the lower access point. By varying these aspects different test methods are used, e.g. the local, distributed, coordinated, and remote test methods [ISO91]. In these methods, the lower access point of the IUT is always accessible via an underlying service; the upper access point may be hidden. There exist variants of these methods that can be applied to a multi-layer IUT where an IUT is tested layer by layer. An embedded method is a special case of multi-layer testing where one layer can be tested embedded in the other layers.

The principle of test case adaptation that takes a given test method into consideration is illustrated in Figure 5. Starting point is a set of test cases and a test method. These are represented as classes, *testcases* and *testmethods*. The class *testmethods* is characterized by some *properties* such as *identname*, *uppertester*, *lowertester*, *sut*, *testsystem*. *Identname* indicates the test method name (e.g. DS, distributed single-layer). *Uppertester* and *lowertester* contain the PCOs names associated to upper and lower tester, respectively. *Sut* states the components of the test system. Note that for certain test methods some of these properties are not defined, e.g. the value of *uppertester* is undefined for the coordinated test method. Test coordination procedures have not been considered here for their irrelevance to our purpose.

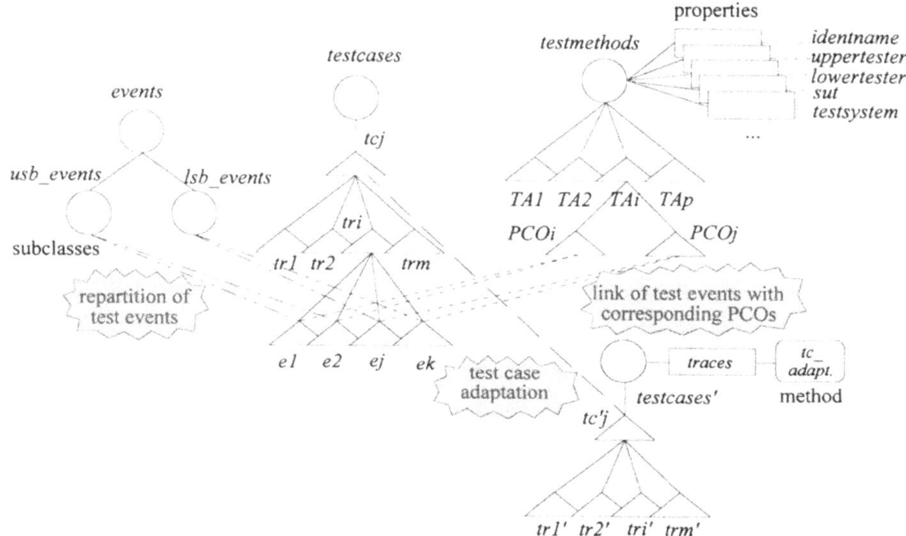

Figure 5 Test case adaptation based on the object paradigm.

There are three main steps for the test case adaptation. First, the test events of the traces of a test case (belonging to the class *testcases*) are identified. They are linked to the corresponding subclasses, *usb_events* or *lsb_events*, according to their occurrence at the upper service boundary or at the lower service boundary of the IUT. The access service boundaries are then linked to their corresponding PCO_i that are associated with the upper tester or to the lower tester depending on the given test method TA_i (cf. Figure 5). Finally, a class of test cases *testcases'* is deduced. There is a one-to-one relation between its test cases and the input test cases belonging to the class *testcases*. There is also a one-to-one relation between a trace of a test case of the class *testcases'* and a trace of a corresponding test case of the class *testcases*. The property *traces* is associated to the class *testcases'* that has to be determined for each member (i.e. a test case) under consideration the given test method. For this purpose, a *method* is attached to this property to determine the value of the property *traces*, i.e. to compute all traces composing the given test case. The *method* implements an algorithm *tc_adaptation* that is not given here.

7 FINAL REMARKS

The automation of the test case derivation in respect to given test purposes is a difficult problem. We have shown that this problem can be alleviated by using knowledge-based techniques. Our approach is based on a partial formalization of test purposes. This can be done for important classes of test purposes. The approach was inspired by the experience of testers who are often capable to assess what kinds of test purposes are relevant in practice. The cooperation between a knowledge engineer and a human expert in testing may help to formalize and classify test purposes. In an earlier phase of formalization and identification of test purposes, the assistance of a human expert is useful. His support is less needed in later sessions with the progressively growing of the knowledge-base of test purposes prototypes.

Currently, we are extending the approach to cover a large number of real protocols by investigating test purposes prototypes used in these protocols. Furthermore, we plan to make an empirical study of the approach in order to demonstrate the effectiveness of the approach, i.e. to which extent the approach is able to identify test purposes.

REFERENCES

[Brin88] E. Brinksma. A theory for the derivation of tests. In S. Aggarwal and K. Sabnani, editors, Protocol Specification, Testing, and Verification, VIII, North-Holland, 1988.

[Brin91] E. Brinksma, J. Tretmans, and L. Verhaard. A framework for test case selection. In B. Jonsson, J. Parrow, and B. Pehrson, editors, Protocol Specification, Testing, and Verification, XI, North-Holland, 1991.

[Chan92] S.T. Chanson and Q. Li. On inconclusive verdict in conformance testing. In G.v. Bochmann, R. Dssouli, and A. Das, editors, 5th International Workshop on Protocol Test Systems, North-Holland, 1992.

[DIN91] DIN 19245. Profibus, Process Field Bus. Beuth Verlag GmbH, Berlin, 1991.

[FMCT95] ISO/IEC. Working draft on formal methods in conformance testing. Paris, 1995.

[Guer95] A. Guerrouat, H. König, and A. Ulrich. SELEXPERT - a knowledge-based tool for test case selection. In Proceedings of 8th International Conference on Formal Description Techniques, Montreal, 1995.

[Guer96] A. Guerrouat and H. König. Automatic derivation of test case with respect to test purposes. Technical report, Department of Computer Science, University of Otto von Guericke Magdeburg, Magdeburg, 1996.

[Hogr92] D. Hogrefe. OSI formal specification case study: the INRES protocol and service, revised. Universität Bern, Institut für Informatik, 1992.

[ISO91] Information Technology - Open Systems Interconnection - Conformance testing methodology and framework - Parts 1-5. ISO/IEC 9646, 1991.

[Lai95] R. Lai and W. Leung. Industrial and academic protocol testing: the gap and the means of convergence. Computer Networks and ISDN Systems, 27(1995): 537-547.

[Lunz94] J. Lunze. Künstliche Intelligenz für Ingenieure. Oldenburg Verlag, Munich, 1994 (in German).

[Robl92] T. Robles, J. Manas, and G. Huecas. Specification and derivation of OSI conformance test suites. 3rd LOTOSPHERE-Workshop, Pisa, 1992.

[Rumb91] J. Rumbaugh, M. Blaha, W. Premarlani, F. Eddy, and W. Lorensen. Object-oriented modeling and design. Prentice-Hall, New Jersey, 1991.

[Shen90] Y.-N. Shen, F. Lombardi, and A.T. Dahbura. Protocol conformance testing using multiple UIO sequences. In E. Brinksma, G. Scollo, and C.A. Vissers, editors, Protocol Specification, Testing, and Verification, IX, North-Holland, 1990.

[Ulri93] A. Ulrich and H. König. Test derivation from LOTOS using structure information. In O. Rafiq, editor, 6th International Workshop on Protocol Test Systems, North-Holland, 1993.

[Velt92] R.J. Velthuys, J.M. Schneider, and G. Zörntlein. A test derivation method based on exploiting structure information. In R.J. Linn and M.Ü. Uyar, editors, 2Protocol Specification, Testing and Verification, XII, North-Holland, 1992.

PART SEVEN

ATM Testing

16

Field Trial and Quality Test of ATM Switching System In Korea

Tae-Won Kim
Kyeong-ho Lee, Taek-Won Jeong
Electronics and Telecommunications Research Institude
161 Kajong-Dong Yusong-Gu, Taejon, 305-600, KOREA
FAX : +82-42-860-5410
kimtw@nice.etri.re.kr, khlee@nice.etri.re.kr, twjeong@nice.etri.re.kr

Abstract

This paper describes the characteristics, quality and application of ATM switching system in the Network Testbed which is established to test interoperability between B-ISDN elements. The switching system in the Network Testbed manages connections of PVC type. Test result of the switching system shows cell loss probability of 10^{-10}. The cell loss probability between two terminals that are about 200Km away from each other in the Network Testbed was order of 10^{-9}. The Network Testbed will be upgraded and used until 1998.

Keywords

Quality Test, Network Test-Bed, Throughput, Switching System, Delay,Variation

1 INTRODUCTION

ATM is recommended as the transfer mode for B-ISDN. ATM switching system is one of the core element of B-ISDN. The ATM switching system should have the ability to operate properly under various load conditions. Generally, quality parameters in ATM network are cell loss, cell misinsertion, cell delay and cell delay variation. Therefore, it is required to design systems which fulfill the quality parameters to provide satisfactory services to users. This paper describes the test results of an ATM switching system. Tests are carried out under mixture of various traffic and are summed by quality parameters such as cell interarrival time, cell delay time and cell loss probability. Also described is the operation of Network Testbed which consists of ATM switching systems, B-NT, B-TA, private ATM switches, and terminals.

2 ATM SWITCHING SYSTEM

The ATM switching system used in Network Testbed is developed for use in public B-ISDN by ETRI since 1992. The system, called TDX-ATM switching system, has the switch size of 256x256 which is reconfigurable into 32x32, 64x64 or 128x128 according to the traffic. As shown in Figure 1, the system consists of 3 kinds of subsystems: ALS(ATM Local Switching Subsystem), ACS(ATM Central Switching Subsystem) and IWS(IWS:Interworking Subsystem).

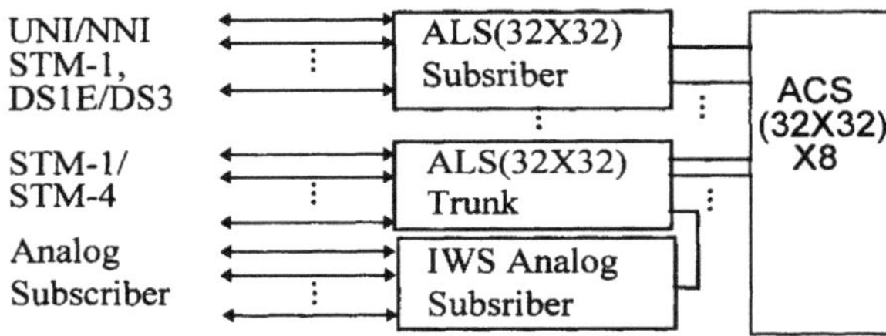

Figure 1 TDX-ATM Switching System

The ALS is a subsystem which consists of a single stage self-routing switch, interfaces for various subscribers and other ATM switching system, and a processor for connection and call control. An ALS itself can operate as a small switching system. The architecture of ALS is shown in Figure 2. The switch network of the ALS is 32x32. Since the link speed of switch is 155.52Mbps, medium(DS3, 45Mbps) or low(DS1E, 2.048Mbps) rate subscribers are multiplexed. The communication between processors in each subsystem is done through the switch network. Thus 32 links of the switch network are used by subscribers, the processor, and the ACS. If the traffic to other ALS is not high, then most of links are allocated for subscriber interfaces and trunk interfaces and only a few links for the connection of ALS and ACS. Thus the switch network in ALS can concentrate input traffic with concentration ratio from 8:7 to 29:2 according to the configuration.

The ASNM(ATM Switch Network Module) is 32x32 folded switch consists of 16x16 switch elements as shown in Figure 3. The 16x16 switch element adopts limited shared buffer type. As shown in Figure 4, each 16x16 switch has many buffers: a common buffer to store cells, 16 cell address buffers to store cell addresses from 16 links, a buffer to store multicasting address, a buffer to store multicasting routes, and a buffer to store usable addresses of the common buffer. The common buffer can store 4,096 cells from 16 links. Each link of 16x16 switch element is transformed serial-to-parallel at the input and parallel-to-serial at the output to lower the operating speed of the switch.

ACS is a subsystem which connects ALSs and IWSs. The ALS consists of switch network and a processor(OMP) for the operation and management of the switching system. The switch of ACS is the same as that of ALS. The processor(OMP) also communicates through the switch network with other processors. IWS is the subsystem for PSTN subscribers. An IWS has the switching capacity of 1,024 time slots. Hence, an IWS can accommodate 8,192 subscribers if concentration ratio of 8:1 is assumed.

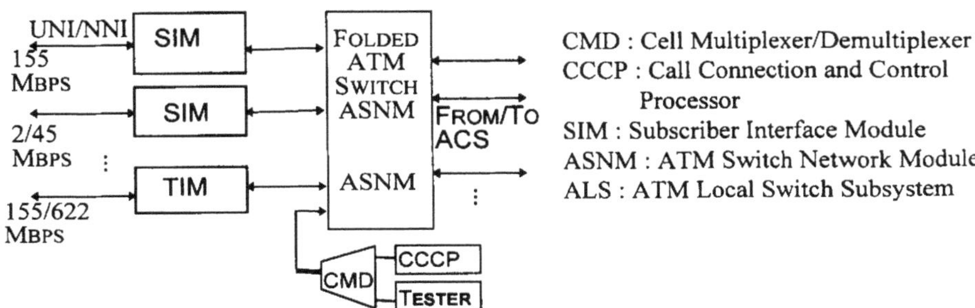

Figure 2 The Structure of ALS(ATM Local Switching Subsystem)

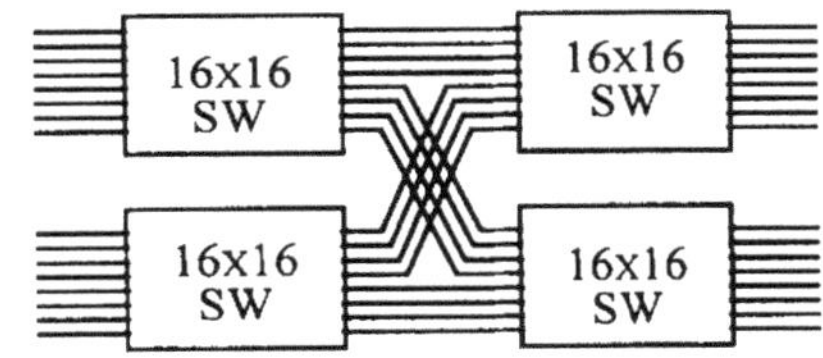

Figure 3 The Structure of ASNM

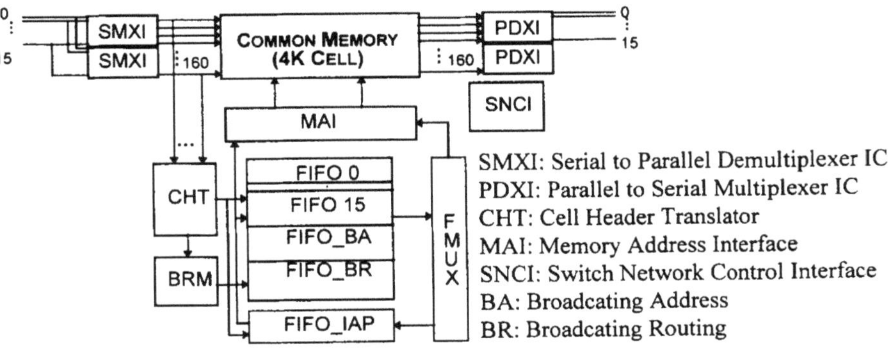

Figure 4 The structure of 16x16 switch

3 TEST OF TDX-ATM SWITCHING SYSTEM

TDX-ATM switching system is tested under various mixture of loads. Tests are carried to find QoS parameters such as cell loss ratio, cell delay and cell interarrival time. In addition to these QoS parameters, we also tested the switch overflow and the stability of the switching system.

3.1 QoS test of TDX-ATM switching system

ATM cells that pass the switch are user cells, signaling cells or OAM cells. User cells may be switched in one stage or 3 stages according as whether the calling subscriber and the called subscriber are accommodated at the same ALS. Several tests are carried for the connection of subscribers of the same ALS or different ALSs. For all of tests, we used test system which can generate traffic and analyze the test result by comparing sent traffic and received traffic.

QoS for the connection of subscribers of the same ALS

The input and output of the test system are connected to the same ALS of the switching system as shown in Figure 5. Two tests are carried our for the connections: first test with CBR load of single channel of 7.488Mbps and second test with load of 100 channels of 1.4976Mbps.

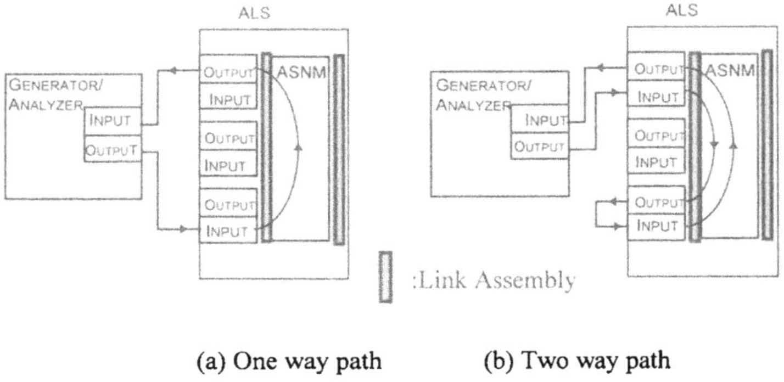

(a) One way path (b) Two way path

Figure 5 Connection of the test system and TDX-ATM switching system

Test results in Table 1 shows average cell delay time of 53.3 μs and maximum cell delay variation of 6.1μs for one way path. Most of the delay is caused by the switch and the link between subscriber interfaces and the switch. This delay may be said as the fixed delay in the reference [1]. In case of two way path, cell delay times are doubled approximately as expected.

Table 1 Cell interarrival time and cell delay time (single 7.488Mbps/100 channels of 1.4976Mbps)

	two way path		*one way path*	
	(single/100 channels)		*(single/100 channels)*	
	CTD(μs)	*CIT(μs)*	*CIT(μs)*	*CTD(μs)*
Minimum	52.6/52.9	53.8/280.2	50.6/280.2	103.1/120.5
Maximum	55.9/55.8	59.9/286.2	62.6/286.1	109.1/121.1
Mean	53.3/54.0	56.62/283.1	56.6/283.1	104.4/120.8

CIT: Cell Interarrival Time CTD: Cell Transfer Delay

QoS for the connection of subscribers of different ALSs

The input and output of test system are connected to the same ALS of the switching system as shown in Figure 6. Test cases and the load are the same as before. The path consists of three switches ALS-ACS-ALS. The central switch, ACS, connects subscribers of different ALSs. Test results in Table 2 shows almost the same cell interarrival time. However, cell delay time is doubled approximately as expected

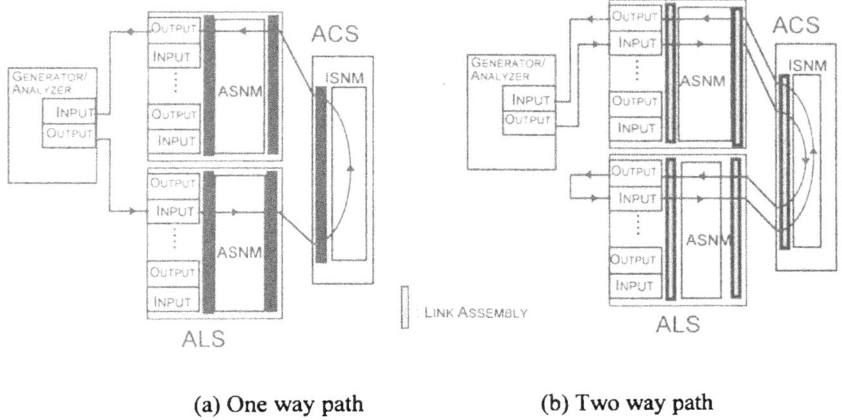

(a) One way path (b) Two way path

Figure 6 Connection of the test system and TDX-ATM switching system

Table 2 Cell interarrival time and cell delay time (single 7.488Mbps/100 channels of 1.4976Mbps)

	one way path		*two way path*	
	(single/100 channels)		*(single/100 channels)*	
	CIT(μs)	*CTD(μs)*	*CIT(μs)*	*CTD(μs)*
Minimum	53.8/280.1	106.7/106	50.6/280.1	216.5/216.5
Maximum	59.8/286.2	110.0/109.4	62.7/286.2	219.8/222.5
Mean	56.62/283.1	107.6/54.0	57.1/283.1	218.7/220.1

QoS for multiple paths of subscribers of different ALSs

In this test, two test systems are connected to TDX-ATM switching system. Two paths are established to know whether there is influence between two paths.

(a) In case of two independent paths

Two test systems are connected to TDX-ATM switching system as shown in Figure 7. There are 8 connections switched through the path for test system 1. The bandwidths of 8 connections are: 45, 40, 30, 15, 7.488, 5, 4, 3.272 Mbps. The total load of 8 connections is 149.760Mbps which is full load for the STM-1 link. The other path switches CBR load of 100Mbps which is generated by the test system 2. As shown in table 3, test results are similar to those of previous tests. The result may be interpreted that there is not any noticeable influence between two independent paths.

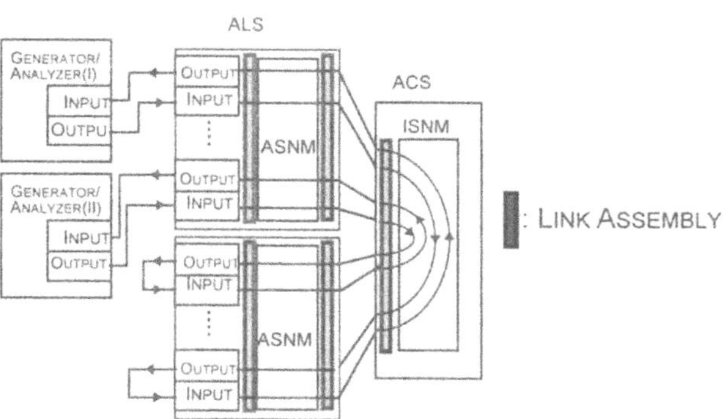

Figure 7 The connection of test systems and TDX-ATM switching system

Table 3 Cell interarrival time and cell delay time for CBR traffic of 7.488Mbps(independent paths)

	cell interarrival time(μs)	*cell delay time(μs)*
Minimum	56.5	213.5
Maximum	57.1	214.1
Mean	56.6	213.7

(b) In case of two dependent paths

Two test systems are connected to TDX-ATM switching system as shown in Figure 8. The external connections are the same as before. But internal paths are arranged so that the traffic from two test systems are mixed internally. Two test systems generate traffic loads different from the previous test. Generated traffic loads are: 45Mbps CBR traffic by test system 1 and VBR traffic with average of 86.79Mbps and peak of 149.76Mbps by test system 2. As shown in the Figure, since part of the path are shared by traffic load from two test systems, cells are buffered when total traffic is more than 149.76Mbps. The buffering of cells means delay of the buffered cells. In this test, we get cell delay times of minimum 207.6μs, maximum 211.0μs, and average 209.9μs if only the test system 1 generates traffic. When both of two test systems generate traffic, there is an increase of maximum cell delay time by 54μs. This test shows that the system switches mixed traffic of CBR and VBR well with a little increase of cell delay.

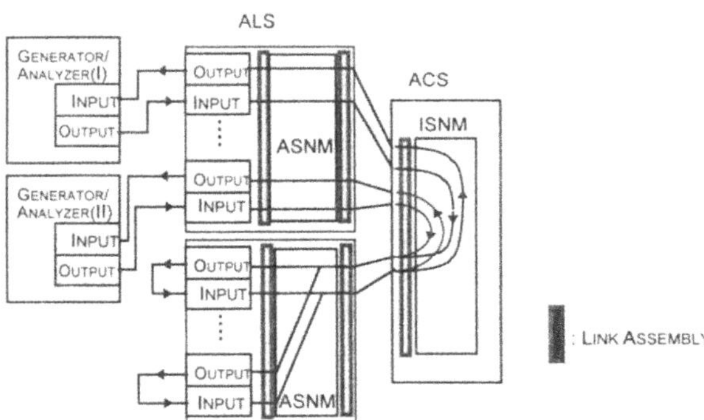

Figure 8 The connection of test systems and TDX-ATM switching system

3.2 Test of performance and stability of the switching system

Test of performance and stability of the switching system is carried by the long-term operation of the switching system and by test of influence from the variation of traffic and overload. The test is performed by stability test of switch and performance test.

(a) Stability test of switch

Figure 9 shows the configuration to test the stability of 32x32 switch fabric in ALS. All of the subscriber boards are looped together to form a path. The path switches CBR traffic of 149.76Mbps generated by the test system. There is no cell loss during the test which continued 24 hours. From this result, it may be said that TDX-ATM switching system has stable switch.

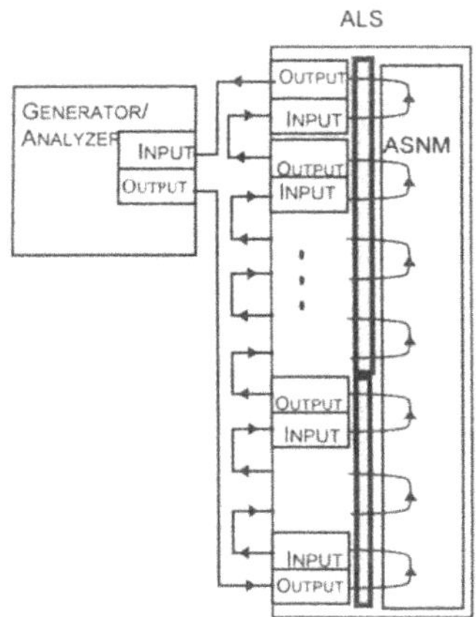

Figure 9 Stability test of the switch

(b) Performance test

In this test, test systems and the switching system are so configured as to test the performance of the switching system. Thus 149.76Mbps CBR stream or the mixture of CBR stream and bursty stream is applied to every port of the switch. In the first test, test systems and the switching system are connected as shown in Figure 10. Each path is independent that traffic stream is not mixed with the stream from other paths. Each of two test systems generates CBR stream of 149.76Mbps. After the test of 38 hours, we get cell loss ratio of 1.1E-11.

Now, test systems and the switching system are connected, as shown in Figure 11, differently from that of previous test. The traffic streams are distributed and mixed together in the switching system. During this test, traffic streams are generated by two test systems: CBR stream of 0.7(=0.7x149.76Mbps) by test system 1 and VBR stream of maximum 149.76Mbps and average 0.0625(=9.36Mbps), or 0.1875(=28.08Mbps), or 0.2875(=43.056Mbps) by test system 2. If CBR traffic and VBR traffic are applied together, there happen 7 hot spots as shown in the figure. After 38 hours of test, we get the test result shown in table 4. According to the result, the switching system operates with load of 0.7625 and 0.8875)

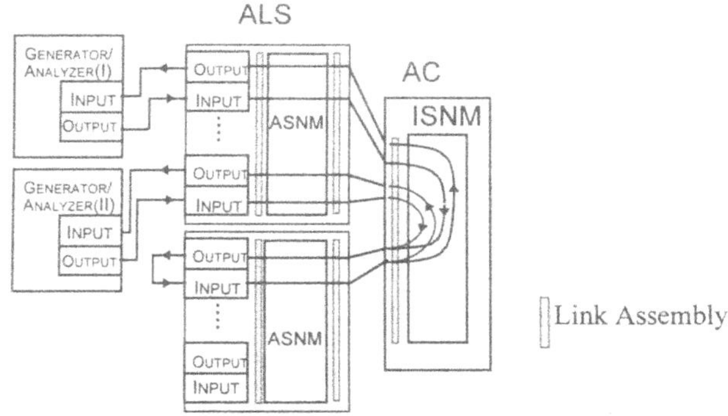

Figure 10 The connection of test systems and the switching system

Table 4 Cell loss ratio(independent paths)

	Load(0.7625)		*Load(0.8875)*		*Load(0.9875)*	
	CBR(0.7)+VBR(0.0625)		*CBR(0.7)+VBR(0.1875)*		*CBR(0.7)+VBR(0.2875)*	
	CBR(0.7)	VBR(0.0625)	CBR(0.7)	VBR(0.1875)	CBR(0.7)	VBR(0.2875)
cell loss ratio	3.3E-10	1.0E-9	2.3E-10	5.0E-10	3.9E-7	9.5E-7

* VBR Traffic applied in Baseline Document MPEG-II Core Service in Table1/Annex2/ IVS /Baseline Document/ITU-SG13/19-94

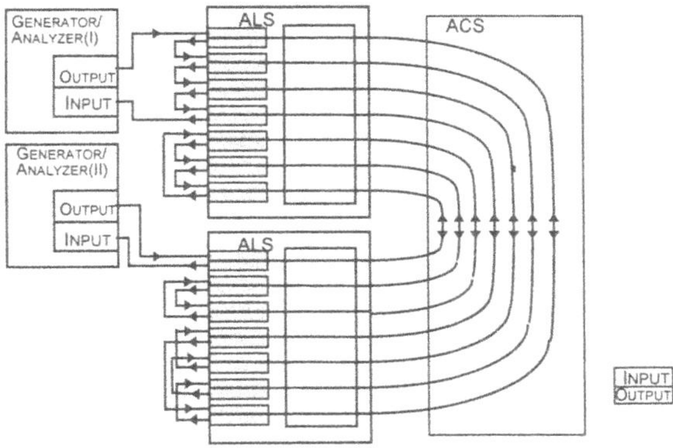

Figure 11 The connection test systems and the switching system

Table 4 Cell loss ratio(independent paths)

	Load(0.7625)		*Load(0.8875)*		*Load(0.9875)*	
	CBR(0.7)+VBR(0.0625)		*CBR(0.7)+VBR(0.1875)*		*CBR(0.7)+VBR(0.2875)*	
	CBR(0.7)	*VBR(0.0625)*	*CBR(0.7)*	*VBR(0.1875)*	*CBR(0.7)*	*VBR(0.2875)*
cell loss ratio	3.3E-10	1.0E-9	2.3E-10	5.0E-10	3.9E-7	9.5E-7

4 APPLICATION OF TDX-ATM SWITCHING SYSTEM

Network Testbed[2] is established mainly to test the interoperability between B-ISDN systems such as public ATM switching system, B-NT, B-TA, private ATM switch, etc. As shown in Figure 12, there are 4 public ATM switching systems in the Network Testbed: 3 systems in Seoul and 1 system in Taejon which is about 200Km away from Seoul. Currently, users are restricted to government organizations, Universities, and libraries. Up to now, most of traffic is for VOD, video telephone, and data communication between LANs. In the connection of Figure 12, we get the cell loss ratio of 2.04E-9 for the end to end connection between subscribers of 200Km away.

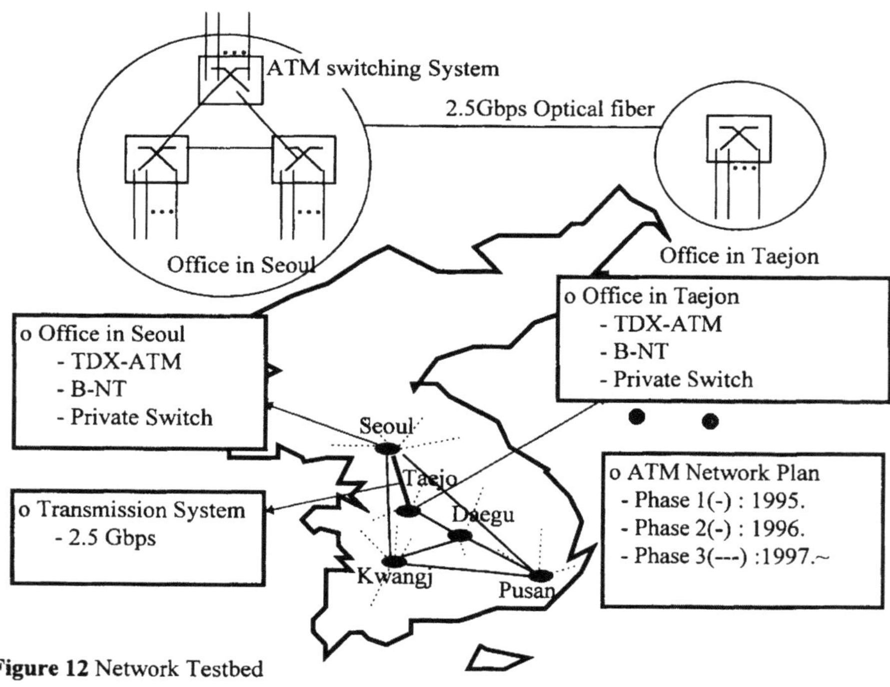

Figure 12 Network Testbed

In addition, The following equipments are operating on service in the Network Test Bed

(a) Video-TA

- Features: STM-1, DS3, VC Connection, DPCM(Differential Pulse Code Modulation)
- Function: Services that let you converse with high-qulaity voice and video through the ATM Network

(b) LAN-TA

- Features: STM-1, SVC Connection, LAN-Router
- Functions: Terminal adaptor that connects LANs with ATM Network

(c) B-NT(CANS:Centralized Access Node System)

- Features: CANS have 1 line of STM-1 at Tb interface point, and 4 lines of STM-1 at Sb interface point(Concentration Ratio(Tb/Sb) = 1/4)
- Functions:Network termination that connects subscribers, terminals to ATM network

(d) Work-Station and Personal Computer with ATM Card

- Features: STM-1, DS1E, MPEG, JPEG
- Functions:
  - High-speed data transmission service between LANs via ATM network
  - Access to multimedia DBs will be available via Internet
  - Transfer the video programs in the Video server via the ATM network
  - Multiparty conferences in which you can exchange data and voice while seeing one one another via ATM network

(e) Private switch :

- Products: Fore Co. switch, 3Com. Co. switch, Private ATM Switch made in korea.
- Features: STM-1, DS3, E1, LAN-router
- Functions: Private network that connect subscribers, terminals or local area network to ATM network

5 CONCULSION

This paper presents test results for the ATM switching system developed by ETRI. According to the results, the switching system shows maximum cell delay of 110μs under normal operation. Also, the test results show the stability of the ATM switching system. The ATM switching system, called TDX-ATM, will be improved in quality, features, and economy by software upgrade and hardware upgrade using ASIC chips and new commercial chips.

6 REFERENCES

Thomas R. Banniza, J.(Oct.1991) Design and Technology of VLSI's for ATM Switches:

IEEE J. on Selected Areas in Communication, Vol.9 No.8.
M. Parker, ISS'95(Apr. 1995) The European ATM Pilot Vol.1

6 BIOGRAPHY

Tae Won Kim was born in South Korea in 1959. He graduated from Taejon University with a B.S. degree in electrical engineering in 1982. Since 1983, he has been in Switching Technology Division of ETRI where he is working as a senior reseracher of Quality Assurance Section. His research interests include B-ISDN communication systems for public network and ATM(Asynchronous Transfer Mode) communication system engineering, especially system modeling and performance evaluation of telecommunication.

Kyeong Ho Lee* was born in South Korea in 1956. He received B.S. and M.S. degrees in electrical engineering from Kwangun University, Seoul, Korea in 1980 and 1982, respectively. Since 1981, he has been in switching Technology Division of ETRI where he is working as the head of Quality Assurance Section. His research interests include network design, telecommunication system modeling and performance evaluation

Taek Won jeong** was born in South Korea in 1955. He received B.S. and M.S. degrees in electrical engineering from Seoul University, Seoul, Korea in 1979 and 1981, respectively, and Ph.D. degree from University of Florida, USA, in 1990. Since 1983, he has been in Switching Technology Division of ETRI, where he is working as the head of ATM Service Section. His research interests include wide band services and B-ISDN switching systems.

17
Conformance testing of ATM Adaptation Layer protocol

Sangjo Yoo *, *Leslie Collica* **, *and Myungchul Kim* *

**Korea Telecom Research and Development Group*
Sochogu Unyundong 17, Seoul, 137-795, Korea
{sjyoo, mckim}@sava.kotel.co.kr

***National Institute of Standards and Technology*
Building 223/ Room B364, Gaithersburg, MD 20899, USA
collica@isdn.ncsl.nist.gov

Abstract

This paper presents an appropriate testing methodology to check the conformity of an implementation following ATM Adaptation Layer protocol in ATM/B-ISDN communication systems. In this paper, we analyze restrictions of the AAL Type 5 protocol to develop Abstract Test Suites and a test system, and we compare test coverage of applicable test methods to choose the best test method for AAL Type 5 Common Part protocol testing. We decide to apply Distributed Test Method for testing of the protocol because of its larger test coverage. To synchronize the execution of test cases between Upper Tester and Lower Tester in distributed testing environment, we propose a simple and efficient timing synchronization mechanism. To illustrate the effectiveness of our test structure, we give an example of test cases described in Tree and Tabular Combined Notation using the proposed timing synchronization mechanism.

Keywords

Conformance Testing, ATM Adaptation Layer, Distributed Testing, TTCN, Upper Tester

1 INTRODUCTION

Asynchronous Transfer Mode (ATM) and Broadband Integrated Services Digital Network (B-ISDN) are rapidly becoming acknowledged as a base technology for the next generation of global communications supporting a variety of applications. ATM/B-ISDN protocols have very complex structures compared with narrowband ISDN or packet networks.

Of the ATM/B-ISDN protocols, ATM Adaptation Layer (AAL) (ITU-T COM SVIII-R 116, 1992)is used to enhance the services provided by ATM layer to support the functions required by the layer above. Therefore, the functions performed in the AAL depend upon the higher layer requirements.

Especially, the AAL Type 5 protocol (ITU-T TD-XVIII/10, 1993) is defined not only for user data transfer in the user plane, but also for signalling in the control plane. To ensure that an implementation conforms to its standards or specifications and then to increase the probability of interoperability between ATM/B-ISDN products from multiple venders, protocol conformance testing in telecommunication has become an important area in the latest years. In order to assure the sufficient test coverage and to perform the conformance testing in accordance with the test requirements, first of all, we should decide the most appropriate test method and test configuration. In this paper, we present an appropriate test methodology using the proposed timing synchronization mechanism for the AAL Type 5 protocol conformance testing.

In Section 2, we analyze the restrictions of AAL Type 5 protocol in developing Abstract Test Suites (ATSs) and a test system for conformance testing. In Section 3, we compare test coverage for test methods, and we list advantages and disadvantages of each method. In Section 4, we show applicable test configuration for AAL Type 5 Common Part protocol conformance testing. In Section 5, we propose a simple and efficient timing synchronization procedure in Distributed Test Method (DTM). The procedure has no additional requirements or management protocols for synchronizing between Upper Tester (UT) and Lower Tester (LT). In Section 6, we show the requirement and implementation structure of the UT. In Section 7, we give an example of test cases, described in Tree and Tabular Combined Notation (TTCN) (ISO/IEC 9646-3, 1993), using the proposed timing synchronization mechanism for AAL Type 5 Common Part protocol. Finally, in Section 8, we conclude this paper.

2 RESTRICTIONS OF AAL PROTOCOL FOR CONFORMANCE TESTING

In order to develop test suites for conformance testing, we should consider test requirements that specify which protocol characteristics should be tested. Once we have defined test requirements including PICS (Protocol Implementation Conformance Statement) proforma, we need to decide what kind of test methods can be used to specify the abstract test suites. Then we can specify test scenarios describing test behaviour between tester and Implementation Under Test (IUT) as an ATS, and we can implement a test system based on the test scenarios. Before we decide a test method, we should carefully review all the restrictions in the protocol under test. In this section, we analyze the restrictions on conformance testing of AAL Type 5 protocol.

Major factors in AAL Type 5 specification [2] that restrict conformance testing are listed as follows:

- various profiles,
- no Service Access Point (SAP) between sublayers,
- simple Protocol Data Unit (PDU) transfer procedure, and
- absence of description for control and management planes.

2.1 Various Profiles

As shown in Figure 1 [2], the AAL Type 5 protocol consists of CPAAL 5 (Common Part AAL Type 5), that is independent of AAL service user, and Service Specific Convergence Sublayer (SSCS), that is dependent on the characteristics of AAL service user. This means that a variety of AAL Type 5 profiles may exist according to the kind of SSCS protocols (including a case that SSCS is null) that support specific AAL user services or groups of services. Examples of possible AAL Type 5 structures are shown as follows:

- AAL Type 5 structure consisting of only CPAAL 5 (i.e., SSCS is null),
- Signalling AAL consisting of CPAAL 5, Service Specific Connection Oriented Protocol (SSCOP), and Service Specific Coordination Function (SSCF). SSCOP provides assured data delivery between AAL connection endpoints, and SSCF performs a coordination function between the service required by user of signalling layer 3 (Q.2931) and the service provided by SSCOP, and
- Other AAL Type 5 structures depending on SSCS protocol (e.g., FR-SSCS over CPAAL 5 for Frame Relay service).

The fact that we have these various profile structures means that several kinds of test suites are necessary. If we want to use UT for conformance testing on AAL Type 5 protocol, we have to define and implement UT functions according to the profile of AAL Type 5.

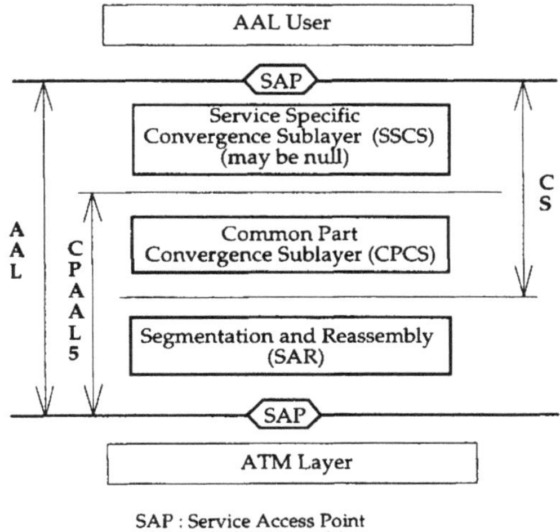

Figure 1 Structure of AAL Type 5

2.2 No SAP Between Sublayers of AAL Type 5

AAL Type 5 protocol is subdivided into multiple sublayers as shown in Figure 1. According to ATM Adaptation Layer Type 5 specification in ITU-T I.363 draft recommendation [2], there exist SAPs between the AAL and the ATM layer, and between the AAL and the AAL user. However there exist no SAPs between sublayers of AAL Type 5. This means that exchanges of parameters (e.g., AAL SDU (Service Data Unit), loss priority, and congestion indication) between sublayers can be implemented by internal interface method of each implementation.

In protocol conformance testing, the fact that there exist no SAPs between sublayers implies that it is very difficult for UT and LT of a test system to access interfaces between sublayers directly. Thus, this gives us restrictions in choosing a test method and implementing a test system.

2.3 Simple PDU Transfer Procedure of CPAAL 5

CPAAL 5 protocol has a very simple protocol behaviour and does not have any PDU (Protocol Data Unit) exchanges between the peer entities for connection establishment or flow control to transfer CPCS-SDUs (Common Part Convergence Sublayer - Service Data Units). Both CPCS and SAR (Segmentation And Reassembly) have only one state (IDLE state). Thus there exists no state transition except a self transition. However, the simple procedure of the protocol gives us difficulty to observe the behaviour of the protocol entity from the outside on the viewpoint of black box testing.

CPAAL 5 does not send any CPCS and SAR PDUs and does not have state transitions when it receives SAR-PDUs and CPCS-PDUs from the peer entity. Thus, testing of the CPAAL 5 protocol for the error handling and appropriate parameter passing at the receiver side has to be dependent on the direct observation of the service primitives from the CPAAL 5 upper interface or of the behaviour of the CPAAL 5 user layer entity.

In a strict sense, SAR does not have its Protocol Control Information (PCI) and error handling procedures. Therefore, if CPCS entity receives an errored CPCS-PDU, it is actually impossible to know whether the error is caused by the fault on SAR or the fault on CPCS entities. This means that when we perform CPAAL 5 conformance testing, separate single-sublayer testing can not satisfy the test requirements including error localization.

2.4 Absence of Descriptions for Control and Management Plane

The descriptions for the control and management plane of CPAAL 5 are not included in ITU-T draft recommendation I.363 [2] and they remain for further study. Thus the possible test coverage for CPAAL 5 is restricted to the user plane specification, and additional study on the control and management plane is required in future.

3 TEST COVERAGE ANALYSIS

To choose the most efficient test method for AAL Type 5 protocol conformance testing, in this section, we compare the test coverage of applicable test methods. In fact, test coverage is largely affected from the test method itself. For more reliable and precise testing, we should use a test method that has larger test coverage than those of others.

In this section, we analyze test methods in the viewpoint of advantages and disadvantages of each method for AAL Type 5 conformance testing. And we present a result on comparing test coverage of each test method.

3.1 Comparison of Test Methods for AAL Type 5 Conformance Testing

For protocol conformance testing, four abstract test methods have been standardized by ISO and ITU-T as ISO/IEC IS 9646 part 1 (ISO/IEC 9646-1, 1993) and ITU-T X.290 (ITU-T X.290, 1992) respectively. These methods are Local Test Method, Distributed Test Method, Coordinated Test Method, and Remote Test Method.

Among four standardized test methods, Local Test Method is the simplest form. The tester in Local Test Method surrounds an IUT, and it is able to control and observe the interfaces of a layer above and below. However this test method is only a conceptual architecture because most implementations do not provide both interfaces externally. For most commercial network equipments, AAL Type 5 protocol is implemented as an embedded layer between ATM layer and AAL user. Therefore the application of

Local Test Method to AAL Type 5 protocol conformance testing is not feasible. In case of Coordinated Test Method, a standardized Test Management Protocol (TMP) is required for coordinating between UT and LT. But the TMP requires extra cost for developing the protocol.

Because of above reasons, Local Test Method and Coordinated Test Method are not considered in AAL testing. Remote Test Method and Distributed Test Method are candidates as applicable test methods for AAL Type 5 protocol testing.

1) Remote Test Method

Remote Test Method is the simplest one to describe a test specification and to implement a test system. The advantages and disadvantages of Remote Test Method for AAL Type 5 testing are as follows:

Advantages

- It is simple to design and to implement the test suite and the test system because we need to consider only LT functions to observe and control IUT through the lower interface of the LT.
- This test method makes no special demands on SUT (System Under Test) for testing (i.e., Remote Test Method makes no assumptions about the internal design of SUT or IUT, and treats SUT as a black box).

Disadvantages

- The control and observation of IUT is performed only through the lower interface of LT. Therefore there are **limitations on test coverage**. Especially, in case of the common part of AAL Type 5, a main function at receiver mode is to pass correct CPCS user data and parameters to the CPCS user after error handling. And, as presented in Section 2.3, CPAAL 5 protocol does not have any PDU exchanges between the peer entities for connection establishment or flow control. Therefore, if AAL Type 5 consists of CPAAL 5 only, it is impossible to test the receiver mode of IUT using Remote Test Method which can not control and observe ASPs (Abstract Service Primitives) from IUT upper interface.
- There are limitations on AAL user commands for testing. In AAL Type 5 sender mode testing, tests for various AAL user data conditions (e.g., testing under different AAL-SDU length sizes or parameter values) may be needed. But in case of Remote Test Method, it is difficult to control AAL user data conditions that we want to set.
- Manual operations may be required because there are no demands on SUT to implement UT functions and to perform the test coordination between UT and LT.

2) Distributed Test Method

Distributed Test Method uses PCOs (Point of Control and Observations) at LT and UT, thus we are able to control and observe IUT through PCOs. The advantages and disadvantages of Distributed Test Method are as follows:

Advantages

- This test method has a larger test coverage than that of Remote Test Method because we can control and observe IUT using LT and UT.

- The test coordination procedures are made more rigorously in a sense that it can include control and observation instructions related directly to ASPs across AAL and AAL user layer boundary.

Disadvantages

- The direct access to the upper boundary is needed to implement UT in SUT.
- Extra cost is needed to design a complex test suite and to implement special requirements for UT.

3.2 Comparison of Test Coverage

AAL Type 5 Common Part Protocol is independent of AAL service user, and it is used as a common protocol for all AAL Type 5 profiles. Before we develop an abstract test suite for CPAAL 5 conformance testing, we compared the test coverage of different types of test methods.

Much related work has been done in deriving test sequence directly from Finite State Machine (FSM) and formal specifications described in formal description techniques such as SDL, ESTELLE, and LOTOS (Krishan Sabnani and Anton Dahbura, 1988)(Tsun S. Chow, 1978)(G. Gonenc, 1970)(Ana R, Cavalli, Jean-Philippe Favreau, and Marc Phalippou, 1994). However, AAL5 Common Part (CPAAL5) protocol specified in ITU-T I.363 does not have state transitions listed as one of the testing restrictions in Section 2.3. Therefore, it is difficult to generate test sequences automatically for all instances of specification.

For testing AAL5 Common Part, the ATM Forum Test Working Group developed test purposes (Nouni, 1995) manually. There are thirty four test purposes, with fourteen test purposes in PDU Format, eight in Sender Mode, and twelve in Receiver Mode. These test purposes are considered to have covered all major functions of the protocol specification. Because of absence of descriptions for the control and management plane (see Section 2.4), test purposes for management and control functions have not been defined. For comparing test coverage for each test method to be used in developing ATS, the test purpose document (Nouni, 1995) is used for the reference target. Test coverage is presented in terms of percentage of realizable tests in the test purposes document.

1) Chosen Test Methods

This section presents differences between two kinds of Distributed Test Method and assumptions in using Remote Single Layer Test Method. We assume that SUT having CPAAL 5 protocol supports direct access to the upper boundary of CPAAL 5 to implement an UT.

Type 1 of Distributed Test Method (D1)

- UT can control and observe only CPCS-SDU parameter among CPCS-primitives.
- UT has minimal functions required for conformance testing. Therefore it is easy to implement UT functions in SUT. This kind of test method can be used more generally in testing AAL Type 5 because the only UT functions required are normal sending and receiving of AAL User Data.

Type 2 of Distributed Test Method (D2)

- UT can control and observe all parameters of CPCS-primitives in addition to D1 capabilities listed above.

Remote Single Layer Test Method (R)

- It is assumed that UT supports implicit sending of CPCS-PDUs for testing (if IUT can not support this sending for testing, test coverage will be decreased).

2) Comparison of Test Coverage

A summary of test coverage comparison for the three applicable test methods is presented in Table 1. While all test purposes are testable in Type 2 of Distributed Test Method (D2), only 47% (16 test purposes) of test purposes (34 test purposes) are testable in Remote Test Method, and 82% (28 test purposes) are testable in Type 1 of Distributed Test Method (D1). Furthermore the test coverage for Remote Test Method may be decreased if testing client can not answer some of PIXIT (Protocol Implementation eXtra Information) questions (for example, can the content of CPCS-SDU (payload) be known by the tester/test operator prior to testing?).

The 18% difference between D1 and D2 test methods comes from the fact that UT in D1 can not handle CPCS-UU, Congestion Indication, and Loss Priority parameters.

It is impossible to test the receiver mode of IUT using Remote Test Method because it can not control and observe the abstract service primitives of IUT upper interface. Only by observing the lower interface of IUT in Remote Test Method, a test system (or test operator) can not know whether IUT discards CPCS-PDUs and parameters or delivers them to CPCS because CPAAL 5 protocol does not send any PDUs to respond upon reception of SAR-PDUs or CPCS-PDUs from the peer entity, as shown in Section 2.3.

Table 1 Summary of test coverage comparison

	Distributed Test (D1)		Distributed Test (D2)		Remote Test (R)	
	Testable	Non-testable	Testable	Non-testable	Testable	Non-testable
PDU Format	13	1	14	0	10	4
Sender Mode	6	2	8	0	6	2
Receiver Mode	9	3	12	0	0	12
Total number of Test Purposes (%)	28 (82%)	6 (18%)	34 (100%)	0 (0%)	16 * (47%)	18 (53%)

* Note : does not require UT function.

4 TESTING CONFIGURATION FOR CONFORMANCE TESTING OF AALTYPE 5 COMMON PART

As explained in Section 2.2, there are a number of AAL Type 5 profile structures that consist of different sublayers depending on AAL user characteristics. In this section, we present an appropriate test configuration for AAL Type 5 profile consisting of CPAAL5 only.

If AAL Type 5 consists of CPAAL 5 only, the profile structure of AAL Type 5 has SAR and CPCS protocols only. In this case, as analyzed in Section 3, Remote Test Method can not cover receiver mode tests for AAL Type 5 protocol. To test the receiver mode of IUT, a tester needs UT function. Even in the sender mode testing, test requirements for various AAL user conditions may be needed. However, if we use Remote Test Method, it is difficult to control AAL user commands that stimulate a variety of AAL user conditions. For these reasons, we chose **Distributed Layer Test Method** that allows access to the lower and upper boundaries of IUT as shown in Figure 2.

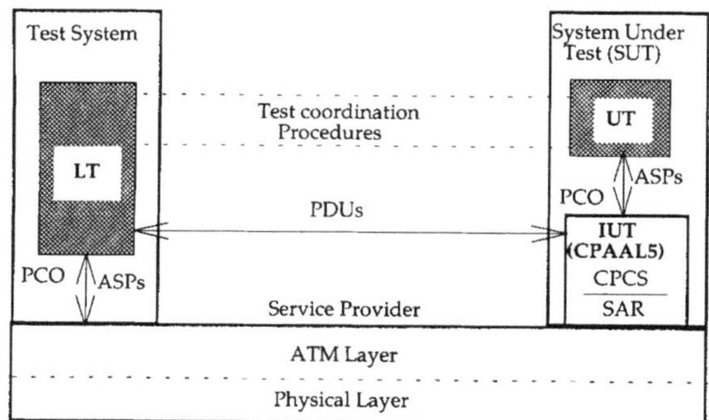

Figure 2 Test configuration for conformance testing of CPAAL 5 protocol

5 TIMING SYNCHRONIZATION AND TEST COORDINATION BETWEEN LT AND UT FOR CPAAL5 TESTING

Timing synchronization and test coordination between LT and UT are very important issues in the design and implementation of LT and UT. Especially in case of Distributed Test Method, synchronization and coordination procedure between LT and UT are practical issues, because there is no standardized test management protocol as in Coordinated Test Method. For synchronization and coordination, we can use the ferry principle (Zeng H.X, Chanson S.T., and Smith B.R., 1989)(Marc F. Witteman and Ronald C. van Wuijtswinkel, 1993) or define a user-defined protocol for testing. However, both of these approaches require extra costs to develop a test suite and a test system.

Fortunately, CPAAL 5 protocol is a simple protocol, and there is no requirement for complex protocol timers. Therefore, the synchronization problem becomes a coordination problem between LT and UT. In this section, a simple and efficient timing synchronization procedure is presented without any additional requirements or protocol for timing synchronization between UT and LT.

5.1 Timers

For conformance testing of AAL Type 5 Common Part protocol using Distributed Test Method, we have defined three timers to achieve timing synchronization and coordination between UT and LT during testing.

1) Preamble Timer : T_PRE

This timer is used by LT to make a verdict as an "INCONCLUSIVE" when the LT does not receive the expected response from IUT before this timer expires after the LT sent an ATM_DATA.request primitive [1] to notify a test case starting in a preamble.

2) **Sender Mode (of the IUT) Timer : T_SEND**

This timer is used by LT to define a period of time which the LT has to wait to guarantee receiving all ATM_DATA.indication primitives that were sent by UT as AAL user data from IUT.

3) **Receiver Mode (of the IUT) Timer : T_REC**

This timer is used by UT to define the time that UT has to wait to guarantee receiving the expected AAL user data as an AAL_UNITDATA.indication primitive that was sent by LT as pre-defined ATM_DATA.request primitives.

To determine the timer durations for three synchronization timers listed above, we have defined several parameters (e.g., Tr, Td1, and Ts which were presented in Table 2, 3, and 4) as PIXIT (Protocol Implementation eXtra Information for Testing) proforma questions. The values for these parameters can be determined experimentally during the testing preparation step by the test operator and the test client.

5.2 Timing Synchronization and Test Coordination Method

Proposed timing synchronization and test coordination procedure between LT and UT starts with sending an ATM primitive that includes the start message and the test case identification to be executed by LT. When UT receives the start message, it sends an acknowledge message to the IUT. Immediately after sending and receiving the acknowledge message, the UT and LT perform predefined test procedure using the proposed timers(Sangjo Yoo and Leslie Collica, 1995).

1) Preamble Timer : T_PRE

For execution of a test case, we need a preamble procedure so that LT gives UT the test case identification to be executed. In the preamble, "*Start (test_case_id)*" ATM_DATA.request primitive and "*Start (test_case_id)*" AAL_UNITDATA.indication primitive are used to notify the starting of a specific test case "*test_case_id*" to the UT. "*Ack (test_case_id)*" ATM_DATA.indication primitive and "*Ack (test_case_id)*" AAL_ UNITDATA.request primitive are also used for starting point (time) of other timers such as T_SEND and T_REC by LT and UT, respectively. Figure 3 shows an example of T_PRE timer operation. The parameters for the preamble timer operation are shown in Table 2.

After sending "Start" ATM primitive, LT starts T_PRE timer. If LT receives an "Ack" ATM primitive from IUT, LT stops timer T_PRE, then executes the test body of the LT part. On receipt of "Start" AAL primitive containing the test case start message and the test case number, UT sends "Ack" AAL primitive to IUT, and then executes the test body of UT part. When timer T_PRE expires prior to receipt of an "Ack" ATM primitive from IUT, LT stops the test case and assign a verdict as an "INCONCLUSIVE".

---

[1] An ATM_DATA.request or ATM_DATA.indication primitive can be implemented as a corresponding ATM cell which has correct primitive parameters (i.e., Cell Loss Priority, Congestion Indication, and ATM Payload) in the ATM cell header and payload.

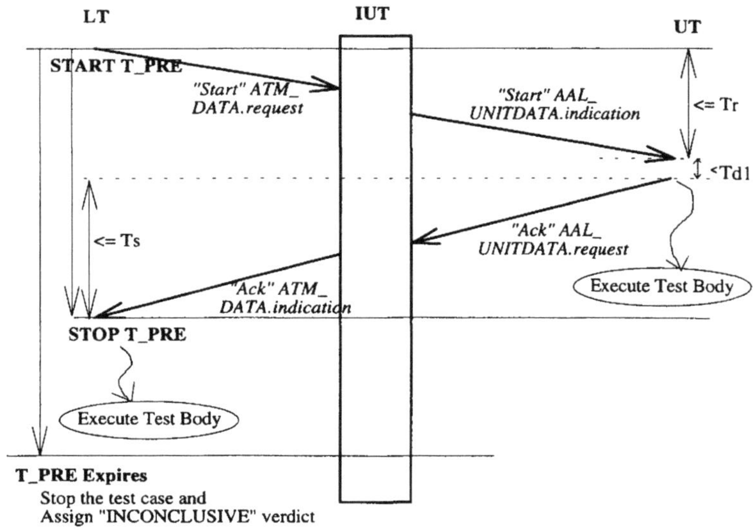

Figure 3 Preamble T_PRE timer operation

Table 2 Parameters for preamble timer operation

Timer/Parameter Name	Timer Duration
T_PRE	Tr + Td1* + Ts
Tr	The maximum time to guarantee that UT receives the corresponding AAL User Data after LT sends ATM primitive (containing 39 octet AAL User Data). The value of Tr is determined by PIXIT.
Td1	The maximum processing time of UT to check the data receiving, generate "Ack" AAL primitive, and send it to IUT. The value of Td1 is determined by PIXIT.
Ts	The maximum time to guarantee that LT receives the corresponding ATM primitive after UT sent the 39 octet AAL User Data. The value of Ts is determined by PIXIT.

* Note : Td1 may not be included, if Td1 is significantly smaller than both Tr and Ts.

2) Sender Mode Timer : T_SEND

T_SEND timer is used by LT to define the period of time which the LT has to wait to guarantee receiving all ATM primitives that were sent by UT as an AAL User Data from IUT. Tms duration is defined to guarantee that LT receives corresponding ATM primitives after UT sent AAL User Data. In IUT sender mode testing, UT shall send immediately pre-defined AAL User Data as an AAL_UNITDATA.request primitive after sending "Ack" AAL primitive. Upon receiving the "Ack"

ATM primitive from IUT, LT starts the T_SEND timer. If the LT receives the last ATM primitives corresponding to AAL User Data that UT sent, the LT stops the timer T_SEND. When timer T_SEND expires prior to receipt of the last ATM primitive from IUT, the LT stops the test case and assigns a verdict as "FAIL". Figure 4 shows an example of T_SEND timer operation.

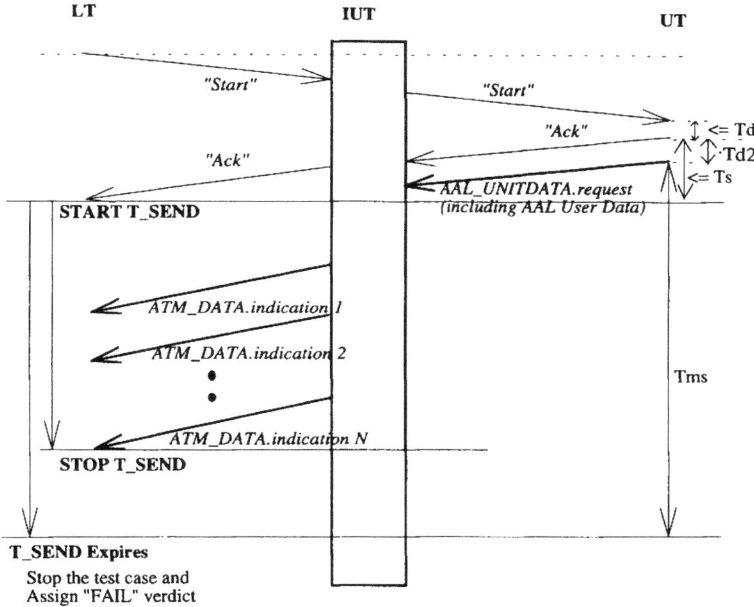

Figure 4 Example of T_SEND timer operation

Table 3 Parameters for sender mode timer operation

Timer/Parameter Name	Timer duration
T_SEND	Tms + Td2* - Ts
Tms	The maximum time to guarantee that LT receives corresponding ATM primitives after UT sent AAL User Data of Max_Deliver_Length size. The value of Tms is determined by PIXIT.
Td2*	The maximum processing time of UT to generate AAL User Data of Max_Deliver_Length size and to send it to IUT. The value of Td2* is determined by PIXIT

* Note : Td2 may not be included, if Td2 is significantly smaller than both Tms and Ts separately.

3) Receiver Mode Timer : T_REC

In IUT receiver mode testing, LT sends "Start" ATM primitive; IUT then sends "Start" AAL primitive to UT. Upon receiving the "Start" AAL primitive, the UT sends "Ack" AAL primitive to the IUT and starts T_REC timer. After receiving "Ack" ATM primitive from IUT, the LT shall immediately send the pre-defined ATM primitives which include a pre-defined AAL User Data. If the UT receives AAL User Data corresponding to ATM primitives that the LT sent, the UT stops timer T_REC. When timer T_REC expires prior to receipt of the AAL User Data from the IUT, the UT stops the test case and assigns a verdict as "FAIL". Figure 5 shows an example of T_REC timer operation.

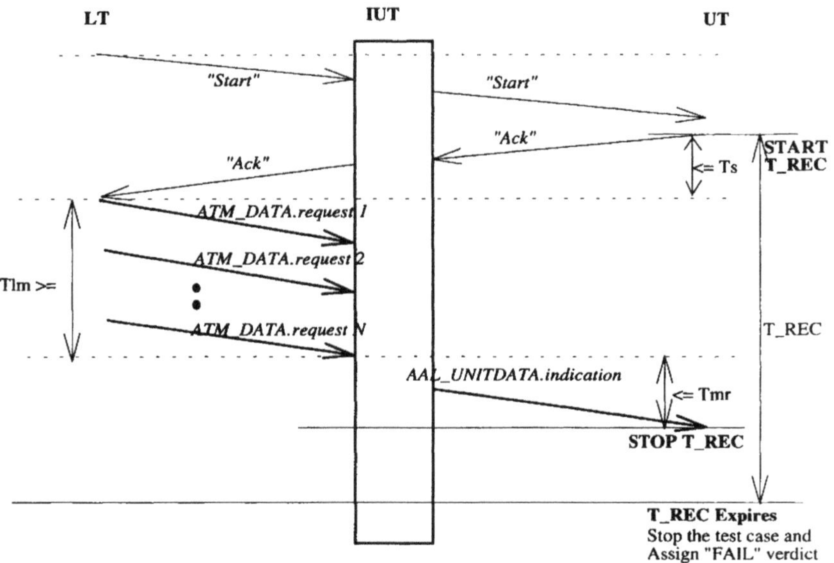

Figure 5 Example of T_REC timer operation

Table 4 Parameters for receiver mode timer operation

Timer/Parameter Name	Timer duration
T_REC	Ts + Tlm + Tmr
Tlm	The maximum time to guarantee that the IUT receives all of the pre-defined ATM primitives (containing Max_Deliver_Length size AAL User Data) sent by the LT. The value of Tlm is determined by PIXIT.
Tmr	The maximum time to guarantee that UT receives corresponding reassembled AAL User Data after LT sent pre-defined ATM primitives that contained Max_ Deliver_Length size AAL User Data. The value of Tmr is determined by PIXIT.

6 UPPER TESTER FOR CPAAL5 TESTING

In ISO/IEC 9646 Part 1 (ISO/IEC 9646-1, 1993) and ITU-T Recommendation X.290 (ITU-T X.290, 1992), Upper Tester has been defined as "the means of providing, during test execution, control and observation of the upper boundary of the IUT, as defined by the chosen Abstract Test Method". However, implementing and using the UT have been recognized that it is not easy to apply in real testing environment because direct access to the upper boundary of the IUT is not always supported and coordination between UT and LT requires a complex procedure. Therefore, we propose an UT that has a simple structure. As shown in Figure 6, if UT manager receives "Start" message, then the manager sends "Ack" and calls the indicated test case execution file. According to the test purpose, the UT sends a predefined AAL User Data or waits an expected AAL User Data from the IUT.

The only support expected by UT in SUT is the sending and receiving of AAL User Data and the timer operation management presented in Section 5. Such minimum capabilities are normally supported by API (Application Programming Interface) which venders provide. Even though some vender's APIs can not handle primitive parameters such as loss priority and congestion indication, recently, standardization bodies including ATM Forum are trying to develop standardized APIs for ATM/B-ISDN implementation. If the standardized APIs are developed, the UT will be able to be implemented and used more easily.

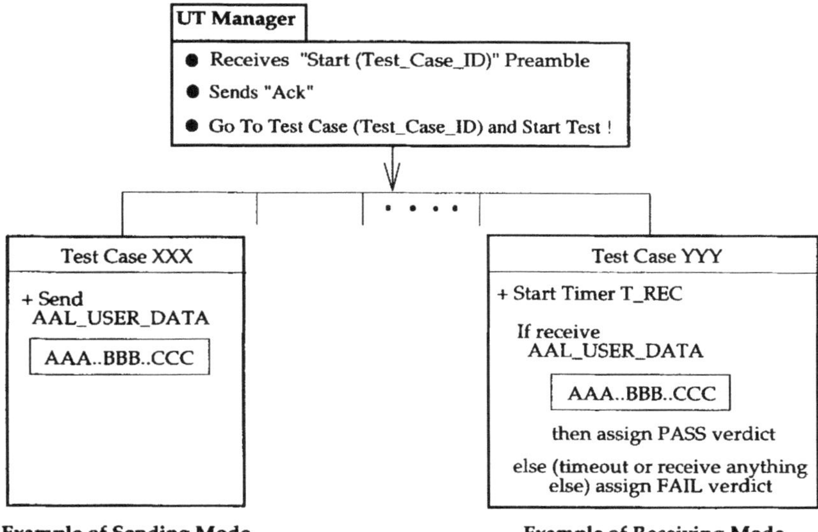

Figure 6 The structure of Upper Tester

7 EXAMPLE TEST CASE FOR CPAAL5 TESTING

An example test case described in TTCN using the proposed test method for CPAAL 5 is given in Table 5. In this test case, the test body calls the test step PREAMBLE presented in Table 6 to initiate the test case execution. In the preamble of Table 6, LT sends "Start" ATM primitive, that contains the test case identification (Test_Case_ID) and the mode flag (R:=1) to indicate receiver mode testing, to IUT. After sending the "Start" ATM primitive, LT starts timer T_PRE. When UT receives "Start" primitive, the UT sends "Ack" primitive to LT and starts timer T_REC. Upon receipt of the "Ack" primitive the LT stops timer T_PRE and starts test body. If timer T_PRE expires prior to receipt of the primitive, the LT stops timer T_PRE and assigns a verdict "INCONCLUSIVE". Since the test case shown in Table 5 is for the IUT's receiver mode testing, the UT expects to receive a corresponding AAL User Data to the ATM primitives sent by the LT. When the UT receives a correct AAL User Data, the UT assigns a "PASS" verdict for the test case. If the UT receives an unexpected AAL User Data or timer T_REC is expired, then a "Fail" verdict is assigned. This test case shows the timing synchronization procedure between UT and LT presented in Section 5.

Table 5 Example of test body for the receiver mode test case

Test Case Dynamic Behaviour

Test Case Name : rec_1
Group : AAL_CP/RECEIVER/
Objective : Verify that the IUT recovers and delivers the CPCS-SDU to the upper layer upon reception of an ATM-SDU with the AUU parameter in ATM_DATA.indication primitive equal to 1.
Configuration :
Default :

Nr	L	Behaviour Description	Constraints Ref	V	Comments
1		(Test_Case_ID:="rec_1",R:=1)			*Test case I.D.=rec_1, and receiver mode testing.*
2		+PREAMBLE(Test_Case_ID,R)			*Go to PREAMBLE test step.*
3		LT_PCO!ATM_D.req	ATM_D_R_1_1		*AUU='0'*
4		LT_PCO!ATM_D.req	ATM_D_R_1_2		*AUU='0'*
5		LT_PCO!ATM_D.req	ATM_D_R_1_3		*AUU='1'*
6		UT_PCO?AAL_UD.ind	AAL_UD_I_1	P	*Corresponding AAL User Data*
7		UT_PCO?OTHERWISE		F	
8		?TIMEOUT T_REC		F	*Timer T_REC expires.*

Table 6 Example of preamble

Test Step Dynamic Behaviour					
Test Step Name : PREAMBLE(Test_Case_ID, R) **Group** : AAL_CP/ **Objective** : Procedure used to notify the start of the test case to the Upper Tester. **Default** :					
Nr	**L**	**Behaviour Description**	**Constraints Ref**	**V**	**Comments**
1		LT_PCO!ATM_D_req	ATM_PRE_Start (Test_Case_ID,R)		*LT sends "Start" to the IUT.*
2		START T_PRE			
3		UT_PCO?AAL_UD_ind	AAL_PRE_Start (Test_Case_ID,R)		*UT receives "Start" from the IUT.*
4		UT_PCO!AAL_UD_req	AAL_PRE_Ack		*UT sends "Ack" to the IUT.*
5		[R=1]			*(Receiver mode testing)*
6		START T_REC			*Timer T_REC is started by UT.*
7		+ LT_PRE			
8		[R=0]			*(Sender mode testing)*
9		+ LT_PRE			
10		UT_PCO?OTHERWISE		I	
11		?TIMEOUT T_PRE		I	*Timer T_PRE expires.*
12		LT_PRE			
13		LT_PCO?ATM_D_ind	ATM_PRE_Ack		*LT receives "Ack" from the IUT.*
14		LT_PCO?OTHERWISE		I	
15		?TIMEOUT T_PRE		I	*Timer T_PRE expires.*

8 CONCLUSION

In this paper, we analyzed test requirements for AAL Type 5 protocol conformance testing. To choose the best test method, we compared test coverage of four applicable test methods, i.e., Local Test Method, Remote Test Method, Distributed Test Method, and Coordinated Test Method after analyzing the restrictions of AAL Type 5 protocol with respect to conformance testing. We found that the Distributed Test Method is the most efficient method in the view point of test coverage and real test environments. To provide necessary timing synchronization and test coordination between LT and UT, we proposed a simple mechanism that uses only three timers. We developed an abstract test suite for AAL Type 5 Common Part protocol testing using proposed mechanism. Finally, we presented an example of test case to show how we described behaviour of the test specification using TTCN.

This paper shows a part of our ongoing research in broadband network protocol testing project. Currently, we are developing test suites for AAL and signalling protocols to prepare testings for broadband network services and to contribute to ATM Forum testing specification works. For further research, we have plans to implement an UT using standardized API and to generate test cases using SDL automatically.

9 REFERENCES

Ana R, Cavalli, Jean-Philippe Favreau, and Marc Phalippou (1994), "Formal Methods in Conformance

Testing: Results and Perspectives", Protocol Test Systems, VI.

G. Gonenc (1970), "A Method for the Design of Fault-Detection Experiments", IEEE Trans. Comput., vol. 19.

ISO/IEC 9646-1 (1993), "Information Technology - Open Systems Interconnection - Conformance Testing Methodology and Framework - Part 1 : General Concepts".

ISO/IEC 9646-3 (1993), "Information Technology-Open Systems Interconnection-Conformance Testing Methodology and Framework-Part 3:Tree and Tabular Combined Notation (TTCN)".

ITU-T COM SVIII-R 116 (1992), "B-ISDN ATM Adaptation Layer (AAL) functional description".

ITU-T TD-XVIII/10 (1993), "AAL Type 5, Draft Recommendation text for section 6 of I.363".

ITU-T X.290 (1992), "Information Processing System - OSI Conformance Testing Methodology and Framework - Part 1 : General Concepts".

Krishan Sabnani and Anton Dahbura (1988), "A Protocol Test Generation Procedure", Computer Networks and ISDN Systems, vol. 15.

Marc F. Witteman and Ronald C. van Wuijtswinkel (1993), "ATM broadband network testing using the ferry principle", Protocol Test Systems, V.

Nouni (1995), "Test Purposes for the AAL Type 5 Common Part Conformance Testing", ATM Forum 95-0033.

Sangjo Yoo and Leslie Collica (1995), "Timing Synchronization for the CPAAL5 Conformance Testing using Distributed Test Method", ATM Forum 95-0309, April 1995.

Tsun S. Chow (1978), "Testing Software Design Modeled by Finite-State Machines", IEEE Trans. on Software Engineering, vol. SE-4.

Zeng H.X, Chanson S.T., and Smith B.R. (1989), "On Ferry Clip approaches in protocol testing", Computer Networks and ISDN Systems 17.

10 BIOGRAPHY

Sangjo Yoo received BA in electronic communication engineering from Hanyang Univ. in 1988, MS in electrical engineering from the Korea Advanced Institute of Science and Technology in 1990. Currently he is with the Korea Telecom Research and Development Group as a member of technical staff and serving as an editor in the ATM Forum. His research interests include protocol testing and formal description technique.

Leslie A. Collica is a Computer Scientist working at the National Institute of Standards and Technology since 1988 on ISDN and Broadband ISDN protocols, conformance and interoperability tests. She is Editor of the Test Working Group in the ATM Forum and is Chairman of the North American ISDN User's Forum. She has a B.S. degree from the University of Maryland.

Myungchul Kim received BA in electronics engineering from Ajou Univ. in 1982, MS in computer science from the Korea Advanced Institute of Science and Technology in 1984, and Ph.D in computer science from the Univ. of British Columbia in 1992. Currently he is with the Korea Telecom Research and Development Group as a managing director, Chairman of Profile Test Specification-Special Interest Group of Asia-Oceania Workshop, and will be Chair of the 10th IWTCS'97. His research interests include protocol engineering on telecommunications and multimedia.

18
Test case generation for ATM protocols using high-level Petri net models

R. Bechtold [a], *G. Gattung* [b], *O. Henniger* [b], *C. Paule* [b]

[a] *Deutsche Telekom, Niederlassung Wiesbaden, Projekt Roland K.-Adenauer-Ring 33, 65187 Wiesbaden, Germany*

[b] *GMD – German National Research Center for Information Technology, Rheinstr. 75, 64295 Darmstadt, Germany*

Abstract

This paper discusses an experience in computer-supported test case generation for ATM protocols, viz. the AAL3/4 protocol (ATM Adaptation Layer Type 3/4) and CLNAP (Connectionless Network Access Protocol). The test cases were generated by means of the test generation tool *GeneTest* based on protocol specifications as product nets, i.e. as a special kind of high-level Petri net.

Keywords

ATM (Asynchronous Transfer Mode), Protocol specification, Test case generation

1 INTRODUCTION

Quality and reliability are factors of growing importance in the highly competitive European market area of communication systems. Only well-tested products can claim to provide the required behaviour and can take over their tasks in a complex communication system. To reach the goal of reliability, a sequence of steps is required, from an unambiguous definition of requirements via conformance tests to interoperability and performance tests.

In the classic area of communication protocols, like X.25, X.400, FTAM, X.75 etc., conformance test suites are available and widely used by manufacturers, users, and independent test laboratories. For new protocols, like the ATM protocols, conformance test suites are hardly available and still need to be developed. Computer-supported test case generation is expected to save time and to cut costs in the development of test suites.

This paper reports about the application of the test generation method presented in [Bau90] to real protocols. The method has been implemented in a tool [Bau91] and has been applied to generate conformance test cases for ATM protocols, in a first stage for the AAL3/4 protocol [I363] and for CLNAP [I364].

The starting point for the test case generation method is a specification of the complete test architecture as a product net [Bur89], i.e. as a special kind of high-level Petri net with n-tuples as tokens and with inhibitor arcs and erase arcs beside the regular arcs. A Petri net is a mathematical model to describe distributed discrete event systems with the advantage to allow very compact descriptions of systems with a large set of states. The states of the system are described implicitly by means of the net model, the initial markings, and the firing rules.

Especially high-level Petri nets with mathematical objects as tokens carrying complex information allow compact system descriptions and are very suitable for specifying communication protocols.

The test case generation method is a method with explicit test purposes, i.e. beside the actual specification additional information about test purposes is required as input to the method. The test specifier has to decide on the test purposes and to translate them manually into initial markings of the net. Then the method ensures that test cases consistent with both the specification and the test purposes are generated. Methods with explicit test purposes offer much flexibility and allow to focus on relevant conformance requirements. On the other hand, they take manual efforts to determine test purposes and to ensure a systematic fault coverage.

The generated test cases are represented in the standardised test description language TTCN [ISO9646].

In Section 2, we shortly discuss the AAL3/4 protocol and CLNAP and their product net specifications that are used as starting point for the test case generation. In Section 3, the test case generation method is explained. Section 4 gives a summary and concluding remarks.

2 PROTOCOL SPECIFICATION

2.1 Functions of AAL3/4 and CLNAP

The AAL of the B-ISDN protocol reference model provides for the adaptation of services provided by the ATM layer to the requirements of higher layers. Its main function is mapping higher layer PDUs into the information fields (of 48 octets length) of ATM cells and reassembling these cells. Depending on the service required by the higher layer, different types of AAL have been defined (types 1, 2, 3/4, and 5). AAL3/4 is for the connection-oriented and the connectionless data transfer service classes. Our study is confined to the message mode service of AAL3/4. The AAL is subdivided into the Convergence Sublayer (CS) and the Segmentation and Reassembly (SAR) Sublayer. The CS in its turn is subdivided into the Common Part Convergence Sublayer (CPCS) and the Service Specific Convergence Sublayer (SSCS). The SSCS protocol depends on the specific AAL user. In our case, with CLNAP above AAL3/4, the SSCS is empty.

The CLNAP provides the connectionless data service in B-ISDN.

2.2 Test architecture

We modelled the test architecture that is applied by Deutsche Telekom for acceptance testing of terminal adapters employed in the area of SMDS/CBDS (Switched Multi-Megabit Data Service/Connectionless Broadband Data Services). The AAL3/4 and CLNAP implementations under test reside in a terminal adapter for connecting LANs to an ATM network (using the HSSI/DXI interface, High Speed Serial Interface/Data eXchange Interface). In the test architecture, the terminal adapter is connected via an ATM network to an ATM test system acting as the lower tester. There is no upper tester in the test architecture. In order to gain some indirect controllability and observability of the upper service boundaries of the IUT, the terminal adapter's LAN output line is looped to its LAN input line.

Depending on the focus of testing, the PCO is placed at different SAPs or interfaces in the test system: immediately beneath the SAR Sublayer for SAR tests, immediately beneath the CPCS for CPCS tests, or immediately beneath the CLNAP for CLNAP tests. As an example, Figure 1 outlines the test architecture for CPCS tests.

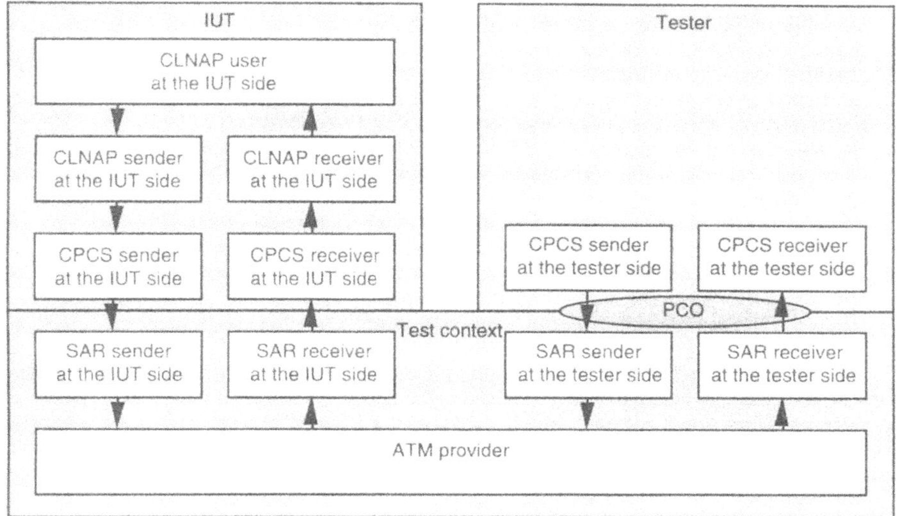

Figure 1 Test architecture for CPCS tests and division into subnets

2.3 Product net model

The complete test architecture including a correct IUT, a test context, and the tester was modelled as a product net. This net is divided into subnets as depicted by the boxes in Figure 1. A subnet is just a graphical unit showing a selected part of the overall product net. Subnets are glued together at places they have in common. All subnets together form the overall product net.

The subnets for SAR, CPCS, and CLNAP sender and receiver at the IUT side model the correct behaviour of the protocols as specified in [I363] and [I364] respectively.

In the CPCS test architecture, the behaviour of the subnets for SAR sender and receiver at the tester side is identical with that of the corresponding subnets at the IUT side, however, supplemented by modelling TIMEOUT and OTHERWISE events. The subnet for the CPCS sender at the tester side models the sending of valid as well as of invalid PDUs by the tester. In the subnet for the CPCS receiver at the tester side, the received PDUs are collected.

The subnet for the CLNAP user at the IUT side is a simplified substitute net modelling only the behaviour visible at the interface. Likewise, the subnet for the ATM provider is a simplified substitute net modelling only the behaviour visible at the ATM SAPs.

We save details of the product net model for the ATM protocols because of limited space. An introduction to product nets as well as a short product net example explaining the applied test case generation method has been given in [Bau90].

3 TEST CASE GENERATION

3.1 Overview

The starting point of the test generation method *GeneTest* is the product net comprising the complete specification of the test architecture, i.e. the protocol to be tested, the test context, and the tester. To generate a test case, the following steps are carried out :

1. The transitions of the product net that are controllable or observable at the PCOs of the chosen test architecture are determined, and the extra information needed for representing these transitions in TTCN is provided.
2. An initial marking for the product net is determined according to a given test purpose.
3. Starting with the initial marking, a complete reachability analysis of the product net is carried out by means of the product net analysis tool *PNM* [GMD94].
4. By means of the test generation tool *GeneTest* [Bau91] the reachability graph is restricted to the occurrences of the controllable or observable transitions. This restriction results in an automaton describing only the behaviour visible at the PCOs. This automaton is converted into TTCN format. At the same time, *GeneTest* generates most of the relevant parts of the declarations and constraints parts of the resulting test suite in TTCN format.
5. The generated test case is edited in order to add verdicts, which are not yet generated automatically. The determination of appropriate verdicts is not defined clearly in [ISO9646]. [Bau95] deals with the formalisation of the verdict assignment.

Figure 2 illustrates the steps of the applied test case generation method. Though generally leading to combinatorial explosion, reachability analysis is feasible in most cases since a separate reachability graph for each initial marking, not a single reachability graph comprising every possible behaviour of the whole system, is computed. However, for some initial markings modelling transmission of many PDUs reachability analysis may take a rather long time.

3.2 Extra information for TTCN representation

According to the chosen test architecture, the test specifier has to determine the transitions that are visible at the PCOs. Occurrences of these transitions are mapped to TTCN behaviour lines in the generated test cases. In *GeneTest*, each transition visible at a PCO is annotated by information necessary for its representation in TTCN:

- the type of TTCN events to which the transition occurrences correspond (SEND, RECEIVE, TIMEOUT, or OTHERWISE events);
- in case of SEND or RECEIVE events: the PCO identifier (if any), the ASP or PDU identifier to be used in the TTCN test suite, a prefix for the TTCN constraint identifiers, the ASP parameter or PDU field identifiers and the way their values are derived from the occurrence variables of this transition (if any);
- in case of TIMEOUT events: the timer identifier;
- in case of OTHERWISE events: the PCO identifier (if any).

3.3 Test suite structure, test purposes, and initial markings

The behaviour of the product net depends on the initial marking of the net. In the sender subnet of the tester, an initial marking models the data to be sent by the tester. In other subnets, some places modelling options are set to an appropriate initial marking.

We manually developed the test suite structure and test purposes for the protocols under consideration according to the guidelines given in [ISO9646]. Care was taken to ensure an adequate coverage of the conformance requirements. Then the test purposes were manually translated into the appropriate initial markings of the product net. The need to define test purposes manually is a weak point of the test case generation method. However, so far methods for automatic generation of test purposes seem to be of limited applicability in practice.

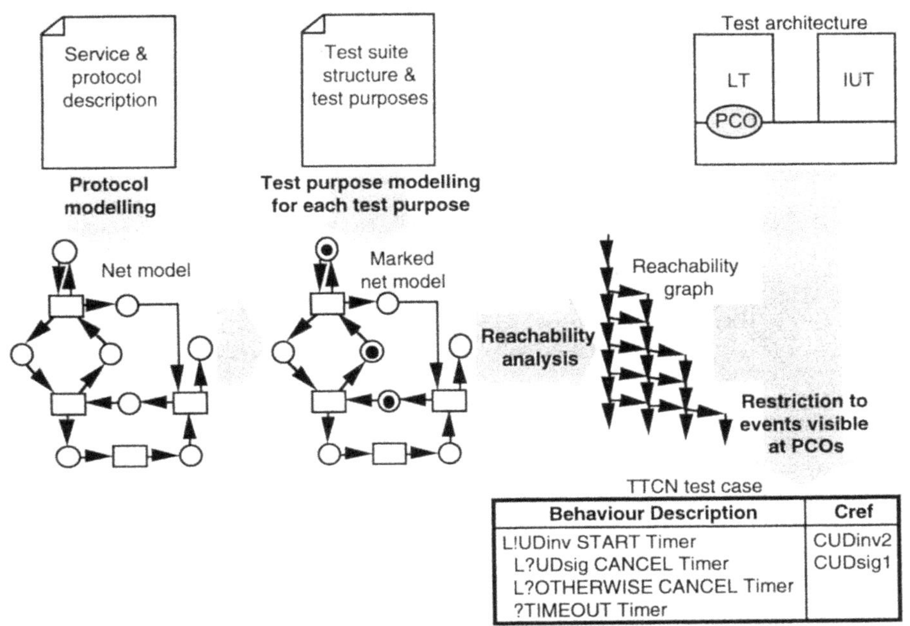

Figure 2 Test case generation from net models

3.4 Manual revision of the test suites

Not all parts of a test suite can be generated automatically from the formal specification, either because relevant information is not contained in the net model or because of limitations of the test generation tool. Therefore, the generated test cases have to be revised manually. The following steps have to be carried out, preferably by means of a TTCN editor tool:

- addition of test suite type definitions, test suite parameter declarations, test case selection expression definitions, test suite constant declarations, test case variable declarations, PCO declarations, timer declarations, test suite operations definitions, and test suite variable declarations (if any) to the declarations part;
- transformation of some constraint values to the correct TTCN format (due to limitations of the current version of the test generation tool);
- assignment of verdicts to the leaf nodes of the TTCN behaviour trees.

3.5 Resulting test suites

The SAR part of the AAL3/4 test suite we produced consists of 104 test cases; the CPCS part has 38 test cases. The CLNAP test suite consists of 76 test cases. Because of the well-known limitations of manually defined test purposes, we cannot declare a certain fault coverage.

To judge the resulting test suites, we compared them with the test suites for the AAL3/4 protocol and for CLNAP that had been developed before by the Roland team of Deutsche Telekom in the classic, manual way. The results of the computer-supported test case generation turned out to exceed our expectations. The resulting test suites meet all requirements applied by Deutsche Telekom and revealed some flaws in the manually developed test suites. The resulting test suites are available in TTCN format; thus, they allow processing by other tools to create executable test suites.

4 CONCLUSIONS

By means of the test generation tool *GeneTest*, which operates on product net specifications, test suites have been generated for the AAL3/4 protocol and for CLNAP. While still requiring considerable manual efforts to develop the product net specification, to determine initial markings for the net, and to manually revise the generated test suite, the test generation method effectively supports the generation of correct test cases for realistic protocols. The tool-supported test generation takes less time compared to the classic, manual test suite development, delivers correct test cases provably consistent with the specification and the specified test purposes, and thus helps to cut the development costs for test suites.

In the course of the AAL3/4 and CLNAP test generation project, some desirable improvements and extensions have been identified, which shall be taken into consideration for a new version of *GeneTest,* such as parameterisation of test cases and constraints, assignment of verdicts, incorporation of ASN.1 definitions, and TTCN presentation of data values.

It is planned to apply the *GeneTest* tool to other ATM protocols in subsequent projects. The generated test suites will be fed into the standardisation area, as a first step within the ATM Forum.

ACKNOWLEDGEMENTS

The authors would like thank the other participants in the test case generation project, B. Baumgarten, A. Giessler, K. Stöttinger, and H. Wiland, for their helpful comments on this experience report.

REFERENCES

[Bau90] B. Baumgarten, A. Giessler, R. Platten. Test derivation from net models. In J. de Meer, L. Mackert, and W. Effelsberg (editors), *Protocol Test Systems, II,* pp. 141-159, Berlin, Germany, 1989. Elsevier Science Publishers B.V., 1990.

[Bau91] B. Baumgarten, A. Giessler, C. Paule. Testgenerierung mit Produktnetzen – Algorithmus, Implementierung und Benutzeranleitung. Arbeitspapiere der GMD, Nr. 605, GMD, Germany, 1991. The user-manual part is updated regularly.

[Bau95] B. Baumgarten. Timed systems behaviour and conformance testing – a mathematical framework. In A.R. Cavalli and S. Budkowski (editors), *Protocol Test Systems, VIII,* Evry, France, 1995. Participants' proceedings.

[Bur89] H.J. Burkhardt, P. Ochsenschläger, R. Prinoth. Product nets: a formal description technique for cooperating systems. GMD-Studien, Nr. 165, GMD, Germany, 1989

[GMD94] Produktnetzmaschine. User manual, GMD, Germany, 1994.

[ISO9646] International Standard ISO/IEC 9646: Information technology – Open Systems Interconnection – Conformance testing methodology and framework. 1991.

[I363] Revised ITU-T Recommendation I.363: B-ISDN ATM Adaptation Layer (AAL) Specification. 1993.

[I364] ITU-T Recommendation I.364: Support of broadband connectionless data service on B-ISDN. 1994.

PART EIGHT

Contributions to Theory

19
Formal methods in conformance testing : a probabilistic refinement

Lex Heerink and Jan Tretmans
Tele-Informatics and Open Systems group, Dept. of Computer Science
University of Twente, 7500 AE Enschede, The Netherlands
{heerink,tretmans}@cs.utwente.nl

Abstract

This paper refines the framework of 'Formal Methods in Conformance Testing' by introducing probabilities for concepts which have a stochastic nature. Test execution is refined into test runs, where each test run is considered as a stochastic process that returns a possible observation with a certain probability. This implies that not every possible observation that could be made, will actually be made. The development process of an implementation from a specification is also viewed as a stochastic process that may result in a specific implementation with a certain probability. Together with a weight assignment on implementations this introduces a valuation measure on implementations. The test run probabilities and the valuation measures are integrated in generalized definitions of soundness and exhaustiveness, which can be used to compare test suites with respect to their ability to accept correct, and to reject erroneous implementations.

Keywords

Conformance testing, test framework, test selection, formal methods, probabilities.

1 INTRODUCTION

Conformance testing is a way to assess the correctness of an implementation with respect to its specification by means of performing experiments on the implementation and observing its responses. In case the specification is given as a formal description we need formal definitions of testing concepts, such as correctness of an implementation with respect to a formal specification, a test purpose, a sound test case, test execution, test generation, etc. Currently, the standardization group on 'Formal Methods in Conformance Testing' (ISO/IEC JTC 1/SC 21/WG 7 project 54, ITU-T SG 10/Q 8) develops a framework for conformance testing based on formal methods defining these concepts [ISO95]. The framework defines terminology, abstract concepts, and minimal requirements on, and relations between these concepts. Since it is defined at a high level of abstraction, e.g., it abstracts from specific test generation algorithms, even from a specific formal description technique, use of the framework requires instantiating these concepts with specific choices for test generation algorithms, for the formal description technique, etc.

This paper builds on the framework of 'Formal Methods in Conformance Testing'. Its goal is to refine this framework by adding probabilities for those concepts which have a stochastic nature. The refinement concerns the testing process; probabilistic extensions of specification languages or models are not considered. The first addition is to consider test execution in a probabilistic setting. In the framework, test execution of a test case against an implementation under test is assumed to yield a unique observation. This assumption is replaced by refining test execution into a number of test runs. Each test run yields an observation of the implementation, but a number of test runs does not necessarily yield all possible observations [LS89]. A probability distribution is added to express which observations are likely to be obtained.

A second refinement of the framework concerns the extension of the soundness and the exhaustiveness of a test suite. Soundness refers to the property of a test suite to accept all conforming implementations, and exhaustiveness indicates that all nonconforming implementations are rejected [ISO95]. These predicates are generalized to measures in the vein of [BTV91, Bri93], which take into account the probability of the occurrence of implementations, the gravity of errors in implementations, and the above mentioned probability on observations made during test runs. It is indicated how these soundness and exhaustiveness measures can then be used to compare test suites, in order to select a good, or the best one.

The outline of this paper is as follows. In section 2 an overview of the framework 'Formal Methods in Conformance Testing' is given, as far as it is relevant for this paper, and some of the assumptions underlying this framework are discussed. Section 3 refines test execution into test runs, and it adds probabilities to the observations. In section 4 a valuation on implementations is defined, which assigns a value based on their probability of occurrence and on their weight. Section 5 uses the probability on test-run observations and the valuation on implementations to define soundness and exhaustiveness as two measures on test suites. Comparison of test suites based on these measures is briefly discussed. In section 6 the concepts are illustrated for labelled transition systems with inputs and outputs; this section may be read in parallel with the sections 2 till 5. Section 7 presents concluding remarks and items for further work.

2 FORMAL METHODS IN CONFORMANCE TESTING

The emerging international standard 'Formal Methods in Conformance Testing' defines a framework for the use of formal methods in conformance testing [ISO95]. It is intended to guide the testing process of an implementation with respect to a formal specification. In this section the main concepts of [ISO95], such as conformance, testing, and conformance testing, are presented, as far as they are needed for the subsequent sections. They are followed by a discussion of some of the explicit and implicit assumptions on which the framework is based.

Conformance The definition of conformance concerns implementations under test (IUT) and specifications, so a universe of implementations $IMPS$, and a universe of formal specifications $SPECS$ are assumed. Implementations are concrete, informal objects, such as pieces of hardware, or pieces of software. In order to reason formally about them, it is assumed that each implementation $IUT \in IMPS$ can be modelled by a formal object i_{IUT} in a formalism $MODS$, which is referred to as the universe of models. This hypothesis is referred to as the test assumption. Note that a model i_{IUT} is only assumed to exist; it is not known apriori.

Conformance is expressed by means of an implementation relation $\mathbf{imp} \subseteq MODS \times SPECS$. Implementation $IUT \in IMPS$ is considered **imp**-correct with respect to $s \in SPECS$ if the model $i_{IUT} \in MODS$ of IUT is **imp**-related to s: i_{IUT} **imp** s. We use $I_s =_{\text{def}} \{i \in MODS \mid i \ \mathbf{imp}\ s\}$ for the set of **imp**-correct implementations, and $\overline{I_s} =_{\text{def}} MODS \backslash I_s$ for its complement.

Testing The behaviour of concrete implementations is investigated by performing experiments on the implementations and observing the reactions that the implementations produce to these experiments. Such experiments are called tests, and they are formally specified as elements of a universe of test cases $TESTS$. A set of test cases is called a test suite. The process of running a test against a concrete implementation is called test execution. Test execution leads to an observation in a domain of observations OBS. To each observation a verdict is assigned by a verdict assignment function $verd_t : OBS \rightarrow \{\mathbf{pass}, \mathbf{fail}\}$. It is said that a concrete implementation $IUT \in IMPS$ **passes** a test suite $T \subseteq TESTS$ if the test execution of all its test cases lead to an observation with verdict **pass**:

$$\begin{array}{ll} IUT \ \mathbf{passes}\ T =_{\text{def}} & \forall t \in T:\ IUT\ \mathbf{passes}\ t \\ IUT\ \mathbf{passes}\ t =_{\text{def}} & \text{test execution of } t \text{ against } IUT \text{ gives } \sigma \in OBS, \\ & \text{such that } verd_t(\sigma) = \mathbf{pass} \end{array} \tag{1}$$

An implementation fails test suite T if it does not pass: $IUT\ \mathbf{fails}\ T =_{\text{def}} \neg(IUT\ \mathbf{passes}\ T)$.

The interpretation of test execution, i.e., of IUT **passes** T, is given by modelling the process of test execution on models of implementations. By comparing the concrete observations made of IUT with the calculated observations of the model of test execution, conclusions can be drawn about the model i_{IUT}, in particular, a set of candidate models for i_{IUT} can be calculated. Let test execution be modelled as a function $exec : TESTS \times MODS \rightarrow OBS$, such that for each test case $t \in TESTS$ and each model $i \in MODS$, $exec(t, i)$ calculates the observation in OBS that results from executing t with the model i. If $exec$ indeed faithfully models concrete test execution, then it can be concluded from successful test execution, IUT **passes** t, that the model of IUT is in the subset P_t of models for which a **pass**-verdict is calculated:

$$\begin{array}{ll} \text{let} & P_t =_{\text{def}} \{\ i \in MODS \mid verd_t(exec(t,i)) = \mathbf{pass}\ \} \\ \text{then} & IUT\ \mathbf{passes}\ t \iff i_{IUT} \in P_t \end{array} \tag{2}$$

and moreover for a test suite T:

$$\begin{array}{ll} \text{let} & P_T =_{\text{def}} \bigcap_{t \in T} P_t \\ \text{then} & IUT\ \mathbf{passes}\ T \iff i_{IUT} \in P_T \end{array} \tag{3}$$

In this way, for each test suite T, the universe of models of implementations $MODS$ is partitioned into models in P_T, which model passing implementations, and models not in P_T. The set P_T is called the formal test purpose of T.

Conformance testing In order to judge whether a concrete implementation IUT conforms to its specification $s \in SPECS$ by means of testing, the notion of conformance, i.e., the set I_s, and test execution, i.e., the set P_T (3), have to be linked, so that from test execution an indication can be obtained whether $i_{IUT} \in I_s$, i.e., whether IUT conforms. A test suite is complete if it can distinguish exactly between all conforming and non-conforming implementations: $I_s = P_T$. Unfortunately, this is a very strong requirement for practical testing: complete test suites are usually infinite, and consequently not practically executable. Hence, [ISO95] poses a weaker requirement on test suites: they shall be sound, which means that at least all correct implementations (and possibly some incorrect implementations) will pass them:

$$\forall i \in MODS:\ i\ \mathbf{imp}\ s \Longrightarrow i \in P_T \tag{4}$$

In case all incorrect implementations (and possibly some correct ones) do not pass the execution of test suite T, the test suite is called exhaustive:

$$\forall i \in MODS:\ i\ \mathbf{imp}\ s \Longleftarrow i \in P_T \tag{5}$$

To quantify the error-detecting capability of a sound test suite a coverage measure can be defined, which expresses the extent to which a sound test suite is exhaustive ($\mathcal{P}(TESTS)$ is the powerset of $TESTS$, i.e., the set of sets of test cases, so the set of test suites):

$$cov : \mathcal{P}(TESTS) \longrightarrow [0,1] \quad \text{satisfying} \quad P_{T_1} \supseteq P_{T_2} \Longrightarrow cov(T_1) \leq cov(T_2) \tag{6}$$

Assumptions In the definition of the formal testing framework of [ISO95] some explicit and implicit assumptions are made. The first assumption, which is explicitly stated, is the test assumption: any concrete implementation IUT can be modelled by an $i_{IUT} \in MODS$. In order to do any formal reasoning about implementations the existence of such a model should be assumed. However, the test assumption is unclear about unicity of i_{IUT}, i.e., about fully-abstractness with respect to observable behaviour, and about consequences of non-unicity.

A second assumption has to do with practical test execution. Execution of one single test case usually consists of several test runs, where each test run consists of applying the test case once to the implementation under test. Due to nondeterminism in the implementation under test each test run may lead to a different outcome, and the outcomes of several, independent test runs make up one observation. Let the class of possible test-run outcomes be $\mathcal{O}$, then an observation is a set of outcomes: $OBS = \mathcal{P}(\mathcal{O})$. Now $exec : TESTS \times MODS \rightarrow \mathcal{P}(\mathcal{O})$ calculates all possible test outcomes of given t and i. However, in the concrete test execution of t against an IUT it is difficult, or impossible, to be sure that all possible outcomes have really been obtained. Concrete test execution consists of performing a finite number of test runs, during which certain nondeterministic behaviours of the implementation may never be encountered. After any finite number of test runs it cannot be known whether all possible outcomes have been obtained or not. Consequently, if concrete test execution gives us an observation $O \subseteq \mathcal{O}$, we cannot conclude that $O = exec(t, i_{IUT})$ as above, and in [ISO95], but only that $O \subseteq exec(t, i_{IUT})$.

But even the conclusion $O \subseteq exec(t, i_{IUT})$ is not always valid; it depends on another assumption, viz. that the concrete observations obtained from test runs are always those which can be calculated from the model of test execution. This only holds if we assume that test cases are correctly implemented, i.e., that each test case is a valid model of its own implementation, and that $exec$ correctly models the concrete observations that can be made during test runs. If the first assumption cannot be assumed to hold then we should verify or test the implementations of our test cases, for which we would have to derive and implement test cases again, which should also be tested, etc. Usually this assumption can be made since test cases are assumed to be an order of magnitude simpler than the implementations that are the aim of our testing. The second assumption requires an accurate modelling of the concrete test execution process, for any test case and any IUT, by means of the function $exec$, which is not always easy. Consider as an example the observation of a time-out. Time-outs are used to detect that an implementation does not react to a given stimulus. However, if no real-time requirements are specified for the reaction, then the time-out value should theoretically be infinite, which is practically infeasible. So it might be that a time-out is observed where only the implementation under test is slow. The observation of this time-out will usually not be modelled in $exec$, where the theoretically infinite timer is considered.

In the probabilistic additions to the formal framework in the next sections we will not challenge the test assumption, but we will reconsider the assumption that all test-run outcomes can be obtained during test execution: we decompose test execution into test runs, taking into account that during concrete test execution some outcomes might be missed; this will be discussed in section 3. Implemented test cases and test execution are, without change, assumed to be correctly modelled by test cases and a function $exec$, respectively.

3 TEST EXECUTION AS PROBABILISTIC TEST RUNS

It was argued in section 2 that the application of a single test case to a concrete implementation usually consists of multiple test runs, where the outcome of each test run may be different, due to nondeterminism in the implementation itself (e.g., a gambling machine modelling the tossing of a coin), or due to nondeterminism introduced by interaction of the implementation with its environment (e.g., a fileserver writing a file to disk that is nondeterministically interrupted by the operating system when the disk is full). Each time a test-run experiment is repeated, it may result in another outcome, and it is never known when all possible outcomes of the *IUT* have been obtained. Consequently, test execution results in a set of test-run outcomes, where some outcomes might be more likely to occur than others. This leads to considering the occurrence of an outcome as a stochastic experiment, in which a single outcome from the set of possible outcomes is drawn. The probability of an outcome to occur can be thought of as depending on the frequency with which the implementation resolves the nondeterministic choices leading to the different outcomes.

Let $\mathcal{O}$ be the class of outcomes, and let test execution be correctly modelled by $exec : TESTS \times MODS \to \mathcal{P}(\mathcal{O})$ (cf. section 2), then in each test run of test case t against implementation *IUT*, modelled by i, an outcome σ from sample set $exec(t,i)$ is drawn, where each outcome in $exec(t,i)$ has a nonzero probability to occur. This can be described by viewing a test-run as a stochastic experiment that produces an outcome $\underline{\sigma}(t,i)$ that takes its value in $exec(t,i)$. Consequently, for every t and i a probability distribution $P_o^{t,i} : \mathcal{P}(exec(t,i)) \to [0,1]$ is assumed, indicating the probability that testing implementation i with test t leads to an outcome in $O \subseteq exec(t,i)$, viz.

$$P_o^{t,i}(O) =_{\text{def}} Pr\{\underline{\sigma}(t,i) \in O\} \tag{7}$$

Note that the distribution of the random variable $\underline{\sigma}(t,i)$ depends on the implementation i and test t; for each i and t, $\underline{\sigma}(t,i)$ may have another distribution.

Instead of combining outcomes into one observation $O \subseteq \mathcal{O}$, and then assigning the verdict $verd_t(O)$, we will assign verdicts to the outcomes, and then combine these outcome-verdicts into one verdict for the observation. The verdict for the observation will be **pass** if all outcome-verdicts are **pass**. Note that by combining outcome-verdicts into an observation-verdict instead of combining outcomes into an observation, we loose some observation power, e.g., consider a system which shall produce either always x or always y. Two test runs yield the outcomes x and y, respectively. Since they are both valid outcomes the test-run verdicts assigned will be both **pass**, whereas the verdict assigned to the observation $\{x,y\}$ would be **fail**.

Let $v_t : \mathcal{O} \to \{\mathbf{pass}, \mathbf{fail}\}$ be a verdict assignment to outcomes, then the probability measure $P_o^{t,i}$ can be used to induce a probability measure on sets of verdicts in $\{\mathbf{pass}, \mathbf{fail}\}$. As each outcome $\sigma \in exec(t,i)$ uniquely leads to a verdict $v_t(\sigma)$ this holds in particular for the outcome $\underline{\sigma}(t,i)$. As $\underline{\sigma}(t,i)$ is stochastically determined, it follows that the verdict assignment v_t is a stochastic function, and hence a probability distribution $P_v^{t,i} : \mathcal{P}(\{\mathbf{pass}, \mathbf{fail}\}) \to [0,1]$ is obtained. The probability that a test run of t with i results in verdict **pass** is the cumulative probability that an outcome in $exec(t,i)$ occurs that leads to a **pass**-verdict:

$$P_v^{t,i}(\{\mathbf{pass}\}) =_{\text{def}} Pr\{v_t(\underline{\sigma}(t,i)) = \mathbf{pass}\} = P_o^{t,i}(\{\sigma \mid v_t(\sigma) = \mathbf{pass}, \sigma \in exec(t,i)\}) \tag{8}$$

A straightforward extension of the measure $P_v^{t,i}$ on test-run verdicts to test-case and test-suite execution verdicts can be given. To integrate test-case and test-suite execution we consider a test suite as a multi-set of test cases rather than a set as in section 2. Multiple test runs of a single test case t can then be represented by multiple occurrences of that test case in the multi-set, written as t^n, so a test suite specifies the test cases it contains together with the number of test runs for each test case. For test suite $\langle t_1, \dots, t_n \rangle$ we have the random variables $\langle \underline{\sigma_1}(t_1, i), \dots, \underline{\sigma_n}(t_n, i) \rangle$, where each $\underline{\sigma_k}(t_k, i)$ $(1 \leq k \leq n)$ takes its value in $exec(t_k, i)$ according to the probability distribution $P_o^{t_k, i}$. According to (1,3) a test suite is passed by i if all the test cases in the test suite are passed. Analogously, the probability that implementation i passes test suite $T = \langle t_1, \dots, t_n \rangle$ equals the probability that all tests $t_1, \dots, t_n$ are passed by i:

$$P_v^{T,i}(\{\mathbf{pass}\}) =_{\text{def}} Pr\{v_{t_1}(\underline{\sigma_1}(t_1, i)) = \mathbf{pass}, \dots, v_{t_n}(\underline{\sigma_n}(t_n, i)) = \mathbf{pass}\} \tag{9}$$

Under the assumption that the random variables $\underline{\sigma_1}(t_1, i), \dots, \underline{\sigma_n}(t_n, i)$ are independent, that is, the outcome of a test run does not depend on previous (or future) test runs, it follows that

$$P_v^{T,i}(\{\mathbf{pass}\}) = \prod_{k=1}^{n} Pr\{v_{t_k}(\underline{\sigma_k}(t_k, i)) = \mathbf{pass}\} = \prod_{k=1}^{n} P_v^{t_k, i}(\{\mathbf{pass}\}) \tag{10}$$

The probability to fail T immediately follows: $P_v^{T,i}(\{\mathbf{fail}\}) =_{\text{def}} 1 - P_v^{T,i}(\{\mathbf{pass}\})$. It is evident that $P_v^{T,i}$ denotes a probability measure on $\mathcal{P}(\{\mathbf{pass}, \mathbf{fail}\})$, which has the obvious property, expressed in the next proposition, that for non-zero probability of **fail** in a test run, test case execution will finally result in **fail** if enough test runs are performed.

Proposition 1 If $P_v^{t,i}(\{\mathbf{pass}\}) < 1$ then $\lim_{n \to \infty} P_v^{t^n, i}(\{\mathbf{pass}\}) = 0$ □

4 A VALUATION ON IMPLEMENTATIONS

By considering test runs as stochastic experiments, section 3 formalized the notion of passing a test suite in the form of the probability that that test suite is passed, when a particular model of an implementation $i \in MODS$ is given. However, in testing we do not have one particular, given model, but we need to reason about classes of possible models. Some of these possible models are more likely to occur than others, and some are more important than others.

In this section we define a valuation measure on models of implementations. This valuation assigns a value to a set of models of implementations, i.e., to subsets of *MODS*. This value gives an indication about the importance of the subset as a possible class of correct implementations with respect to a particular specification s and implementation relation **imp**. The valuation is defined analogous to [Bri93] as a measure-theoretic integral, which takes into account both the likeliness of occurrence of the implementations and the importance of the individual implementations, expressed by a probability and a weight on models of implementations, respectively.

Weight of implementations An implementation relation distinguishes between correct and incorrect models of implementations. However, to express the importance of each implementation, more discriminating power has to be added. This is done by assigning a weight to each implementation.

Let $s \in SPECS$ be a specification, and $\mathbf{imp} \subseteq MODS \times SPECS$ an implementation relation, then a function $w : MODS \to \mathbb{R}\backslash\{0\}$ is a weight assignment function on *MODS* with respect to s and **imp**, if for all $i \in MODS$:

$$w(i) > 0 \Longleftrightarrow i \ \mathbf{imp} \ s \tag{11}$$

A weight assignment assigns a positive real number to each conforming implementation, and a negative number to each erroneous implementation. For conforming implementations the weight can express that one implementation is better than another; negative weights express the gravity of errors in erroneous implementations: if $w(i_1) < w(i_2) < 0$ then both i_1 and i_2 are not correct, but the errors of i_1 are more severe than those of i_2. Note that the weight is defined with respect to a specification s and an implementation relation **imp**: it assigns a weight to each implementation as a candidate for an implementation of the particular s and **imp**.

Probability on implementations Given a specification $s \in S$ and an implementation relation **imp** implementers will start developing a concrete implementation $IUT \in IMPS$. Many different implementations, modelled by different models, may occur as the result of the implementation process, some of them conforming, and others nonconforming. Not every possible implementation will have the same chance to occur as the result of implementing a given specification, e.g., when designing a coffee machine which is supposed to serve coffee and tea it is less likely to end up with a completely different kind of machine (e.g., a gambling machine) than to end up with a slightly different, but possibly incorrect, coffee machine. Moreover, assuming that an implementer can make several independent mistakes with non-zero probability it is less likely for the implementer to make all possible mistakes than to make only a small number of mistakes.

Similar to [Bri93] we view the design and implementation process of a complex system as a stochastic experiment that draws a concrete implementation IUT modelled by the stochastic function $\underline{i} \in MODS$ from the sample set of all possible implementations $MODS$ according to a certain probability distribution. Let $MODS$ be discrete, and let the probability density function be given by p_s, i.e., $p_s(i)$ denotes the probability that implementation i occurs as the result of the design and implementation process, then

$$P_s(I) =_{\text{def}} \sum_{i \in I} p_s(i) \tag{12}$$

denotes the probability that the activity of implementing specification s produces an implementation that is modelled by a member of I.

Valuation Using the probability density p_s and the weight assignment w it is possible to define a valuation on discrete sets of implementations, which captures the importance of a set of implementations in terms of their weight and their probability of occurrence:

$$\mu(I) =_{\text{def}} \sum_{i \in I} w(i)\ p_s(i) \tag{13}$$

The expression $\mu(I)$ denotes the value of implementations in $I \subseteq MODS$. It expresses, in some sense, the aggregate correctness of the candidate implementations in I with respect to the given s and **imp**. Note that correct implementations may compensate incorrect implementations (and vice versa). Consequently, $\sum_{i \in I} w(i)\ p_s(i) < 0$ does not imply that the probability of obtaining an erroneous implementation in I is high: the negative value might occur because of one unlikely, but very negatively weighted erroneous implementation. The expectation value of the weight of an arbitrarily designed implementation in $MODS$ is expressed by $\sum_{i \in MODS} w(i)\ p_s(i)$.

Although in most practical cases it suffices to consider the valuation as a, possibly infinite, summation, it is more abstract and general to rewrite (12) and (13) as measure-theoretic integrals, which also encompass continuous domains of implementations. Consider a probability measure P_s on sets of implementations $I \subseteq MODS$ defined by

$$P_s(I) =_{\text{def}} Pr\{\underline{i} \in I\} \tag{14}$$

then, under sufficient assumptions of neatness of the underlying sets (they should be Borel), the valuation μ can be expressed as the measure-theoretic integral

$$\mu(I) =_{\text{def}} \int_I w(i)\, dP_s \tag{15}$$

Proposition 2

1. $\int_\emptyset w(i)\, dP_s = 0$
2. If $I_s \neq \emptyset$ then $\int_{I_s} w(i)\, dP_s > 0$
3. If $\overline{I_s} \neq \emptyset$ then $\int_{\overline{I_s}} w(i)\, dP_s < 0$

□

5 COMPARING TEST SUITES

The purpose of conformance testing is to increase the confidence in the correct functioning of an implementation by detecting whether it conforms to its specification or not. Test-suite execution is expected to reject, i.e., yield the verdict **fail** with, nonconforming implementations, and to accept, i.e., yield the verdict **pass** with, conforming ones. Since a perfect test suite exactly doing this is not likely to be encountered in practice (cf. section 2), a method for comparing the quality of test suites is needed in order to select the best test suite for a given conformance testing problem. A comparison of test suites can be made if a quantitative measure can be assigned to each test suite. Such a measure should quantify the ability of a test suite to reject nonconforming implementations and to accept conforming ones, and it should take into account the probability of occurrence of implementations (a test suite that detects errors which are very likely to occur has higher quality), the weight of implementations (a test suite that detects a nonconforming implementation with severe errors, i.e., with very negative weight, has higher quality), and the probability that an erroneous implementation indeed yields the verdict **fail** (a test suite that rejects an erroneous implementation with higher probability has higher quality).

Soundness and exhaustiveness The probability of occurrence of implementations was expressed by the probability measure P_s on sets of implementations (14), and the probability that an implementation i yields the verdict **pass** with test suite T was expressed by the probability measure $P_v^{T,i}$ (9,10). Under the assumption that the probability of occurrence of implementations (the measure P_s) is independent of the probability of yielding **pass** (the measure $P_v^{T,i}$) we can integrate these measures to obtain the probability measure on $MODS \times \{\mathbf{pass}, \mathbf{fail}\}$:

$$\pi_T(I, V) =_{\text{def}} \int_I \int_V dP_v^{T,i}\, dP_s \tag{16}$$

$\pi_T(I, V)$ expresses the probability that an implementation $i \in I \subseteq MODS$ occurs, and that i, when tested with T, yields a verdict in $V \subseteq \{\mathbf{pass}, \mathbf{fail}\}$. So, $\pi_T(I, \{\mathbf{pass}\})$ denotes the probability that an implementation $i \in I$ occurs which passes test suite T. Note that the integrals cannot be interchanged, since the inner integral depends on $i \in I$.

Taking also the weight of implementations (section 4) into account we obtain analogously the valuation measure λ:

$$\lambda_T(I, V) =_{\text{def}} \int_I \int_V w(i)\, dP_v^{T,i}\, dP_s \tag{17}$$

Given a test suite T, the valuation λ assigns a value to each pair of $I \subseteq MODS$ and $V \subseteq \{\textbf{pass}, \textbf{fail}\}$. Of course, the interesting values of λ are the combinations of conforming and nonconforming implementations with the verdicts **pass** and **fail**:

	$V = \{\textbf{pass}\}$	$V = \{\textbf{fail}\}$
$I = I_s$	$\lambda_T(I_s, \{\textbf{pass}\})$	$\lambda_T(I_s, \{\textbf{fail}\})$
$I = \overline{I_s}$	$\lambda_T(\overline{I_s}, \{\textbf{pass}\})$	$\lambda_T(\overline{I_s}, \{\textbf{fail}\})$

For a good test suite, i.e., one with a large ability to reject nonconforming implementations and to accept conforming ones, the values $\lambda_T(I_s, \{\textbf{pass}\})$ and $\lambda_T(\overline{I_s}, \{\textbf{fail}\})$ are optimized, or, equivalently, the values $\lambda_T(I_s, \{\textbf{fail}\})$ and $\lambda_T(\overline{I_s}, \{\textbf{pass}\})$ are minimized.

We define two measures on test suites, soundness *snd*, quantifying the ability to accept conforming implementations, and exhaustiveness *exh*, quantifying the ability to reject nonconforming implementations, by normalization of the valuation λ with respect to all conforming implementations, and all nonconforming implementations, respectively. The soundness of test suite T with respect to $I_s \subseteq MODS$ and weight assignment w is

$$snd(T) =_{\mathrm{def}} \frac{\lambda_T(I_s, \{\textbf{pass}\})}{\lambda_T(I_s, \{\textbf{pass}, \textbf{fail}\})} \tag{18}$$

Similarly, the exhaustiveness of test suite T with respect to I_s and w is

$$exh(T) =_{\mathrm{def}} \frac{\lambda_T(\overline{I_s}, \{\textbf{fail}\})}{\lambda_T(\overline{I_s}, \{\textbf{pass}, \textbf{fail}\})} \tag{19}$$

Of course, these definitions are only valid if $\lambda_T(I_s, \{\textbf{pass}, \textbf{fail}\}) \neq 0$ and $\lambda_T(\overline{I_s}, \{\textbf{pass}, \textbf{fail}\}) \neq 0$, but these requirements are easily satisfied, since they correspond to the existence of conforming and nonconforming implementations, respectively. If they would not exist, there were no need for testing at all.

Proposition 3

1. $0 \leq snd(T) \leq 1$
2. $0 \leq exh(T) \leq 1$
3. T is sound following (4) if and only if $snd(T) = 1$
4. T is exhaustive following (5) if and only if $exh(T) = 1$
5. T is complete if and only if $snd(T) = 1$ and $exh(T) = 1$

□

As noticed in section 4 it mostly suffices to consider summations instead of integrals. Let $p_v^{T,i}$ be the density function corresponding to $P_v^{T,i}$, i.e., $p_v^{T,i}(v) = P_v^{T,i}(\{v\})$, then we can use

$$\lambda_T(I, V) = \sum_{i \in I} \sum_{v \in V} w(i)\; p_v^{T,i}(v)\; p_s(i) \tag{20}$$

Comparison In the approach of section 2 [ISO95], test suites are compared using the coverage function, which is defined to quantify the extent to which sound test suites are exhaustive. Moreover, test suites are trivially compared with their soundness: test suites which are not sound are simply not considered, i.e., they are worse than any sound test suite. So, in fact, test suites are ordered lexicographically with the pair $\langle \mathit{soundness}, \mathit{coverage} \rangle$, where *soundness* can only take two possible values, viz. sound or not sound.

The two measures *snd* (18) and *exh* (19) generalize the original definitions in (4) and (5): soundness and exhaustiveness are not logical properties anymore, but they are continuous measures on test suites with values between 0 and 1. Similar as above we can now consider an ordering on pairs of the form

$$\langle \mathit{snd}(T), \mathit{exh}(T) \rangle \tag{21}$$

for comparing test suites. The rôle of *exh(T)* is analogous to the coverage function above: it defines a measure of the extent to which a test suite is exhaustive. And indeed, it can be shown that *exh* is a coverage satisfying (6), if *exec* faithfully models concrete test execution (cf. section 2), i.e., if $P_v^{T,i}(\{\mathbf{pass}\}) = 1 \Leftrightarrow i \in P_T$ and $P_v^{T,i}(\{\mathbf{pass}\}) = 0 \Leftrightarrow i \notin P_T$. If this assumption does not hold then *exh* is not a coverage, and it is difficult to define one which satisfies (6).

There are various ways to compare test suites by ordering pairs of the form $\langle \mathit{snd}(T), \mathit{exh}(T) \rangle$: the lexicographical ordering, projections on one of the constituents, addition, vector addition or multiplication of both constituents, comparing the maxima or minima, etc. The actual way of ordering the tuples (21) will depend on the application. In testing the software of a nuclear power plant the exhaustiveness will be the most important: not finding an error if there is one has much more disastrous consequences than finding an error where there is none. If, on the other hand, testing is expensive, then detecting errors where there are none is costly, and should be avoided: soundness will prevail.

6 PROBABILISTIC TESTING WITH INPUTS AND OUTPUTS

In this section we instantiate the framework discussed in the previous sections and illustrate its applicability by a running example. The structure of this section follows the structure of the framework. First, we present some elementary notation for the description of system behaviour as labelled transition systems. Next, we instantiate the universes *SPECS*, *MODS* and *TESTS* as (restricted sets of) labelled transition systems. Then the concepts of conformance and testing (section 2) are instantiated in such a way that they are applicable for the systems under consideration, followed by the concepts defined in section 4 (valuation on implementations). Finally, we discuss how section 5 (comparison) can be instantiated.

Preliminary definitions The behaviour of systems is modelled by means of labelled transition systems. We use the standard definitions of labelled transition systems as can be found in, e.g., [Tre95]: a transition system p is a quadruple $p = \langle S, L, \rightarrow, s_0 \rangle$, where S is a (countable) set of states, L is a (countable) set of observable actions, $\rightarrow \subseteq S \times (L \cup \{\tau\}) \times S$ is a set of transitions, and $s_0 \in S$ is the initial state. The special action $\tau \notin L$ denotes the unobservable action. The universe of labelled transition systems over L is denoted by $\mathcal{LTS}(L)$. A trace σ is a sequence of observable actions ($\sigma \in L^*$), and $\Longrightarrow$ denotes the transition relation between states s, s' when performing trace σ, i.e., $s \stackrel{\sigma}{\Longrightarrow} s'$ indicates that s' can be reached by performing the observable sequence of actions $\sigma \in L^*$. Furthermore, we define the set of traces by $\mathit{traces}(s) =_{\mathrm{def}} \{\sigma \in L^* \mid \exists s' : s \stackrel{\sigma}{\Longrightarrow} s'\}$, the set of reachable states from s by $\mathit{der}(s) =_{\mathrm{def}} \{s' \mid \exists s' : s \stackrel{\sigma}{\Longrightarrow} s'\}$, the set of initial observable actions from $S' \subseteq S$ by $init(S') =_{\text{def}} \{x \in L \mid \exists s' \in S', s \in S : s' \stackrel{x}{\Longrightarrow} s\}$, the set of reachable states after $\sigma \in L^*$ by s **after** $\sigma =_{\text{def}} \{s' \mid s \stackrel{\sigma}{\Longrightarrow} s'\}$, and a boolean predicate on states, s **after** σ **refuses** $A =_{\text{def}} \exists s' : s \stackrel{\sigma}{\Longrightarrow} s'$ and $\forall x \in A : s' \stackrel{x}{\not\Longrightarrow}$, where $A \subseteq L$. As usual, we will not distinguish between a labelled transition system and its initial state.

Specifications We instantiate the formalism *SPECS* with the set of all labelled transition systems over L, that is, we take $SPECS = \mathcal{LTS}(L)$. Figure 1 depicts a specification $s \in \mathcal{LTS}(L)$ for $L = \{sh, cb, tb, cof, tea\}$. Here, it is assumed that *sh* models the insertion of a shilling, *cb* and *tb* model the pressing of a coffee button and a tea button, respectively, and *cof* and *tea* model the provision of coffee and tea, respectively.

Models of implementations As many systems communicate by means of clearly distinguishable input actions and output actions [Pha94, Tre95], we assume that the set of actions L that a system can perform can be partitioned in a set of input actions L_I and a set of output actions L_U. Furthermore, (viz. [Pha94, Tre95]) we assume that implementations, unlike specifications, never can refuse any input in each reachable state, i.e.,

$$\forall p' \in der(p), \forall a \in L_I : p' \stackrel{a}{\Longrightarrow} \tag{22}$$

Such systems are called input-output transition systems, and the universe of such systems is denoted by $\mathcal{IOTS}(L_I, L_U)$. We take $MODS = \mathcal{IOTS}(L_I, L_U)$. Note that $\mathcal{IOTS}(L_I, L_U) \subseteq \mathcal{LTS}(L_I \cup L_U)$, thus all input-output transition systems are also labelled transition systems. Figure 1 shows implementations $i_1, \ldots, i_4$ for $L_I = \{sh, cb, tb\}$ and $L_U = \{cof, tea\}$.

Tests A test t over L_U and L_I is a 6-tuple $\langle S, L_U, L_I, \rightarrow, \nu_t, s_0\rangle$, where $\langle S, L_I \cup L_U, \rightarrow, s_0\rangle \in \mathcal{LTS}(L_I \cup L_U)$ is a deterministic, finite-behaviour labelled transition system over L_U and L_I, and $\nu_t : der(s_0) \rightarrow \{\mathbf{pass}, \mathbf{fail}\}$ is a function assigning verdicts to the reachable states of t. When a test is run against an implementation, it is either able to receive all outputs from the implementation, or provide a single input to the implementation, or deadlock. That is, any test satisfies

$$\forall s' \in der(s_0) : init(s') = \{a\} \subseteq L_I \text{ or } init(s') = L_U \text{ or } init(s') = \emptyset \tag{23}$$

The universe of tests over L_I and L_U is denoted by $\mathcal{T}(L_I, L_U)$. We take $TESTS = \mathcal{T}(L_I, L_U)$. Cf. section 3, a test suite is modelled as a multiset of tests. Figure 1 shows tests t_1, t_2, t_3 for $L_I = \{sh, cb, tb\}$ and $L_U = \{cof, tea\}$.

Implementation relation An implementation relation is a relation between the set of specifications *SPECS* and the set of models of implementations *MODS* (section 2). In [Tre95] a family of implementation relations $\mathbf{ioconf}_{\mathcal{F}}$ (for $\mathcal{F} \subseteq (L_I \cup L_U)^*$) is defined that are applicable to the kind of systems we are considering. Informally, an implementation i is a correct implementation with respect to specification s and implementation relation $\mathbf{ioconf}_{\mathcal{F}}$ if for every trace $\sigma \in \mathcal{F}$ each output, or absence of outputs, that the implementation i can perform after having performed sequence σ is specified by s. Let $i \in \mathcal{IOTS}(L_I, L_U), s \in \mathcal{LTS}(L_I \cup L_U)$ and $\mathcal{F} \subseteq (L_I \cup L_U)^*$, then

$$i\ \mathbf{ioconf}_{\mathcal{F}}\ s =_{\text{def}} \forall \sigma \in \mathcal{F} : out(i\ \mathbf{after}\ \sigma) \subseteq out(s\ \mathbf{after}\ \sigma) \tag{24}$$

where $out(p\ \mathbf{after}\ \sigma) =_{\text{def}} (init(p\ \mathbf{after}\ \sigma) \cap L_U) \cup \{\delta \mid p\ \mathbf{after}\ \sigma\ \mathbf{refuses}\ L_U\}$ for $\sigma \in traces(p)$, and $out(p\ \mathbf{after}\ \sigma) =_{\text{def}} \emptyset$ otherwise. Here, the special action $\delta \notin L_I \cup L_U$ models absence of

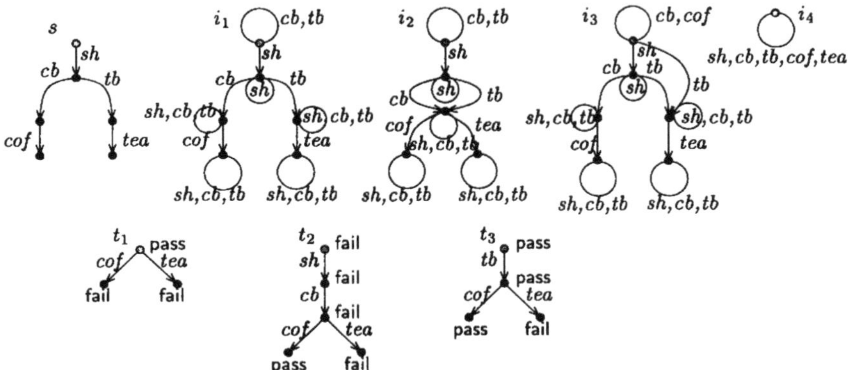

Figure 1: Examples of specifications, implementations and tests.

output actions. If $out(i \textbf{ after } \sigma) \subseteq out(s \textbf{ after } \sigma)$ then we write $i \models_s \sigma$ (and $i \not\models_s \sigma$ for its negation, in which case we say that i has fault σ with respect to s).

Example 4 For specification s (figure 1) we instantiate the implementation relation **imp** (section 2) with $\textbf{ioconf}_{\{\epsilon, sh.cb\}}$. For implementation i_1 we have $out(i_1 \textbf{ after } \epsilon) = \{\delta\} \subseteq \{\delta\} = out(s \textbf{ after } \epsilon)$, and $out(i_1 \textbf{ after } sh.cb) = \{cof\} \subseteq \{cof\} = out(s \textbf{ after } \epsilon)$. Hence $i_1 \textbf{ ioconf}_{\{\epsilon, sh.cb\}} s$. However, $out(i_2 \textbf{ after } sh.cb) = \{cof, tea\} \not\subseteq \{cof\} = out(s \textbf{ after } sh.cb)$, i.e. *not* $(i_2 \textbf{ ioconf}_{\{\epsilon, sh.cb\}} s)$. Here, i_1 is a correct implementation ($i_1 \in I_s$), and i_2, i_3, i_4 are incorrect ($i_2, i_3, i_4 \in \overline{I_s}$). □

Test execution The running of test $t \in \mathcal{T}(L_I, L_U)$ against implementation $i \in \mathcal{IOTS}(L_I, L_U)$ is modelled by the LOTOS synchronization operator $||$. The execution of a test against an implementation returns a set of observations that may occur, defined by

$$exec(t, i) =_{\text{def}} \{\sigma \in (L_I \cup L_U)^* \mid (t||i) \textbf{ after } \sigma \textbf{ refuses } (L_I \cup L_U)\} \tag{25}$$

(viz. the deadlock traces of $(i||t)$). The set $exec(t, i)$ denotes the set of possible test run outcomes.

Verdict assignment To each set of observations $O \subseteq exec(t, i_{IUT})$ that can be made when a test $t \in \mathcal{T}(L_I, L_U)$ is executed against implementation IUT a verdict has to be assigned (cf. (1)). We instantiate the verdict assignment function by

$$verd_t(O) =_{\text{def}} \begin{cases} \textbf{pass} & \text{if } \forall \sigma \in O : v_t(\sigma) = \textbf{pass} \\ \textbf{fail} & \text{if } \exists \sigma \in O : v_t(\sigma) = \textbf{fail} \end{cases} \tag{26}$$

where v_t denotes the verdict assignment on outcomes $\sigma \in exec(t, i)$ given by

$$v_t(\sigma) = \begin{cases} \textbf{pass} & \text{if } \nu_t(t \textbf{ after } \sigma) = \textbf{pass} \\ \textbf{fail} & \text{if } \nu_t(t \textbf{ after } \sigma) = \textbf{fail} \end{cases} \tag{27}$$

i.e., the verdict **pass** is assigned to a set of outcomes O if and only if each individual test outcome leads to a **pass**-state in test t.

Example 5 Consider test t_2 and implementation i_2, then there are two possible test outcomes: $exec(t_2, i_2) = \{sh.cb.cof, sh.cb.tea\}$. Each outcome leads to a different verdict (viz. (27)) $v_{t_2}(sh.cb.cof) = \nu_{t_2}(t_2 \textbf{ after } sh.cb.cof) = \textbf{pass}$ and $v_{t_2}(sh.cb.tea) = \nu_{t_2}(t_2 \textbf{ after } sh.cb.tea) = \textbf{fail}$. Consequently, if test execution gives us an observation $O = \{sh.cb.cof\}$ then (according to (26)) $verd_{t_2}(O) = \textbf{pass}$, whereas if test execution gives us observation $O = \{sh.cb.cof, sh.cb.tea\}$ then (according to (26)) $verd_{t_2}(O) = \textbf{fail}$. □

Probabilistic test runs If test execution is considered as probabilistic test runs (section 3), a probability distribution $P_o^{t,i}$ is assumed. Consequently, according to (8) a probability distribution $P_v^{t,i}$ on verdicts follows. For simplicity we assume that the probability distribution $P_v^{t,i}$ is given for $t \in \mathcal{T}(L_I, L_U)$ and $i \in \mathcal{IOTS}(L_I, L_U)$. In the next table the set $exec(t, i)$ for $t \in \{t_1, \ldots, t_3\}$ (figure 1) and $i \in \{i_1, \ldots, i_4\}$ (figure 1), together with the set of verdicts $\{v_t(\sigma) \mid \sigma \in exec(t, i)\}$ is given. For each set of verdicts consisting of more than one element, the probability $P_v^{t,i}(\{v\})$ on the occurrence of v is assumed to be known, and denoted behind the verdict. If the set of verdicts contains one element v, then evidently $P_v^{t,i}(\{v\}) = 1$.

		t_1	t_2	t_3
i_1	*exec*	$\{\epsilon\}$	$\{sh.cb.cof\}$	$\{tb\}$
	verd	{**pass**}	{**pass**}	{**pass**}
i_2	*exec*	$\{\epsilon\}$	$\{sh.cb.cof, sh.cb.tea\}$	$\{tb\}$
	verd	{**pass**}	{**pass**(1/4), **fail**(3/4)}	{**pass**}
i_3	*exec*	$\{cof\}$	$\{sh.cb.cof\}$	$\{tb.tea\}$
	verd	{**fail**}	{**pass**}	{**fail**}
i_4	*exec*	$\{cof, tea\}$	$\{sh.cb.cof, sh.cb.tea\}$	$\{tb.cof, tb.tea\}$
	verd	{**fail**}	{**pass**(1/4), **fail**(3/4)}	{**pass**(2/5), **fail**(3/5)}

Example 6 The probability that implementation i_4 passes test case t_2 is $1/4$, ($P_v^{t_2,i_4}(\{\textbf{pass}\}) = 1/4$), and the probability that i_4 passes test case t_3 is $2/5$ ($P_v^{t_3,i_4}(\{\textbf{pass}\}) = 2/5$). Under the assumption that the outcome of tests does not depend on the outcome of previous (or future) tests, the probability that i_4 passes test suite $\langle t_2, t_3\rangle$ is given by (10): $P_v^{\langle t_2,t_3\rangle,i_4}(\{\textbf{pass}\}) = 1/4 \times 2/5 = 1/10$, and consequently $P_v^{\langle t_2,t_3\rangle,i_4}(\{\textbf{fail}\}) = 9/10$. □

Weight of implementations A weight assignment (11) quantifies the (in)correctness of implementations and can be obtained from weighing the faults that an implementation possesses. The function $w : \mathcal{IOTS}(L_I, L_U) \to \mathbb{R}\backslash\{0\}$ defined by

$$w(i) = \begin{cases} 1 & \text{if } i \textbf{ ioconf}_{\mathcal{F}}\ s \\ \sum_{\sigma\in\mathcal{F}} \{\!|g(\sigma) \mid i \not\models_s \sigma|\!\} & \text{otherwise} \end{cases} \tag{28}$$

where $g : \mathcal{F} \to \mathbb{R}_{<0}$, and $\{\!| \cdot |\!\}$ denotes a multiset, is a weight assignment function (according to (11)) on $\mathcal{IOTS}(L_I, L_U)$. The function g expresses the gravity of violating $i \models_s \sigma$. Note that the function w is not able to discriminate between correct implementations (all correct implementations are assigned the same value) and that implementations possessing the same faults are assigned the same (negative) weights.

Probability of implementations Let p_σ be the probability that an arbitrary implementation i has fault $\sigma \in \mathcal{F}$, i.e. $i \not\models_s \sigma$, then (under the assumption that all faults are independent)

$$p_s(i) = \prod_{\substack{\sigma \in \mathcal{F} \\ i \models_s \sigma}} (1 - p_\sigma) \times \prod_{\substack{\sigma \in \mathcal{F} \\ i \not\models_s \sigma}} p_\sigma \tag{29}$$

is a probability density function on $\mathcal{IOTS}(L_I, L_U)$. As implementations (dis)satisfying the same requirements are assigned equal probability, p_s can be considered as a probability density function on classes of implementations having the same faults. We shall assume that $i_1, \ldots, i_4$ are representative members of their classes, i.e., it suffices to consider $MODS = \{i_1, \ldots, i_4\}$.

Valuation on implementations Following (13) we can compute the valuation for the set of (in)correct implementations by taking the weights of implementations using (28), and the distribution of the implementations using (29).

Example 7 Let $g(\epsilon) = -5$ and $g(sh.cb) = -3$ then it follows: $w(i_1) = 1, w(i_2) = -3, w(i_3) = -5$ and $w(i_4) = -3 + -5 = -8$. Implementation i_4 violates both requirements imposed by $\mathbf{ioconf}_{\{\epsilon,sh.cb\}}$ and is therefore considered worse compared to all other implementations. Furthermore, assume the probability for arbitrary implementation i to violate the requirement $i \models_s \epsilon$ equals 1/3 (i.e. $p_\epsilon = 1/3$), and the probability to violate the requirement $i \models_s sh.cb$ equals 2/5. Then it follows $p_s(i_1) = (1 - 1/3) \times (1 - 2/5) = 2/5$, $p_s(i_2) = (1 - 1/3) \times 2/5 = 4/15$, $p_s(i_3) = 1/3 \times (1 - 2/5) = 1/5$, and $p_s(i_4) = 1/3 \times 2/5 = 2/15$. From (13) $\mu(\{i_1\}) = 1 \times 2/5 = 2/5$ and $\mu(\{i_2, i_3, i_4\}) = -43/15$. Thus, $\mu(\{i_1, i_2, i_3, i_4\}) = \mu(\{i_1\}) + \mu(\{i_2, i_3, i_4\}) = -37/15$. □

Comparison In the following table we denote the values of $snd(T)$ (18) and $exh(T)$ (19) for some specific T

T	$\lambda_T(I_s, \{\mathbf{pass}\})$	$\lambda_T(\overline{I_s}, \{\mathbf{fail}\})$	$snd(T)$	$exh(T)$
$\langle t_1 \rangle$	2/5	−31/15	1	31/43
$\langle t_2 \rangle$	2/5	−21/15	1	21/43
$\langle t_3 \rangle$	2/5	−41/25	1	123/215
$\langle t_1, t_1 \rangle$	2/5	−31/15	1	31/43
$\langle t_1, t_2 \rangle$	2/5	−40/15	1	40/43

Example 8 Using (20) it follows $\lambda_T(I_s, \{\mathbf{pass}, \mathbf{fail}\}) = 2/5$ and $\lambda_T(\overline{I_s}, \{\mathbf{pass}, \mathbf{fail}\}) = -43/15$ for all test suites T. For test suite $T = \langle t_1, t_2 \rangle$ it follows $\lambda_T(I_s, \{\mathbf{pass}\}) = 1 \times 1^2 \times 2/5 = 2/5$, thus (18) $snd(T) = 1$. Furthermore, $\lambda_T(\overline{I_s}, \{\mathbf{fail}\}) = -3 \times (1 - 1 \cdot (1/4)) \times 4/15 + -5 \times (1 - 0 \cdot 1) \times 1/5 + -8 \times (1 - 0 \cdot (1/4)) \times 2/15 = -40/15$. From (19) it follows $exh(T) = 40/43$. Moreover, $snd(\langle t_2^n, t_3 \rangle) = 1$, and $exh(\langle t_2^n, t_3 \rangle)$ is computed by

$$\frac{-3 \times (1 - (1/4)^n \cdot 1) \times 4/15 - 5 \times (1 - 1^n \cdot 0) \times 1/5 - 8 \times (1 - (1/4)^n \cdot 2/5) \times 2/15}{-3 \times 4/15 + -5 \times 1/5 + -8 \times 2/15}$$

For $n \to \infty$ this yields $\lim_{n \to \infty} exh(\langle t_2^n, t_3 \rangle) = 1$, hence, following proposition (3.5), test suite $\langle t_2^n, t_3 \rangle$ approaches a complete test suite if t_2 is executed infinitely many times for $\mathbf{ioconf}_{\{\epsilon,sh.cb\}}$ with respect to specification s.

Another question might be how often t_2 has to be executed in the test suite $\langle t_2^n, t_3 \rangle$ to reach larger exhaustiveness than $\langle t_1, t_2 \rangle$. This can be calculated by solving the equation $exh(\langle t_2^n, t_3 \rangle) > exh(\langle t_1, t_2 \rangle) = 40/43$: this gives $n > 1$, thus, test suite $\langle t_2, t_2, t_3 \rangle$ has larger exhaustiveness in rejecting erroneous implementations than test suite $\langle t_1, t_2 \rangle$. □

7 CONCLUSIONS

In this paper a probabilistic refinement of [ISO95] with respect to the execution of tests was discussed. In particular, the usually made assumption that every possible outcome that can occur when executing tests against an implementation will be observed (as in, e.g., [ISO95, DNH84]), has been relaxed; it was assumed that outcomes can occur with a certain probability distribution on the set of possible outcomes $exec(t, i)$. This induced a probability distribution on the acceptance or rejection of implementations by test suites.

Furthermore, the notion of correctness has been extended by assigning weights to implementations in order to not only distinguish between correct and incorrect implementations, but also discriminate between different (correct or incorrect) ones by means of their weight (e.g., with respect to the severeness of faults that they possess). Together with a probability distribution on the occurrence of implementations that indicates the probability that a particular (set of) implementation(s) can occur as the result of building a concrete implementation, a valuation on implementations has been obtained. Using such valuations it has been shown that one can quantify the expected correctness value of (sets of) implementations.

By having combined both the probability for a test suite to accept or reject implementations, and the valuation on implementations, normalized measures were obtained, that quantify the extent to which a test suite is able to detect incorrect implementations (19), and the extent to which a test suite is able accept correct implementations (18). Such measures were used to compare, and thus select, test suites with respect to their ability to detect most of the frequently occurring implementations with severe errors, and accept most of the frequently occurring correct systems.

Further work In this paper only one of the assumptions underlying [ISO95] is relaxed. However, [ISO95] states many more assumptions, as stated in section 2, such as the test assumption, the assumption that test cases are correctly implemented, and the assumption that the function *exec* correctly models the process of concrete test execution. Similar as has been done in this paper, one can investigate the consequences when these assumption are relaxed, too.

One of the difficult points in the presented approach is how to obtain the necessary distributions and weight functions. Section 6 gave some possible directions for obtaining them; another way is to use fault domains or fault models. As is indicated in [BDD+92, Bri93] the valuation on implementations (section 4) may be obtained by classifying implementations with respect to a (finite) set of faults or fault models that implementations may possess. In this way the set of implementations $MODS$ is partitioned into classes, such that each class contains all implementations possessing a specific combination of faults. By assigning weights to faults (e.g., by the method proposed by [ACV93]), and assigning probabilities to the occurrence of faults, valuations over classes of implementations can be obtained in a similar way as [BTV91, Bri93]. It needs to be studied how such a partitioning over faults can be used to obtain realistic valuations on implementations.

The applicability of statistical techniques in the current probabilistic extension needs to be investigated. Often the exact distribution with which implementations pass or fail a test is not known apriori because some characteristic parameters of the distribution are not known. In order to estimate the (partially) unknown distribution a sample of the phenomenon under study is taken, and this is used to estimate the unknown parameter(s). Also, the integration of the theory of testing statistical hypotheses such as the acceptance or rejection of the null-hypothesis (H_0 : the system under test is incorrect) and the alternative hypothesis (H_1 : the system under test is correct) and the probabilistic interpretation of test runs as discussed in this paper needs to be explored, in order to ensure that a system is sufficiently tested before considered correct.

In this paper a method is described to obtain a coverage measure before actually executing any test at all. That is, the measures soundness (18) and exhaustiveness (19) can be obtained apriori to the actual test execution process (under the assumption that all relevant parameters, such as distributions P_s and $P_v^{i,T}$ are known in advance). However, one can also measure the assessed soundness (or exhaustiveness) that is obtained after really having executed test suite T. Such measures are called the aposteriori measures. The relation between these two measures has to be investigated, in order to decide whether or not to reject an implementation if the assessed measures are not sufficiently close to the apriori values.

A final remark concerns the continuing discussion about the verdict **inconclusive** in formal testing. Purely semantically, **inconclusive** is equivalent to **pass**: there is no reason to reject the implementation. However, intuitively **inconclusive** refers to not reaching a particular test purpose, if such a test purpose is given apriori. The exact position of **inconclusive**, especially in the context of uncertainty and probability of this paper, remains for further study.

References

[ACV93] J. Alilovic-Curgus and S.T. Vuong. A metric based theory of test selection and coverage. In A. Danthine et al., eds, *Protocol Specification, Testing, and Verification XIII*, pages 289–304. North-Holland, 1993.

[BDD+92] G. von Bochmann, A. Das, R. Dssouli, M. Dubuc, A. Ghedamsi, and G. Luo. Fault models in testing. In J. Kroon et al., eds, *Fourth Int. Workshop on Protocol Test Systems*, pages 17–30. North-Holland, 1992.

[Bri93] E. Brinksma. On the coverage of partial validations. In M. Nivat et al., eds, *AMAST'93*, pages 247–254. Springer-Verlag, 1993.

[BTV91] E. Brinksma, J. Tretmans, and L. Verhaard. A framework for test selection. In B. Jonsson et al., eds, *Protocol Specification, Testing, and Verification XI*, pages 233–248. North-Holland, 1991.

[DNH84] R. De Nicola and M.C.B. Hennessy. Testing equivalences for processes. *Theoretical Computer Science*, 34:83–133, 1984.

[ISO95] ISO/IEC JTC1/SC21 WG7, ITU-T SG 10/Q.8. *Formal Methods in Conformance Testing, working draft.* ISO Project 1.21.54, ITU-T Z.500. ISO, ITU-T, September 1995. Output from ISO/ITU-T Meeting in Geneve.

[LS89] K.G. Larsen and A. Skou. Bisimulation through probabilistic testing. In *Sixteenth Ann. Symp. on the Principles of Programming Languages.* ACM, 1989.

[Pha94] M. Phalippou. *Relations d'Implantation et Hypothèses de Test sur des Automates à Entrées et Sorties.* PhD thesis, L'Université de Bordeaux I, France, 1994.

[Tre95] J. Tretmans. Testing labelled transition systems with inputs and outputs. In A. Cavalli and S. Budkowski, editors, *Participants Procs. of the Int. Workshop on Protocol Test Systems VIII - COST 247 Session*, pages 461–476, Evry, France, September 4-6 1995. Institut National des Télécommunications.

20

TDL: an ASN.1-based tests description language

Paul KABORE, André SCHAFF
CRIN & INRIA Lorraine, Bâtiment Loria, B.P. 239,
54506 Vandoeuvre-les-Nancy Cedex France, Tel: 83.59.20.48,
Fax: 83.41.30.79, e-mail: {kabore,schaff} @loria.fr

Abstract
We present in this paper a language to describe tests for ASN.1 data structures. The main problem in test data generation is to derive from abstract data structures consistently inputs which are effective to find faults in the system under test. The cause is twofold. First, semantic information is not included with the ASN.1 declarations. Second, since the test input space is generally infinite, we must relay on external statements of the properties of the tests we want to use. The intention of this paper is to formalize semantic informations description in ASN.1 and to define a formalism which allow the description of test data. A test description language is then proposed to bridge the large gap between ASN.1 specifications and actual test inputs.

Keywords
ASN.1, tests description, test data generation.

1 INTRODUCTION

ASN.1 [ISO-8824 90, ISO-8824.1 92] (Abstract Syntax Notation One) has been standardized to define the data structures of Protocol Data Units. Techniques have been developed that generate test sequences for the behaviour part of communicating protocols. These test sequences attempt to reveal the presence of specific faults during testing [Bochmann 91]. One of the most difficult parts of testing is the actual generation of test data, which has traditionally been done by hand. The work described in this paper is a first attempt to automatically generate test data following the fault based testing techniques. Basically test data are created by setting parameters in the PDU structure defined in ASN.1. As the complexity of protocols increases, we would like to create these test data automatically or semi-automatically using a test data generator tool. However, parameters included in the PDUs have special meanings and informations about these parameters are described in natural language which is not suitable for computer processing. Thus the problem is how to generate adequate test sets of finite size with respect to a system under test and effective in identifying faults, given the following problems:

1. since formal description of parameters setting conditions is missing, how to automatically generate test data which fulfill these conditions;
2. the input space is generally too large and often infinite, so which range of data must be generated to allow a decision on the reliability of a system under test in a reasonable limit of time.

In most testing environments, problem 1 is solved by getting the test executor to set the parameters interactively. Problem 2 is caused by the nondeterminism in the choice of ASN.1 values similar to nondeterminism in specification languages: the possible choices of a value which obey some ASN.1 type may be too large and it would be very hard to decide which instances can be selected. Some methods have been proposed in software engineering to generate test data. These methods include random tests data generation ([Myers 79],[Duran 84]), partition testing [Ammann 93] ... Random testing should be adequate when a fault-prone region is uniformly distributed across the complete input space. Performance of partition-testing depends on the partition criteria. We propose a formalism which allows test designers to describe test data and then their expertise to an automatic test data generator. Hence, we present a formalism to specify constraints on ASN.1 data structures and a Test Description Language, TDL, which uses this formalism to specify test data. A test description consists in partitioning the input space using constraints.

The remainder of the paper is organised as follows: in the next section we describe the constraints specification formalism. Then we use this formalism to define in section 3 the Tests Description Language. Finally a conclusion is provided in section 4.

2 CONSTRAINTS

This section deals with an extension to the ASN.1 description language with constraints. The objective of this extension is to formalize the informal description of the PDUs parameter settings conditions. Some similar works have been made in [Varvitsiotis 93] and [Van de Burgt 89]. Our constraints definition formalism can be viewed as an extension of these works and we follow the constraints definition syntax of ASN.1 92 [ISO-8824.3 92]. In particular, we introduce operators into the ASN.1 language. These operators have been introduced for the following reasons:

1. specification languages generally use operators to describe data types. When we would like to use ASN.1 as a specification language, it would be necessary to use operators over ASN.1 types (e.g in GDMO [ISO-10165.4 92, Bapat 93]);
2. to describe tests for a given ASN.1 data structure, it is useful to be able to specify some test data by combining specific components of this structure using operators.

ASN.1 92 allows sub-type constraints and general constraints definition [ISO-8824.3 92]. However, the sub-type constraints definition formalism is not general enough (e.g we cannot specify constraints which combine several ASN.1 structure components) and general constraints are still specified informally. Our aim is to formally specify the dependencies between components of an ASN.1 data structure. The constraints description formalism is used to specify dependencies describing either conformance of PDUs instances to their ASN.1 abstract structure or test data.

Definition 21 *Given an ASN.1 type T, the domain of T $\mathcal{D}(T)$, is the set of all ASN.1 values that obey T.*

We specify constraints by defining elements of $\mathcal{D}(T)$, i.e. sets of ASN.1 values. Sets are defined symbolically using predicates, one or more predicates per set. A predicate is a function over $\mathcal{D}(T)$ that returns TRUE or FALSE and denotes the set of values that satisfy the predicate (i.e. values for which the predicate returns TRUE). So, operators over constraints are the usual operators over sets such as the UNION, the INTERSECTION and the COMPLEMENT operators. Relational operators (=, <, <=, >, >=, <>) define the set {TRUE} or {FALSE} and are used over predicates. A constraint specification is an expression with enumerated sets or predicates as terms. Predicates are constructed on attributes as in [Varvitsiotis 93].

2.1 Attributes, predicates and actions

Given an ASN.1 value V of type T, V carries a set of attributes on which we can apply predicates and actions. Attributes carried by V can either satisfy the predicates or not. For a given attribute, we define a set of predefined predicates and actions shown in figure 1. Furthermore, complex predicates and actions can be constructed combining existing predicates (respectively actions) using logical operators (respectively mathematical operators). On figure 1, `X` denotes a value of some ASN.1 type, `encoded(X)` is an action that computes the encoded value of `X` according to some encoding rules (e.g BER [ISO-8825 87]). #`X` is the length of `X` if `X` is a STRING, SEQUENCE OF or SET OF value. `odd()` (respectively `even()`) is a predicate which returns TRUE if its argument is odd (respectively even). `pres(X)` and `abs(X)` indicates if `X` is present or absent for optional types. `X[i]` is the element at position `i` of `X`, for STRING or SEQUENCE OF values. The `sorted()` predicate supposes the existence of a total order over the domain of the ASN.1 type: lexicographic order over characters strings and natural order over integers and reals. `alphabet()` is an action that returns the set of characters or elements of its argument, if this argument is of STRING, SEQUENCE OF or SET OF type. The remainder of this section deals with complex predicates and actions definition. The section also presents the operators used to combine defined predicates and actions into complex ones. For convenience, predicates and actions are defined on types, which means that they have to be applied to any instance of that types.

Attributes	**Predicates**	**Actions**
Value	odd(x)⁽²⁾, even(x)⁽²⁾, uppercase(x)⁽³⁾, lowercase(x)⁽³⁾ sorted(x)⁽³⁾ ⁽⁵⁾	x ⁽¹⁾ encoded(x)⁽¹⁾ x[i]⁽³⁾ ⁽⁵⁾
Length		#x⁽³⁾ ⁽⁴⁾ ⁽⁵⁾
Alphabet		alphabet(x)⁽³⁾ ⁽⁴⁾ ⁽⁵⁾
Option	pres(x)⁽⁶⁾, abs(x)⁽⁶⁾	

(1) Applicable to all types (2) Applicable to Integer type
(3) Applicable to string types (4) Applicable to Set Of type
(5) Applicable to Sequence Of type (6) Applicable to optional types

Figure 1 Attributes for ASN.1 values

We will use the pointing notation defined in [ISO-8824.1 92] to refer to a specific element of an ASN.1 structure:

- the notation `@`*component* refers to the element *component* of the outermost structure of an ASN.1 definition;
- the notation `@.`*component* refers to the element *component* of the innermost structure of an ASN.1 definition.

This notation has been extended as follows:

- the notation `@` refers to the current element;
- the notation `@`$component_1.component_2...component_n$ refers to the $component_n$ assuming that $component_1, \ldots, component_{n-1}$ are structured components.

For example, consider the following ASN.1 definition:

```
T ::= SET {b REAL,                      -- 1
           c SEQUENCE { b BOOLEAN,      -- 2
                        e INTEGER       -- 3
                      }
          }
```

If we write @ or @b on line 1, that points out the component b of this line. On the other hand, to refer to the component b of line 2, we can use the notations @, @.b, or @c.b, written on this line. Writing @b on line 2 or 3 refers to the component b of line 1. The notations @.b and @c.b used on line 3 points out the component b of line 2.

2.2 Constraints specification

The ASN.1 sub-typing mechanism allows the refinement of a parent type. For example, a sub-type can be defined for an OCTET STRING type by either constraining its alphabet or its size:

```
MyType ::= OCTET STRING (FROM ('1'|'2'|'3'))(SIZE(1..50))
```

Constraints are defined using intervals or enumerated sets of literal (integer values or characters). It is possible to combine set and interval expressions in a same constraint definition like this:

```
MyType ::= INTEGER ((5|10|15)|(20..100))
```

However, we cannot define set intersections, set exclusions and constraints using symbolic or computed values. For components of an ASN.1 structure, we would like to define dependencies between these components using suitable operators. For this purpose we have introduced operators in the ASN.1 language:

```
T ::= SET { a BIT STRING,
            b INTEGER (#@.a+10),
            c REAL (10..(@.b+#@.a)/2)
          }
```

This specification means that for an instance of type T, the component b must have a value equal to the length of component a plus 10. And the value of component c must be between 10 and the value computed from the formula (@.b+#@.a)/2. Dependencies between numeric attributes (integer values, string length, etc.) can be expressed using the usual mathematical operators: +, -, *, /, div, mod. Set intersections and set exclusions are expressed as follows:

```
T1 ::= INTEGER ((10..500) AND ODD())
T2 ::= INTEGER ((10..500) AND NOT (5|15|75|115))
```

For values of a STRING type, constraints can be expressed on their length as on numeric values shown above. Constraints can also be expressed using the alphabet() action to define dependencies between the alphabet of the values and alphabets of other components using the from and + (string concatenation) operators. For BIT STRING values, we can use the following additionnal operators: `&&` (bitwise logical *and*), `||` (bitwise logical *or*), `!` (bitwise logical *not*). For string values, we can also do some comparisons as well as define intervals assuming the lexicographic order:

```
T1 ::= SET { b OCTET STRING,
             c OCTET STRING (FROM (('A'..'Z') AND NOT ALPHABET(@.b))),
             d OCTET STRING (@.b+@.c)    -- string concatenation
           }

T2 ::= Set { b BIT STRING,
             c BIT STRING,
             d BIT STRING (!@.b && @.c)
           }

T3 ::= SET {
             a SET OF INTEGER (0..100),
             b SET (10+@.a) OF INTEGER,
             c SEQUENCE OF BIT STRING,
             d SEQUENCE (!@.c) OF BIT STRING
           }
```

Operators used on STRING, SEQUENCE OF or SET OF values are grouping operators and model iterations: these operators have to be interpreted as follows:

```
10+{1,2,3} = {11,12,13}
!{"1001001","1100"} = {"0110110","0011"}
{1,2,3} < {11,12,13}
```

Constraints on SET OF and SEQUENCE OF types are as in ASN.1 92 extended with operators.

```
T1 ::= SET {
        a INTEGER,
        b SEQUENCE,
        c SET (SIZE (@.a+#@.b) OF INTEGER (0..@.a+#@.b)
        }
```

General constraints are constraints which are defined and specified by designers. The syntax used in [ISO-8824.3 92] for general constraints is:

```
ASN.1 type (CONSTRAINED BY
                { -- enbglish text describing the constraints}
            !exceptionSpec
           )
```

So the general constraints specification is a special form of ASN.1 comments. No means are given to describe them formally. It is only stated that one can use a programming language for this purpose. We attempt here to formalize the informal description of general constraints. This formalization is based on constraints specifications as above. The difference from it is the use of conditional constraints, variables and a few more operators that allow the description of more complex constraints. Indeed, every constraint can be defined using the general constraint definition syntax. In sub-typing, the attribute constrained is generally implicit and in the general constraints syntax attributes are explicitly pointed out as we'll see later. Each constraint definition is introduced by the key word `constrained by` as in [ISO-8824.3 92]. Conditional constraints are defined as follows:

```
IF conditions THEN assertions FI
IF conditions THEN assertions ELSE assertions FI
```

where *conditions* and *assertions* have the same syntax. Examples of general constraints are shown below:

```
T1 ::= OCTET STRING (CONSTRAINED BY { variables:
                                        b,c;
                                      constraints:
                                        b FROM('a'),
                                        c FROM('b'),
                                        @ = b+c}
                     !exceptionSpec)


T2  ::= SEQUENCE {
        a INTEGER {un(10), two(20), trois(30)},
        b BIT STRING OPTIONAL
                (CONSTRAINED BY { IF @.a >= 10 THEN PRES(@) FI;
                                  IF @.a = 20 THEN #@.b <= 8 FI
                                }
                 !exceptionSpec),
        c BOOLEAN,
        d SEQUENCE OF INTEGER
            (CONSTRAINED BY { IF @.c THEN #@.d < #@.b FI;
                              IF NOT(@.c) AND @.a > 10 THEN @[2] = @.a*2 FI
                            }
             !exceptionSpec)
        }
```

Note that components are always referred using the pointing notation, while variables are not. Constraints are defined as disjunctions of conjunctions of constraints: if `a`, `b`, `c` and `d` are constraints, the definition

```
(CONSTRAINED BY {a,b,c;
                 a,d}
)
```

is equivalent to:

```
(CONSTRAINED BY {(a AND b AND c) OR
                 (a AND d)}
)
```

3 THE TESTS DESCRIPTION LANGUAGE

Since exhaustive testing is infeasible, the problem of selecting input data is the problem of guessing which tests will provide the most information about the tested system. The iterative aspect of tests production is a mechanical task. However, deciding which test inputs have the right properties requires ingenuity, experience and sometimes intuition. These informations cannot be provided by the ASN.1 structures. The tester's knowledge of potential pitfalls and boundary conditions within the implementation is a guide in that search. The Test Description Language allows test designers to provide a symbolic description of properties of the test inputs over which a given system should be tested. This section describes the TDL language, a test data description language for ASN.1 structures. This language uses the constraints description formalism introduced in the previous section.

3.1 Properties

TDL allows the description of test inputs in a structured and coherent way according to ASN.1 data structures. Test inputs are described symbolically with a collection of properties. A property describes a set of alternatives for some ASN.1 leaf type (e.g INTEGER, STRING, ...). The alternatives of a given property are exclusive choices. The idea behind the notion of properties is the mathematical equivalence classes. A property is a set of equivalence classes, each alternative of the property corresponding to an equivalence class. For some ASN.1 structure, a test template can be designed by selecting exactly one equivalence class from the properties associated to each leaf type of this structure. And a test input is obtained by choosing a representative of each equivalence class selected. So we can create one or many test inputs for a given test template.

A property describes some characteristics a test designer want a test input to have. These characteristics are naturally described using constraints. These constraints define a set of ASN.1 values and a test input must satisfy these constraints. The stronger are the constraints, the more precise are the test input described. We then define a notion of tests refinement: the most precise properties are defined using constants (i.e. ASN.1 values) and the less precise are the empty properties (no more restriction is made on the input domain than the one indicated in the standard document). Suppose the following ASN.1 structure:

```
MyType ::= Sequence {
             a INTEGER,
             b BOOLEAN
           }
```

If no properties are associated to components a and b, we have the possible input sets
`a` $\in$ `]-`∞`, +`∞`[` and `b` $\in$ `{TRUE, FALSE}`
Since this input space is infinite, the test inputs which can be chosen from this set may not be precise. Meanwhile, if we define the properties

```
PROP a {prop1 (100)},
PROP b {prop1 (TRUE)}
```

we would have one possible choice of input

```
{a 100,b TRUE}
```

We can weaken the constraints and then extend the properties as follows

```
(1)  PROP a {prop1 (100)}                            : 2 possible choices
(2)  PROP a {prop1 (1..100)}, PROP b {prop1 (TRUE)} : 100 possible choices
(3)  PROP a {prop1 (1..100)}                         : 200 possible choices
```

Figure 2 shows how tests are refined.

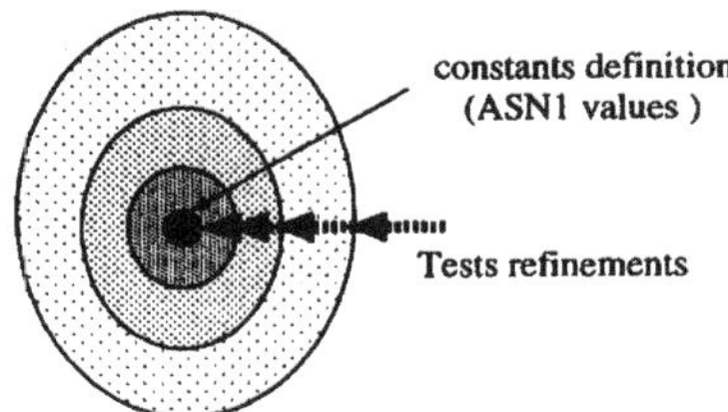

Figure 2 Tests refinement

Properties definition begins with PROP followed by the set of alternatives. Each alternative is given a name and its characteristics are described using constraints. The example properties shown above are defined using one alternative per property. An example with more alternatives is

```
PROP a {small(1..100)|middle(500..1000)|big(1000 ..<5000)}
PROP b {small((1..20) OR (60..100))|big(500..1000)}
```

A test template is constructed by choosing one alternative from each property definition, for example `{(a small), (b small)}` for the above definition. If for a given property, the number of test inputs that have to be created is equal to the number of its alternatives, each alternative can be chosen exactly once. However, if the number of the test inputs is less than the number of alternatives, we must choose an alternative in the set of alternatives. The key question is how to make a choice from a set of alternatives. In the previous definition, an alternative may be chosen at random with an equal probability of choosing any particular alternative. This may not be the best way of choosing alternatives. For many systems, we would like to choose some alternatives more often than others, because generally errors are not uniformly distributed among the alternatives (see figure 3).

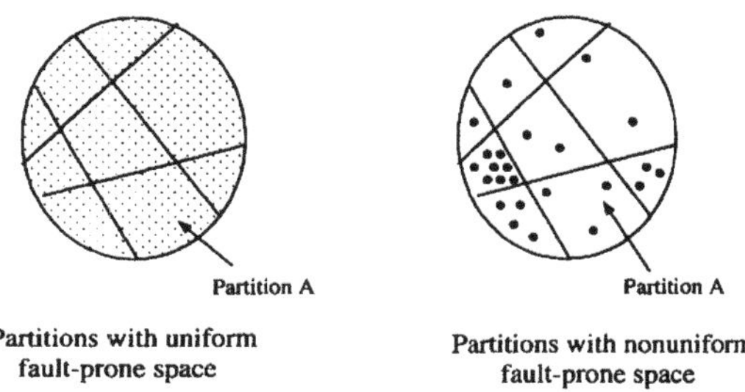

Figure 3 Partitions in a test space

To handle this problem, a solution is to allow weights to be associated to alternatives to force some to be selected more often than others. Indeed, much of the research that will go in the TDL description language concerns the identification of selection rules. Probably the most natural selection rule is specifying the probability of selecting a particular alternative. Probabilities model the repartition of potential faults among the alternatives. Suppose that for four faults detected, the probable repartition according to component a is two inputs with property *small*, one input with property *middle* and one input with property *big*. The test description must be as follows:

```
PROP a {petit%2(1..100)|moyen%1(500..1000)|grand%1(>5000)}
```

Properties for string components are described using the constraints definition formalism on string types, e.g constraints specified using the length and alphabet attributes. For example, we may define the following properties for a component s of OCTET STRING type:

```
PROP s { normal%2 (UPPERCASE(@[1]) AND #@ IN (1..100))|
         error%1 (LOWERCASE(@[1]) AND #@ IN (1..100))
       }
```

Moreover, we would like to describe tests with a particular distribution of the characters in a given string. This is particularly useful for generating tests for arithmetic circuits and network managed objects which model network resources. Some arithmetic circuits and then the managed objects which model these circuits are sensitive to the number of ones and zeros of a bit string. We then extend the manner we associate weights to alternatives to characters of strings. The following definition allows the generation of bit strings with a lots of ones and a few zeros:

```
PROP s {normal%2 (FROM(%25 '1'|%5 '0'))|...
       }
```

The most obvious use of weights is to guide the test generator to favour those test data that the designer feels will detect the most errors.

3.2 Tests description

A test description for an ASN.1 structure is a collection of properties of the components of that structure. A test description is assigned a name which may be the same as the ASN.1 type name. For example, for the ASN.1 type

```
MyType ::= SET { a INTEGER,
                 b BITSTRING
               }
```

we can define the following test

```
MyTest ::= { PROP a { small(1..10) | big(80..100)},
             PROP b { one(FROM ('1')) | zero(FROM('0'))}
           }
```

When defining tests, there are many cases that require the use of a value that must be computed before. Such a value must be generated by the generator in an intermediate stage. Variables allow test designers to do this. Suppose one wants to add another alternative to the above definition in order to create a test input for which the value of component b is an arbitrary number of ones followed by an arbitrary number of zeros. One must use variables as the following example illustrates:

```
MyTest ::= { Variables:
                 x,y;
               Properties:
                 PROP x { one(FROM('1'))},
                 PROP y { zero(FROM('0'))},
                 PROP a { small(1..10) | big(80..100)}
                 PROP b { one(x:one) | zero(y:zero) |
                          one_zero(x+y)}  -- concatenation of x and y
           }
```

The operation ":" models reusability of alternatives already defined in the same ASN.1 structure. To import alternatives, you indicate the name of the property followed by the set of alternatives. Suppose you have defined a property for component a as follows:

```
PROP a { one(1..10) | two(...) | three(...)}
```

You can import the alternatives defined for a to define alternatives for some component b belonging to the same structure, if a and b have compatible types:

```
PROP b { one (@.a:one) | two_three(@.a:one|@.a:three)}
```

So you can group many imported alternatives into one alternative that is the union of these alternatives. It is possible to refine the imported alternatives using the grouping operators (since an alternatives is "set of" values) as follows:

```
PROP b { one (@.a:one+10) | ...}
```

which is equivalent to:

```
PROP b { one(11..20) | ... }
```

Refinement operators are operators over alternatives. To import alternatives defined in another test description you use the "::" operator:

```
MyTest::Myprop
```

which imports all the alternatives defined for property MyProp defined in the test MyTest. You can select the alternatives of Myprop, group some of them into one alternative using the operator ":" as shown above. Operators ":" and "::" are syntactic operators and allow reusability, extention and refinement of tests descriptions.

Dependencies are defined using mathematical or other operators as we have seen in section 2. If you want to define a test data which is the combination of values of other components, in order to specify dependencies of some PDUs parameters or to test some particular combinations, for the ASN.1 structure

```
MyType ::= SEQUENCE { a [0] INTEGER,
                      b [1] INTEGER OPTIONAL,
                      c SET {d BIT STRING,
```

```
                              e BOOLEAN
                          }
                 }
```

you can define it as follows:

```
MyTest ::= { PROP a {small(1..100) | big(500..1000)},
             PROP b {first(@.a+100) | second(ABS(@))},
             PROP c { PROP d { first(IF PRES(@.b) THEN #@ IN (1..50)) |
                               second(IF ABS(@.b) THEN #@ IN (50..100))
                             },
                      PROP e { ... }
                    }
           }
```

4 CONCLUSION

Test descriptions are an important intermediate representation between ASN.1 abstract data structures and actual test data. In general, it is desirable to know the extent to which a test description can be derived from an abstract data specification and the extent to which a tester must rely on information external to the specification. In this paper we have helped to answer this question by presenting a mechanism that will help the tester when creating test data from ASN.1 abstract specifications.

It is unreasonable to expect the derivation of test data from abstract structures without explicit statements of the properties of the tests we want to create. Thus, the role of tests description as identified here is to relieve the burden of routine tasks and free the test engineer to concentrate on other areas, e.g properties he wants to test. Tests description can be useful in early phases of the design process to identify interesting properties of the ASN.1 specification. And we hope to provide test creators with as much automated support as possible to derive consistently test data. In future work, we will try to provide selection procedures that designate which combinations of alternatives should be tested and representatives test inputs for each combination.

REFERENCES

[Ammann 93] Paul Ammann et Jeff Offut. *Using formal methods to mechanize category-partition testing.* Techreport no. ISSE-TR-93-105, George Mason University, Virginia, USA, 1993.

[Bapat 93] Subodh Bapat. *Richer semantics for management information.* Integrated Network Management III, pages 15–28. *Elsevier Science Publishers B.V.*, 1993.

[Bochmann 91] G. v. Bochmann, A. Das, R. Dssouli, M. Dubuc, A. Ghedamsi et G. Luo. *Fault Models in Testing.* J. Kroon, R.J. Heijink et E. Brinksma, éditeurs, *Protocol Test Systems* IV, pages 17–30, October 1991.

[Duran 84] J.W. Duran et S.C. Ntafos. *An evaluation of random testing.* IEEE Trans. on SE, SE-10(4):438–444, 1984.

[ISO-10165.4 92] ISO, *Structure of Management Information - Part 4: Guidelines for the Definition of Managed Objects*, IS, ISO-10165.4, January 1992.

[ISO-8807 87] ISO, *LOTOS: A Formal Description Technique based on the Temporal Ordering of Observable Behaviour*, Draft International Standard, ISO-8807, June 1987.

[ISO-8824 90] ISO, *Specification of Abstract Syntax Notation Number One (ASN.1)*, International Standard, ISO-8824, December 1990.

[ISO-8824.1 92] ISO, *"Abstract Syntax Notation Number One (ASN.1) - Part 1: Specification of Basic Notation"*, Draft International Standard, ISO-8824.1, december 1992.

[ISO-8824.3 92] ISO, *"Abstract Syntax Notation Number One (ASN.1) - Part 3: Constraint specification"*, Draft International Standard, ISO-8824.3, december 1992.

[ISO-8825 87] ISO, *Specification of Basic Encoding Rule for Abstract Syntax Notation Number One (ASN.1)*, International Standard, ISO-8825, december 1987.

[Myers 79] G. L. Myers. *The Art of Software Testing.* John Wiley & Sons Inc., 1979.

[Van de Burgt 89] S.P. Van de Burgt et P.A.J. Tilanus. *Attributed ASN.1. FORTE'89*, pages 298–310, December 1989.

[Varvitsiotis 93] A.P. Varvitsiotis et G.I. Stassinopoulos. *Extending asn.1 into a full-fledged constraint language in the context of osi protocol conformance testing.* Computer Networks and ISDN Systems, 25(11):1243–1263, June 1993.

21
Report on the standardization project "Formal methods in conformance testing"

Dieter Hogrefe, Stefan Heymer
Institut für Telematik

Universität Lübeck
D-23538 Lübeck
{hogrefe,heymer}@itm.mu-luebeck.de

Jan Tretmans
Tele-Informatics and Open Systems group
Dept. of Computer Science
University of Twente
NL-7500 AE Enschede
tretmans@cs.utwente.nl

Abstract

This paper presents the latest developments in the "Formal Methods in Conformance Testing" (FMCT) project of ISO and ITU–T. The project has been initiated to study the role of formal description techniques in the conformance testing process. The goal is to develop a standard that defines the meaning of conformance in the context of formal description techniques. We give an account of the current status of FMCT in the standardization process as well as an overview of the technical status of the proposed standard. Moreover, we indicate some of its strong and weak points, and we give some directions for future work on FMCT.

Keywords

Conformance testing, formal methods, standardization.

1 INTRODUCTION

Formal Methods in Conformance Testing (FMCT) is a project initialized by ITU–T and ISO to study the meaning of conformance in the context of formal description techniques such as SDL, LOTOS, and Estelle. What does it mean that a product conforms to a specification given in such a formal notation? From time to time status reports have been provided at conferences and workshops. The last one has been given in [Hog95]. Please consult [Hog95] for a general introduction to the subject. FMCT is coming to an end now, and it is time to review the current developments.

Given a formal specification, how can one find out whether an implementation conforms to this specification?

The motivation of this activity is the existence of formally described standards and therefore the need to define the meaning of conformance with respect to formal specifications. As the use of formal methods in conformance testing is a very general problem in software engineering, its principles should be applicable to any application area, in particular any application area in telecommunications, e.g., OSI, ODP and IBCN.

Since the Conformance Testing Methodology and Framework (CTMF) [ISO91] defines conformance in a very general, informal sense, it serves as a natural basis for the more formal work on FMCT. By nature there is a close relationship between CTMF and FMCT. In particular, the concepts and their definitions should be compatible between the two documents.

The FMCT project is currently maintaining two working documents:

- Framework for FMCT (CD 13245-1, [ISO95b])
- Guidelines for FMCT (CD 13245-2, [ISO95a])

The framework gives some general information about FMCT and defines the basic concepts and terminology. Annex A, which will be incorporated into the standard, but which will be informative, explains the relationship between the FMCT concepts and the existing FDTs Estelle, LOTOS and SDL. The annex explains, for example, the concept of the implementation relation with respect to the three languages and gives an interpretation of PICS proforma and PICS. The annex will also contain a tutorial.

The guidelines, which will be published separately, deal with test generation methods as they have been developed for finite state machines and other models. A common example, the INRES protocol and service [Hog92], is used to show the applicability of existing test generation methods to an OSI protocol specification. INRES is not a real protocol, but it contains many features of OSI protocols. It is an abridged version of the Abracadabra protocol used in [ISO90] for illustration.

2 CURRENT STATUS

The work on FMCT is done in close cooperation between ITU-T and ISO/IEC. The cooperation was achieved by running a collaborative team. This cooperation proved to be very fruitful and gathered resources from both ITU-T and ISO in order to progress a single project. The "Framework for FMCT" underwent a ballot within ISO until mid of April 1996.

The results of this ballot were forwarded to an editing meeting for the ISO committee draft scheduled for mid May 1996 in Kansas City. The results are the following:

- Out of 23 national bodies that were members of SC21, 11 bodies voted.
- Of these 11, there were the following votes:
  - Abstain: 2
  - In Favour: 8
  - Against: 1

The main causes for the "against" vote (other than typographical errors) were uncertainty about the practical relevance of the proposed standard (cf. section 4 on the weakness of FMCT), and that some of the necessary preconditions for the formal treatment of conformance testing seemed to be dubious. On an editing meeting inside ITU–T in Geneva in April 1996 it was noted, that a tutorial on the use of FMCT would help this. Annex A of the FMCT document could be extended in such a way. This could also help in terms of a wider acceptance of this work.

It was also noted on the same meeting that the relationship between CTMF and FMCT has changed over the study period from formalizing CTMF to a more independent point. Many points in FMCT are not related to CTMF anymore.

Once the comments given with the "against" vote are disposed (i. e. resolved), the Committee Draft on FMCT (document number CD 13245–1) shall be accepted by ISO. It will then be put on ballot as a Draft International Standard (DIS). In the mean time the document will put forward as draft recommendation Z.500 to ITU–T. The current project planning aims at a recommendation and international standard, respectively, in 1998.

The project planning for the "Guidelines for FMCT" is shifted by one year with respect to the framework.

Beyond the work on testing done during the study period from 1993 to 1996, work on verification has been started and will be carried on in the near future.

The editing meeting in Geneva in March 1996 discussed intensively the issue of interoperability testing. It was noted that interoperability testing seems to fit into the FMCT framework. This can be seen in figures 1 and 2, where figure 1 shows the situation encountered in conformance testing, while figure 2 shows the situation for interoperability testing. In both figures *IUT* is the implementation under test, *UT* is an upper tester, and *LT* a lower tester. For interoperability testing additionally the specification *Spec* is taken into account. In both figures the dashed lines enclose the part of the combined specifications for which test cases could be generated.

3 TECHNICAL STATUS OF FMCT

The FMCT framework is presented at a high level of abstraction, e.g., it abstracts from specific test generation algorithms, even from a specific formal description technique. The framework defines terminology, abstract concepts, and minimal requirements on, and relations between these concepts. The main ingredients of the framework are concerned with how to define conformance of an implementation with respect to a formal specification [ISO95b, section 6], how to reason formally about testing concepts, such test architecture and test execution [ISO95b, section 7], and how to use testing concepts to test for conformance [ISO95b, section 8]. These three concepts are briefly described, without being complete; a complete description can, of course, be found in the standard itself.

Conformance Conformance concerns the definition of what is a correct implementation of a formal specification. FMCT only defines what is necessary to define conformance in a formal setting, not how conformance is defined in a particular case. To define conformance

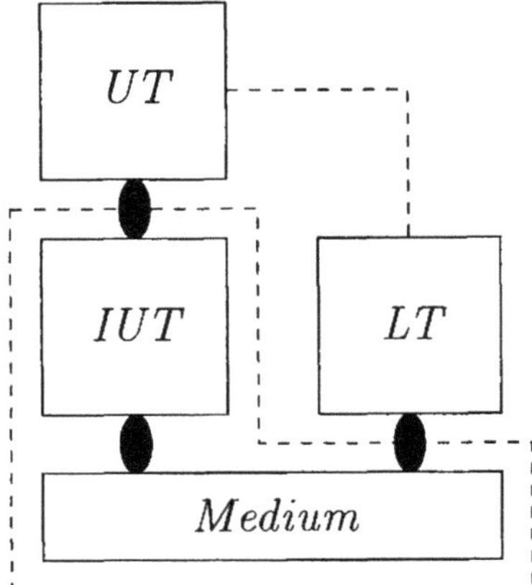

Figure 1: Test architecture for conformance testing

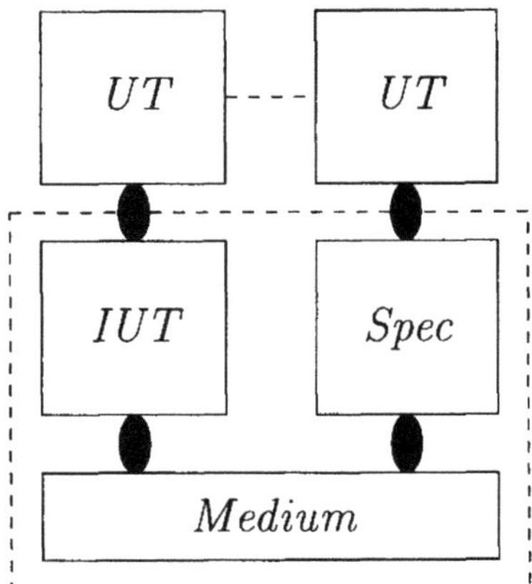

Figure 2: Test architecture for interoperability testing

it is assumed that there is a formal specification of required behaviour, $s \in SPECS$, where $SPECS$ denotes the formal description technique at hand, and a concrete implementation under test $IUT \in IMPS$, where $IMPS$ is the universe of all possible implementations. In order to reason formally about the concrete, physical implementations, it is assumed that each implementation can be modelled by a formal object m_{IUT} in a formalism $MODS$, which is referred to as the universe of models. This hypothesis is referred to as the test assumption. Conformance shall then be expressed by an implementation relation between formal models of implementations and specifications: $\mathbf{imp} \subseteq MODS \times SPECS$.

Testing Testing consists of performing experiments, specified in test cases, against an implementation under test, and observing its responses. Testing occurs in a test architecture, which is an abstract description of the environment in which the IUT is tested. It describes the relevant aspects of how the IUT is embedded in other systems during the testing process, and how the IUT interacts with these systems. The embedding systems are called the test context, and it is formalized as a function $\mathcal{C} : MODS \rightarrow MODS$ mapping behaviour of the IUT to behaviour of the IUT in its test context: $\mathcal{C}(IUT)$.

Test cases are specified in a test notation, referred to as $TESTS$, and a set of them is called a test suite. The process of running a test against a concrete implementation is called test execution. Test execution leads to an observation in a domain of observations OBS. Each observation is interpreted by assigning a verdict to it: $verd_t : OBS \rightarrow \{\mathbf{pass}, \mathbf{fail}\}$. An $IUT \in IMPS$ **passes** a test suite $T \subseteq TESTS$ if test execution of all its test cases leads to an observation with verdict **pass**.

A function $exec : TESTS \times MODS \rightarrow OBS$ is required, which models test case execution, i.e., given a test case $t \in TESTS$ and a model of an implementation $m \in MODS$, $exec(t, m)$ calculates the observation in OBS that results from executing t with the model m. By relating the observations made during the real test execution of the IUT with the tester to the observations calculated with $exec$, it can be concluded whether the model of IUT is in the subset of models for which a **pass**-verdict is calculated. This subset is called the formal test purpose, and it is denoted as P_t:

$$\begin{array}{lll} \text{let} & P_t =_{def} \{ m \in MODS \mid verd_t(exec(t,m)) = \mathbf{pass} \} \\ \text{then} & IUT \ \mathbf{passes} \ t \iff m_{IUT} \in P_t \end{array} \tag{1}$$

and moreover for a test suite T:

$$\begin{array}{lll} \text{let} & P_T =_{def} \bigcap_{t \in T} P_t \\ \text{then} & IUT \ \mathbf{passes} \ T \iff m_{IUT} \in P_T \end{array} \tag{2}$$

Conformance testing In conformance testing the notions of conformance, by means of the implementation relation **imp**, and of testing are linked. This means that a test-suite generation algorithm is required, such that each possible implementation IUT passes the generated test suite (if and only) if it is conforming. Such a test-suite generation algorithm is expressed as a function $gen_{\mathbf{imp}} : SPECS \rightarrow \mathcal{P}(TESTS)$; $(\mathcal{P}(TESTS)$ is the powerset of *TESTS*, i.e., the set of sets of test cases, so the set of test suites):

$$m_{IUT} \ \mathbf{imp} \ s \quad \underset{exhaustive}{\overset{sound}{\underset{\Longleftarrow}{\Longrightarrow}}} \quad IUT \ \mathbf{passes} \ gen_{\mathbf{imp}}(s) \qquad (3)$$

Since testing usually involves only searching for errors (the left-to-right implication in (3)), while it is never certain that all possible errors have been found (the right-to-left implication), FMCT requires that all generated test suites shall be sound.

To quantify the error-detecting capability of a sound test suite a fault coverage measure can be defined, which expresses the extent to which a sound test suite is exhaustive:

$$cov : \mathcal{P}(TESTS) \longrightarrow [0,1] \quad \text{satisfying} \quad P_{T_1} \supseteq P_{T_2} \text{ implies } cov(T_1) \leq cov(T_2) \qquad (4)$$

Moreover, a cost function can be specified to define the costs of generating, implementing, and executing a test suite:

$$cost : \mathcal{P}(TESTS) \longrightarrow \mathbf{R}_{\geq 0} \quad \text{satisfying} \quad T_1 \subseteq T_2 \text{ implies } cost(T_1) \leq cost(T_2) \qquad (5)$$

Apart from the items mentioned above, FMCT distinguishes between parameterized and instantiated specifications, it deals with implementation options, implementation conformance statement (ICS), and static conformance requirements, which state requirements on the sets allowable implementation options (the conformance aspects defined by means of an implementation relation are referred to as dynamic conformance). Moreover, FMCT defines fault models, and it presents several test-suite size reduction strategies: strategies to reduce the size of a test suite while preserving soundness. For details we refer to the (emerging) standard itself [ISO95b].

4 STRENGTH, WEAKNESS, AND FUTURE WORK

The high level of abstractness of the framework for FMCT is its strength, but at the same time it is also its weakness. Making use of this abstract framework requires instantiating its concepts with specific choices for the formal description technique (*SPECS*), for the set of models (*MODS*), for the implementation relation (**imp**), for test generation algorithms, etc. Requirements for such an instantiation are defined in [ISO95b, section 9]. The abstractness allows for very many different instantiations, making the standard applicable in many different areas, and making it easily adaptable to the specific needs of a particular testing problem. It also makes that it can cope easily with changes in technology and with evolving testing methods, since FMCT is not restricted to a particular specification technique, test generation method, or test-suite quality criterion, expressed as a coverage measure.

On the other hand, the abstractness and openness also mean that FMCT is not of direct help to daily testing problems. For any application of the testing framework it first needs to be instantiated, and this is not done in the present FMCT document. Annex A of FMCT presents possible instantiations for the standardized formal description techniques

Estelle, LOTOS, and SDL, but these instantiations are more intended to be illustrative than of practical value.

Hence, the most important next step in the standardization effort for formal protocol conformance testing is filling in the abstract framework with concrete formal description techniques, implementation relations, test contexts, test generation algorithms, test-suite size reduction techniques, coverage measures, etc. Some work in this direction has already been done: the document of Guidelines for FMCT [ISO95a] contains an list of current, state-of-the art techniques in protocol conformance testing, and how they could fit within the framework. This section will discuss some issues which arise when instantiating the framework with concrete implementation relations, etc., thus giving some directions for necessary future work on FMCT. The discussion is structured according to the three main concepts discussed in section 3: conformance, testing concepts, and conformance testing.

Conformance To define conformance by means of an implementation relation, first the specification language *SPECS* and the class of models *MODS* need to be determined. Actually, the precise language is less important than the semantic model on which it is based, as is also apparent from Annex A of FMCT: although many different formal languages exist, the ones that are of interest for protocol conformance testing can usually be expressed semantically in some sort of state machine. Nowadays, mainly two families of state machines prevail: Finite State Machines (FSM) (or Mealy Machines) and Labelled Transition Systems (LTS). The main difference between them is illustrated in figure 3: while an FSM has transitions labelled with pairs of (input,output), an LTS specifies interactions, which can, but need not be interpreted as input or as output. An FSM thus implicitly makes the assumption of synchronicity between input and output: every input or stimulus immediately results in an output or response (possibly the empty output), while in LTS each input or output has its own transition.

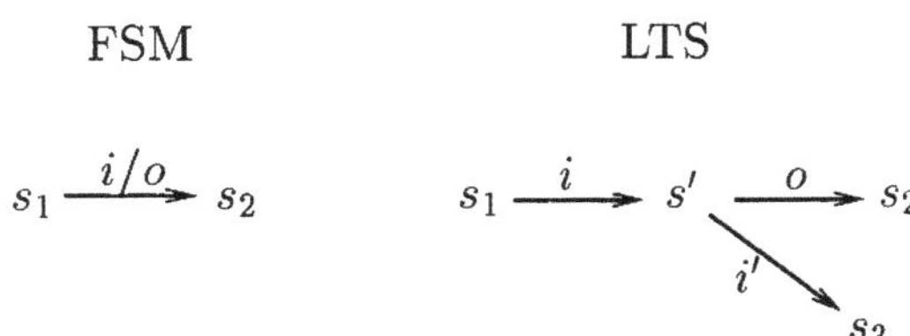

Figure 3: FSM and LTS transitions

Implementation relations for FSM, i.e., $SPECS = MODS = FSM$ and $\mathbf{imp} \subseteq FSM \times FSM$, are usually based on a some kind of relation between the states of the implementation and the states of the specification (isomorphism or homomorphism), where mostly the assumption is made that the number of states in the FSM implementation model is not larger than the number of states in the FSM-specification.

For LTS, i.e., $SPECS = MODS = LTS$ and $\mathbf{imp} \subseteq LTS \times LTS$, many possible implementation relations can be found in the literature, e.g., bisimulation equivalence, trace preorder/equivalence, testing preorder/equivalence, conformance **conf** etc. [Gla93]. Consensus exists that an implementation relation suitable for conformance testing, should be based on a reasonable notion of observation, i.e., it should be possible to detect a non-conforming implementation with a test and a 'reasonable' observation; an implementation relation should not require properties on an implementation which can never be tested with a 'reasonable' testing scenario. Since many of the above mentioned implementation relations were not developed with conformance testing in mind, but applications like verification, model checking, and theorem proving, not all of them are suitable for conformance testing, e.g., bisimulation equivalence is not testable with a realistic testing scenario [Abr87]. Although consensus consists about the non-suitability of bisimulation equivalence, no consensus exists yet about what should be the ideal implementation relation.

A property of the above mentioned relations on LTS is that they do not distinguish between inputs and outputs. Instead, they use abstract interactions for communication, where it is assumed that two communicating partners negotiate to establish an interaction which they will jointly execute. It is questionable whether such a communication scheme coincides with observations made of realistic systems, which usually do distinguish between inputs and outputs, as FSM models do. To overcome this problem, lately implementation relations have been considered for LTS that do distinguish between inputs and outputs, and where input actions are always enabled in implementations, such as the relations quiescent preorder [Vaa91], $R_1, ..., R_5$ [Pha94], and **ioconf** [Tre95].

Consensus exists that a (very) limited set of suitable implementation relations should be defined as a prerequisite for further development of FMCT, however, no consensus exists yet, which these implementation relations are.

Testing Like for *SPECS* and *MODS*, a choice must be made for the test notation *TESTS*. Many languages could be used, the best-known of which is TTCN [ISO91, part 3], however, lack of a completely specified formal semantics for TTCN hinders its application in a formal context. Extensions and modifications have been proposed [FH95, WP96], but they lack the general acceptance and tool support of TTCN. Either TTCN should be better defined, or a reasonable alternative should be proposed.

In defining its general testing concepts, FMCT makes certain implicit assumptions. One of these is that an observation corresponding to the calculated observation $exec(t, m)$ can always be made. This might pose problems, especially when test execution consists of multiple test runs, and when due to nondeterminism it is never certain whether all possible test runs have really been obtained. The current FMCT framework does not have the possibilities to deal with such uncertainties; more on this can be found in [HT96].

Conformance testing For conformance testing we need to generate some test cases. Again, FMCT is very general and abstract by considering test generation as a function from *SPECS* to $\mathcal{P}(TESTS)$ with the only requirement that the produced test suites shall be sound. No clue is given how to obtain such a test generation algorithm, or how to show its soundness.

Depending on the specification formalism and implementation relation many test generation algorithms have been presented in literature, see also the Guidelines document.

unfortunately, not all of them have been shown to be sound. Most test generation algorithms have been developed for FSM: Transition Tours, W-method, Distinguishing Sequences, Characterizing Sequences, Unique Input-Output Sequences (UIO) with some variants (UIOv, UIOg, ...), etc. Mostly they are restricted to work with deterministic FSM, and they can be proved to be sound and exhaustive if the additional assumption is made that the number of states in the IUT is at most equal to the number of states in the specification.

The traditional testing theory for LTS stems from the work on testing equivalences [DNH84]: Canonical Tester, CO-OP, etc., but these use the interaction-based communication scheme as explained above. More recent are test generation algorithms that use the distinction between inputs and outputs [Pha94, Tre95].

A problem in almost all algorithms is how to deal with data parameters in interactions, especially if the existence of transitions depends on predicates over these data. Only first, rather ad-hoc solutions are currently available.

Normative choices for test generation algorithms need, and should not be made, as they follow from choices for specification and modelling formalisms, an implementation relation, and an test architecture. In principle, they are the freedom of test-suite developers or test-tool developers. However, it would help if guidance were provided in how to develop test generation algorithms for the limited set of predefined implementation relations, and in how to show their soundness.

To compare the quality of test suites FMCT proposes to define a coverage function with, again, minimal requirements. In order to do meaningful comparison of test suites, based on well-established and standardized criteria, coverage measures need to be standardized. However, this appears to be a difficult issue, and active research is going on in this direction, which makes it difficult to establish consensus on a suitable definition of such a measure now. Heuristic approaches are expected to prevail for the time being.

Acknowledgements

The contents of this paper are not just the work of the three authors. The ideas were taken from the FMCT working document [ISO95b] which has been developed by the joint ITU/ISO project on FMCT. Numerous people were and are still involved in this work.

References

[Abr87] S. Abramsky. Observational equivalence as a testing equivalence. *Theoretical Computer Science*, 53(3):225–241, 1987.

[DNH84] R. De Nicola and M.C.B. Hennessy. Testing equivalences for processes. *Theoretical Computer Science*, 34:83–133, 1984.

[FH95] L.M.G. Feijs and M. Huizer. TSF: A test specification formalism. In A. Ponse, C. Verhoef, and S.F.M. van Vlijmen, editors, *Proceedings: ACP'95*, number 95-14 in Computing Science Report, pages 347–365, Eindhoven, The Netherlands, 1995. Eindhoven University of Technology.

[Gla93] R.J. van Glabbeek. The linear time – branching time spectrum II (The semantics of sequential systems with silent moves). In E. Best, editor, *CONCUR'93*, Lecture Notes in Computer Science 715, pages 66–81. Springer-Verlag, 1993.

[Hog92] D. Hogrefe. OSI formal specification case study: The INRES protocol and service, revised. Technical Report IAM-91-012, Update 1992, Universität Bern, Institut für Informatik und Angewandte Mathematik, Bern, Switzerland, 1992.

[Hog95] D. Hogrefe. Framework for formal methods in conformance testing. In T. Mizuno, T. Higashino, and Shiratori. N., editors, *Seventh International Workshop on Protocol Test Systems*, IFIP Transactions. Chapman & Hall, 1995.

[HT96] L. Heerink and J. Tretmans. Formal methods in conformance testing: A probabilistic refinement. In B. Baumgarten, editor, *Ninth International Workshop on Testing of Communicating Systems*. Chapman & Hall, 1996.

[ISO90] ISO TC97/SC21. Guidelines for the application of Estelle, LOTOS and SDL. technical report TR 10167, ISO, 1990.

[ISO91] ISO. *Information Technology, Open Systems Interconnection, Conformance Testing Methodology and Framework.* International Standard IS-9646. ISO, Geneve, 1991. Also: CCITT X.290–X.294.

[ISO95a] ISO/IEC JTC1/SC21 WG7. Information Retrieval, Transfer and Management for OSI, FMCT Guidelines on Test Generation Methods from Formal Descriptions, working draft. Committee Draft CD 13245–2, ISO, February 1995.

[ISO95b] ISO/IEC JTC1/SC21 WG7. Information Retrieval, Transfer and Management for OSI, Framework: Formal Methods in Conformance Testing. Committee Draft CD 13245–1, ISO, September 1995.

[Pha94] M. Phalippou. *Relations d'Implantation et Hypothèses de Test sur des Automates à Entrées et Sorties.* PhD thesis, L'Université de Bordeaux I, France, 1994.

[Tre95] J. Tretmans. Testing labelled transition systems with inputs and outputs. In A. Cavalli and S. Budkowski, editors, *Participants Proceedings of the Int. Workshop on Protocol Test Systems VIII — COST 247 Session*, pages 461–476, Evry, France, September 4-6 1995. Institut National des Télécommunications. Extended abstract of Memorandum INF-95-26, University of Twente, Enschede, The Netherlands, 1995.

[Vaa91] F. Vaandrager. On the relationship between process algebra and input/output automata. In *Logic in Computer Science*, pages 387–398. Sixth Annual IEEE Symposium, IEEE Computer Society Press, 1991.

[WP96] T. Walter and B. Plattner. Prospect – a proposal for a new test specification language and its implementation. In A. Cavalli and S. Budkowski, editors, *Eight International Workshop on Protocol Test Systems*. Chapman & Hall, 1996.

PART NINE

Test Practice

22
Network testing strategies for local mobility service

R. Pignatta, D. Turco, P. Marchese
CSELT
Via Reiss Romoli 274, Torino, Italy, Tel. +39 11 228 5555,
Fax +39 11 228 5069, E-Mail Roberto.Pignatta@cselt.stet.it
F. Corno
Italtel
Castelletto di Settimo M.se (MI), Italy, Tel. +39 2 4388 8708
Fax +39 2 4388 8245, E-Mail cornofab@ic8u82.settimo.italtel.it
G. Raneri
Telecom Italia
Via di Valcannuta 250, Roma, Italy, Tel. +39 6 3688 5690
Fax +39 6 3688 5727

Abstract

Cordless Mobility may be regarded as a new PSTN service which allows customers to use their private cordless terminals in a public domain area, typically the urban circle. New functionalities are required to switching systems and to the network itself, as a wireless access, Intelligent Network nodes availability, mobile call management, including location register and terminal authentication capabilities. The paper presents the testing approach chosen to verify in a laboratory environment a prototypal CM solution based on DECT access to Italtel UT100 exchanges and an IN infrastructure, highlighting the necessity of a comprehensive testing strategy (protocol conformance, end to end service testing, system performance evaluation) and new testing tools for network integration activities.

Keywords

DECT, interoperability, cordless mobility.

1 INTRODUCTION

In the international TLC community growing interest exists for Cordless Terminal Mobility (CTM) service. The service allows customers to use advanced cordless systems, such as DECT and CT2, not only in private/residential environments, but also in public or office domains, still preserving a personal profile. Several activities (e.g. within ETSI NA and RES3, Eurescom P507, DECT Operator's Group) and trials by most network Operators and switching manufacturers are in progress to investigate the best way to use radio technologies and the means of connecting cordless terminals into PSTN through Intelligent Network capabilities (see [1] and [2]).

In this framework, a preliminary solution for CTM or Cordless Mobility (CM), service is proposed by Italtel (see [2] and [8]), the UT 100 switching system Italian manufacturer. This solution is based on DECT access to local switches and improved IN capabilities. This paper focuses on describing the testing activities performed by CSELT, the corporate research centre of the STET holding, Italtel and Telecom Italia in order to verify, in a controlled environment (test plant), the correct operation of network nodes, wireless equipment and application programs.

2 A PRELIMINARY SOLUTION FOR CM IMPLEMENTATION

2.1 Network Infrastructure

The preliminary solution for a local mobility service proposed by Italtel (see [1], [2] and [8]) is based on the integration at Local Exchange (LE) level of a DECT access and the use of dedicated intelligent nodes (HDB, Home Data Base) as shown in the Figure 1. The solution relies upon the availability, in the current Italian Intelligent Network, of SSP (Service Switching Point) capabilities on all main access switches.

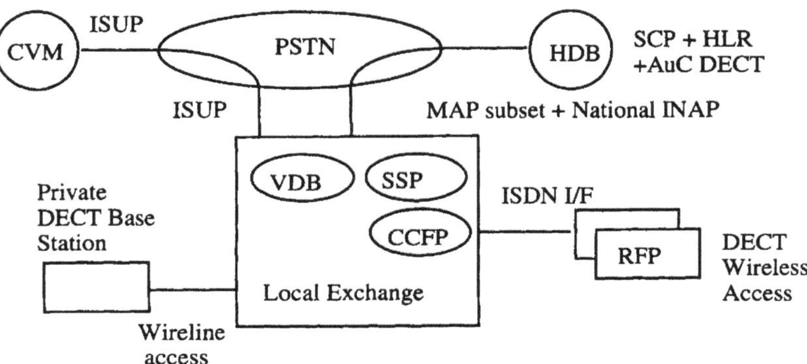

Figure 1 A reference network configuration for CM service.

DECT Radio Fixed Parts (RFPs) are directly connected to the switching system using standard copper pairs and a proprietary interface based on ISDN BRA. The exchange is upgraded with a number of new capabilities (referred to as CCFP, Common Control Fixed Part), including DECT (layers 2 and 3) protocol management, bit rate conversion from ADPCM 32 kbps to standard 64 kbps, remote powering of RFP, Visitor Data Base (VDB) functionalities. In the lack of international standards, a combination of Mobile Application Part (MAP) [9] and the national IN Application Part (INAP) is used to dialogue, using standard CCS7 connections, with the HDB.

The HDB is needed to provide both terminal mobility data and customer profiles. As such, it incorporates Service Control Point (SCP) capabilities, Home Location Register (HLR) capabilities and Authentication Centre (AuC) functionalities. HLR and AuC solutions are derived from the current GSM implementation, but need adaptations to cope with DECT access requirements.

The proposed network configuration may also include a Centralized Voice Mail (CVM) system, which unsuccessful calls to roaming users are directed to.

2.2 Basic Service Features

In the proposed solution, customers use a single portable Generic Access Profile (GAP) [4] DECT handset for both wireless (via public RFP) and wireline (via a private cordless base station) network access. As the customer switches on his handset inside the public radio coverage, or moves from one location area to another one, the handset automatically updates his position in the network by changing his location information in the HDB. When the user reaches the radio coverage of his private base station (normally placed in a residential or office environment), the terminal can be used as a standard cordless, with charging advantages.

The CM customer is associated to a single telephone number, regardless of the underlying access method (wireless or wireline) to the network. Number translation to the roaming number (used when the user is registered in the public domain), or to the wireline telephone number (used when the user is registered in the private home domain), is made by the HDB. In public environment, when the handset is off or out of radio coverage, incoming calls may be routed to a Centralized Voice Mail.

The proposed solution offers handover capabilities as well, to ensure voice connectivity when moving to new RFPs during conversation. In line with ETSI DECT GAP standards, both bearer and connection handover are fully supported.

Security aspects include standard DECT features as handset authentication (with a cryptographic challenge-response mechanism) and data confidentiality (by means of a key stream used for the encryption of data at DECT MAC layer).

The proposed service profile may comprise a number of additional services and features such as Centralized Voice Mail subscription and activation, call barring, call forwarding. They are offered using IN capabilities.

Simple Call Flow

A simple call flow is shown in the Table 1 below to demonstrate the signalling call processing for a mobile terminating call, involving no additional features, to a CM user registered at a public RFP.

In the following each item describe, in the right order, the signalling call procedures involving all network elements (RFP, HDB, Local Exchange, PT) for a mobile terminated call scenario.

1. The calling user dials the telephone number of the called user and the call is routed to the PSTN local exchange (SSP) via a IAM ISUP-S message.
2. The SSP recognizes the CM call and sends an INAP Provide Instruction query to the intelligent node HDB.
3. The HDB sends a MAP Provide Roaming Number operation to the visited Local Exchange (the exchange which covers the geographic area where the user is roaming in) and gets a Roaming Number (an E164 number, dynamically assigned by the visited exchange to reach the CM user) which is transferred to the SSP via an INAP Connection Control Connect operation.
4. The SSP routes the call to the visited local exchange using the Roaming Number as the terminating address.
5. The visited local exchange pages the user. When the CM user's handset is located, the local exchange authenticates the handset, activates ciphering procedures, completes the call setup and provides audible ring-back to the caller.

Additional messages are required if, for instance, the CM user doesn't answer to the incoming call (which results in a call forwarded to the Centralized Voice Mail).

Table 1 A simple Cordless Mobility call flow example

Calling party | PSTN Exch. | HDB | Local Exch. | PT

Incoming Call
--------------------->>

Provide Instructions (INAP operation)
--------------------->

Routing Request (MAP operation)
--------------------->

Response, with Roaming Number (MAP operation)
<---------------------

Connection Control (INAP operation)
<---------------------

Routing to MSRN
--->>

Paging (DECT)
- - - - - - - - - - ->

Paging Response (DECT)
<- - - - - - - - - - -

Ring back
<<--

Connection (DECT)
<- - - - - - - - - - -

Answer
<<--

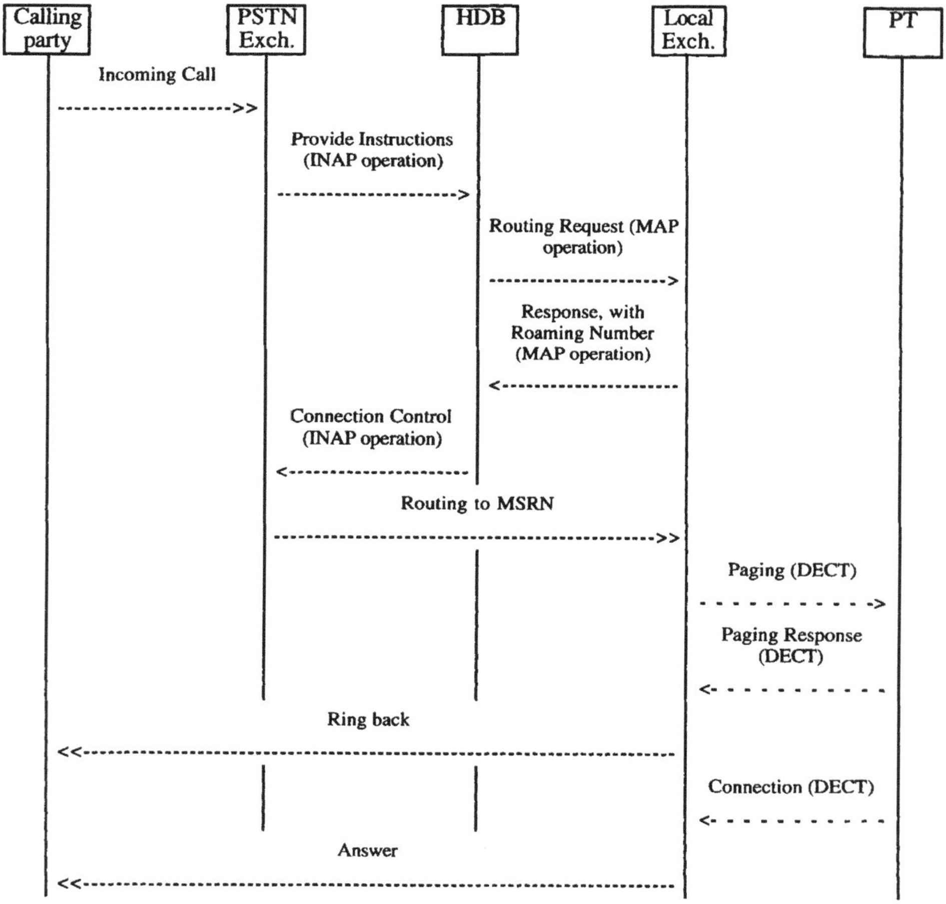

3 CORDLESS MOBILITY TESTING CHALLENGES

The verification, even in a prototypal environment, of a complex service as that described above, requires a comprehensive testing program to inspect each network element, in particular the RFP, the LE and the databases, ensure that systems would interoperate each other, and verify that the users could appropriately access the service using standard terminals. The approach to Cordless Mobility testing must cope with new challenges with respect to "traditional" switch based service testing needs, briefly summarized below.

1 Integration of wireless and wireline access to switching system.
The possibility to use the same terminal in both private (using a wireline network access) and public (using a radio network access) environment increases the complexity in service testing (mobility aspects, compatibility with subscribed service features, environment switch procedures).
2 Interaction between different protocols in the network.
Integration of Mobile Application Protocols (as GSM MAP) and IN Application Protocols (as the CS-1 like INAP protocol used in the Italy IN) in field applications needs appropriate test configurations. The proposed CM service, moreover, as briefly described above, includes interactions between INAP, MAP, ISUP and DECT protocols, with complex interactions which need comprehensive end to end verifications.
3 Use of newly released portable terminal.
The proposed Cordless Mobility service is based on the availability of DECT terminals with GAP profile. However, terminals are still prototypal and not completely certified as the relevant standards are not officially approved yet. This adds complexity in isolating problems in the network.
4 Introduction of new network systems (RFPs and DECT based HDB).
New network systems, as an HDB tailored to DECT authentication procedures, and equipments, as DECT RFPs, are necessary for the deployment of the service. In the lack of standards, this adds interoperability issues.
5 Lack of testing specifications.
Network solutions to offer Cordless Mobility are not yet consolidated, this implies the unavailability of test suites for network service features which have to be developed from scratch, using the service description. Regarding the DECT access, it must be remarked that the existing standardized (e.g. pr-TBR22 [3]) test suites are preliminary and need field verification.
6 Lack of adequate testing tools.
For DECT testing, a number of tools exist for physical level (radio) verification. However they are only partially adequate for network integration testing, which need upper layers monitoring and simulations, as well as terminal emulators and load traffic generators for RFP, HDB and LE in a laboratory environment.

4 TEST ACTIVITIES

To cope with the requirements of Cordless Mobility service testing, a test plan was agreed between Italtel, Telecom Italia and CSELT, in order to share the effort of network integration activities.

The consolidated expertise gained by the partners in IN and switch testing (see [5], [6] and [7]) was useful to formulate an adequate plan, whose objective was to prove the appropriate behaviour of the systems in a controlled environment, before eventual deployment of the nodes in the real network for field experiments.

The testing activities were performed, for a three month period, in test plants located in Italtel, Telecom and CSELT laboratories.

The major items covered in the plan included:

- DECT GAP protocol conformance testing;
- HDB CCS7 and OA&M testing;
- RFP functional testing;

- LE OA&M testing;
- end to end Cordless Mobility service testing;
- HDB verification under traffic conditions;
- LE verification under traffic condition.

Different test configurations were used, depending on the specific testing activity. End to end service verification required, as expected, the most complex testing configuration, shown in the following Figure 2.

For the in-lab testing activities, SSP functions were enabled on a separate UT Exchange (referred to as UT LE in the Figure 2), while CCFP and VDB capabilities on another node (UT DECT in the Figure 2).

Moreover, HDB capabilities were distributed in two physical entities: user profiles were managed by a mirrored national IN SCP node, while mobility management and authentication capabilities were managed on a dedicated "GSM like" HLR/AuC node upgraded with DECT authentication algorithms and CM user management. Without substantial changes to service profile requirements, this configuration added complexity to the tested network.

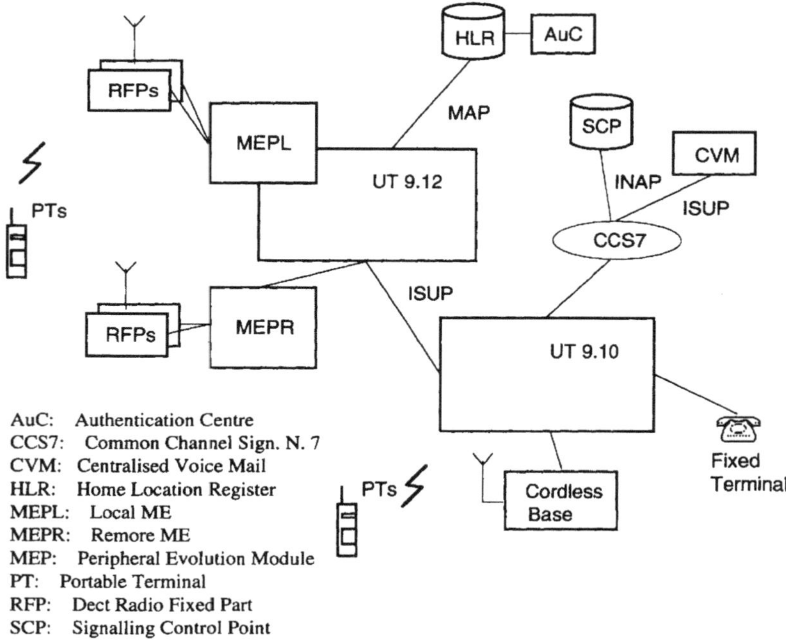

Figure 2 Test configuration for CM service integration testing.

4.1 DECT Protocol Testing

This activity, finalized to verify the conformance of DECT implementation on the network elements, was based on ETSI standard pr-TBR22 [3]. The objective of the TBR (Technical Basis for Regulation) is to ensure interoperability of DECT GAP implementations at the air interface and comprises type approval tests for both FP (Fixed Part) and PP (Portable Part). In particular, only the FP test cases with reference to Network Layer (NWKL) were used in the service validation test session, as the data link layer (DLC) and the layers below (Medium Access Control Layer MAC and Physical Layer PHL) were object of detailed system integration testing performed during the LE and RFP design.

Abstract Test Suite
The NWKL pr-TBR22 test suite for FP is an ISO 9646 conformant ATS (Abstract Test Suite) written in TTCN notation and is composed by 84 test cases. In accordance with the DECT features supplied by UT100 a test case selection was made. In particular the following NWKL features were considered:

- outgoing call;
- off hook;
- on hook;
- dialled digits;
- go to DTMF signalling (defined tone length);
- pause (dialling pause);
- incoming call;
- authentication of PP;
- location registration;
- on-air key allocation;
- identification of PP;
- alerting;
- encryption activation FP initiated;
- access rights;
- terminate access rights FP initiated.

About 50 test cases were selected and executed. The excluded 34 test cases were related to features not supported on UT100 LE (i.e. On-Air Key Allocation and Access Rights procedures). Extra test cases needed to be added for a better coverage of the IUT capabilities in terms of error handling (i.e. test cases conceived for UT100 behaviour verification with DECT NWKL layer messages containing wrong information elements) and DECT NWKL layer procedure coexistence, as required in [4].

Execution Environment
The basic test equipment chosen for DECT GAP protocol conformance testing was a commercially available modular platform for the development and realization of DECT Executable Test Suites (ETS). The tool, through a socket connection mechanism on UNIX environment, communicates with a DECT front-end, and is essentially composed of a test run manager that can be activated via a graphic user interface based on Motif. Moreover the testing system includes all the ETSI pr-TBR22 test cases executable form for DECT Network layer.

The platform supplies the executable form of ETSI pr-TBR22 ATS for DECT in an enhanced

'C' language. In particular, a macro has been defined for each TTCN primitive, in such a way that translation from TTCN to source code is simplified. After compilation, tests can be run directly on the tester and both trace and verdict are stored in a file.

The tests on UT100 LE were performed using a standalone network configuration as shown in the Figure 3.

In this network configuration, to allow optimal use of laboratory resources, instead of a real intelligent node, an HDB Emulator, developed by CSELT, supplied mobility management functions. The emulator reproduces the HDB behaviour at the HDB-VDB interface for a large set of scenarios, and can handle simultaneous transactions originated by up to 4 VDBs.

Moreover, a spectrum analyzer was used to monitor radio bearers during a DECT communication.

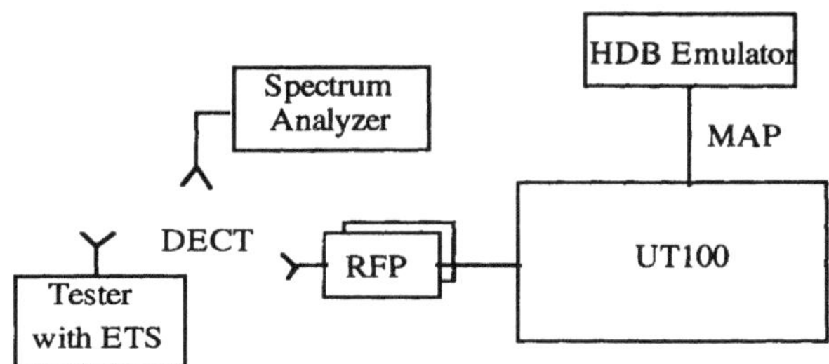

Figure 3 UT100 LE DECT protocol testing: network configuration.

The test session was useful to prove both the ATS and the ETS, at their first field usage. In particular some flaws were discovered on the ATS, promptly forwarded to ETSI through the test tool supplier. Moreover, as ATS coverage was considered not optimal, especially for protocol error handling, coexistence of CC (Call Control) and MM (Mobility Management) procedures, MM procedures nesting, additional test cases were added to the initial plan.

Solution for data link layer (DLC) testing

As DLC is deeply involved in connection handover procedure, a special testing configuration was set-up by the Italtel system integration department. This configuration, in order to assure a reliable and deterministic reproduction of handover and verify switch behaviour, bypasses the air interface by means of a RFP emulator directly connected to the RFP/LE interface. The DLC features under observation in this configuration were:

- basic data link capabilities;
- conformance to the relevant protocol standard (checksum calculation, fragmentation, retrasmission etc.);
- coordination with other functionalities in order to perform connection handover;
- evaluation and tuning of response time.

Only basic DLC capability testing was completed before the network testing took place, while the other specific features (i.e. DLC error handling), with a less impact to network wide functional verifications of all involved systems, were analyzed in parallel with end to end service test session.

4.2 HDB Testing

CM users' profiles are contained in the HDB, including also mobility management functionalities (Location Registration and Authentication). Careful attention was paid to the new capabilities required for mobility management with DECT access, while SCP functionalities were already verified in previous testing sessions for IN deployment. The HDB testing program covered CCS7 protocols, OA&M procedures and HDB evaluation under traffic conditions.

HDB CCS7 Protocol Testing

HDB dialogues with Visitor Data Bases (VDB) through a CCS7 protocol stack: MTP (Message Transfer Part), SCCP (Signalling Connection Control Part), TCAP (Transaction Capabilities Application Part), MAP (Mobile Application Part).

For MTP testing a Test Suite developed by CSELT in a CEC sponsored CTS project was adopted. The Abstract Test Suite (ATS), based on ITU Q.781 and Q.782 and recommendations, written in TTCN notation, includes over 300 tests, 170 of which applicable to the IUT; the Executable Test Suite was obtained using a tool which translates the TTCN MP code into the test source code of a commercial protocol tester.

SCCP and TCAP testing were based on two existing Test Suites, both entirely developed by CSELT using TTCN notation, already used for IN qualification activities. The SCCP Test Suite covers connectionless service functions as well as management aspects, and includes over 100 tests. As for TCAP Test Suite, it includes about 170 test cases.

As for Application layer protocol testing, with respect to the general objective to guarantee interoperability, it was considered sufficiently covered by end to end service testing and consequently no additional tests were specified and executed.

OA&M

The objective was to verify the procedures related to system administration and CM service management; in particular the following aspects were considered:

- configuration management;
- CCS7 management;
- system administration;
- HDB database administration;
- HDB performance statistics;
- event and alarm logging;
- HDB management;
- AuC administration;
- system failures;
- load and save procedures.

About 180 tests cases were specified in textual notation and performed in a stand-alone configuration using the user interface provided by the system under test. For some tests requiring background traffic (HDB performance statistics) a VDB traffic emulator (similar to the already cited HDB emulator) developed by CSELT on a commercial protocol tester was used.

HDB evaluation under traffic condition

HDB evaluation under traffic conditions focused on capacity evaluation for different traffic types, response delay at different traffic levels, impact of administrative activities on call processing and system behaviour under degraded system configurations.

Tests were executed to verify system capacity and CCS7 response time with respect to four different call scenarios: location registration, authentication parameters request, incoming call handling and a combination of all of them (traffic mix). In the lack of other data, two different traffic models were considered, one based on GSM recommendations, and the other based on provisional Cordless Mobility traffic forecasts: they mainly differed for Location registration activity and user's call frequency.

Regarding administrative tests, a number of operator's actions were examined under different traffic conditions, to evaluate the system administration time. As CM is conceived for a large residential market, administration activities as introduction/removal of users from the database, trace activation/deactivation, log files activation/deactivation could be demanded.

Robustness and stability tests allowed to determine the call loss rate and the HLR behaviour in overload conditions and during restoration procedures.

Figure 4 shows the test configuration. A MAP traffic generator and a protocol tester emulated each one a couple of VDBs. The generator was based on a software package commercially available for GSM testing, while the protocol tester, used for sample traffic on which to perform response delay measurements, was based on the cited VDB emulation package developed by CSELT.

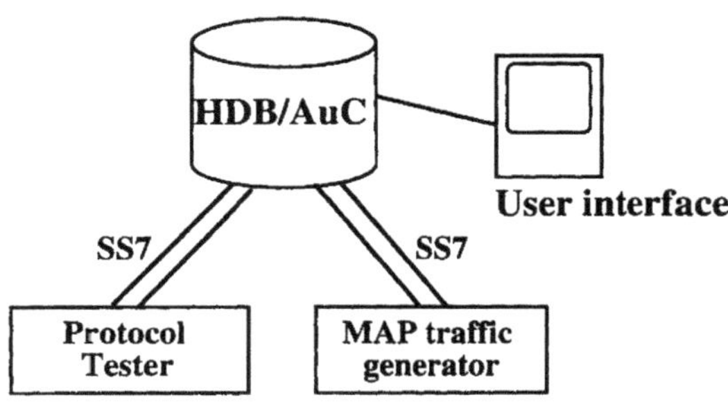

Figure 4 HDB performance test configuration.

4.3 RFP Protocol and Functional Testing

In the experimented architecture, the RFP manages the MAC protocol and controls radio communication with the portable part. In the upstream direction the RFP communicates with the local exchange through an ISDN like interface using a proprietary adaptation layer able to convey MAC abstract primitives on the physical interface. Through the same interface also all the information related to the RFP OA&M procedures are transported.

Due to the complexity of the RFP functionalities, within Italtel and CSELT, an ad-hoc group responsible for the validation of the RFP itself was created. In order to be able to distinguish any problems due to the LE from any problems due to the RFP, a completely stand alone test configuration was setup, with the development of a local exchange emulator and a portable part emulator. This approach was specially realized to reproduce, in a controlled environment, the following procedures:

- basic communication capabilities of the ISDN-like interface;
- downloading of the RFP software and configuration;
- RFP operational state control;
- radio and MAC layer capabilities.

The stand alone configuration for RFP testing was really useful in understanding and solving in a very short time the problems arisen during end to end service validation, allowing for parallel activities.

4.4 Local Exchange OA&M Testing

CM service required the development of a wide set of maintenance and operational procedures on the UT100 local exchange. Some of these new functionalities were tuned in a strict co-operation between RFP and local exchange testing engineers. About 300 test cases were specified and detailed, in a textual notation, on the basis of supplier's documentation, covering various OA&M aspects. In particular, the functions related with RFP software downloading, RFP configuration and RFP operational states management made use of a special testing configuration which adopted the powerful self monitoring capabilities active on both local exchange UT DECT and RFP.

The verification of the management of the new DECT related functionalities inserted in the existing UT100 architecture as charging record generation, traffic measurement control, system and network configuration was performed in the traditional step by step way, with special care in the measurement of the new synchronization capabilities necessary for handover scenarios.

4.5 End to End Service Integration Testing

Objectives and Methodology

Although essential to anticipate problems, standalone test activities does not always reveal all interoperability problems in the network. In fact, with the presence of many network elements (in this case PT and RFP DECT, UT100 LE, network databases HDB and SCP), manufactured by different suppliers, subtle variations in interpreting and implementing service specifications can originate interoperability problems. In addition, use of DECT access technique for public applications is relatively new and, in many cases (i.e. signalling procedures for handover), technical specifications are too loose and imprecise.

To verify system interoperability, an extensive activity was devoted by CSELT to end-to-end CM service integration testing. These tests are conceived as network wide functional verifications of all involved systems before their field application with real users. They are focused on the verification of interoperability between CM nodes, the complex interaction between the different network protocols (interworking between MAP, INAP, ISUP and DECT), and the analysis of DECT impact on the switch and user's service perception.

To achieve the objective, the following methodology has been adopted:

1. thorough analysis of service specifications to identify all relevant user's actions and behaviours, including unexpected events;
2. detailed analysis of all relevant network field configurations: RFP configurations, exchange set-up, location and handover domains;

3. cross analysis between the set of service scenarios and network field configurations, to derive the relevant test cases and choose the appropriate test plant configuration;
4. identification of a test suite structure, test purposes and test details;
5. identification of test suite execution methods at the test plant.

The early definition of the test suite and its discussion within the testing team, was proven very useful to integrate manufacturer's documentation for the LE, the RFP, the HDB, the DECT GAP terminal, highlighting ambiguities and anticipating interoperability problems.

Test Suite and Test Purposes (TP) specification

Based on the previously defined methodology, a Test Suite has been defined using traditional test description. Use of TTCN or other more formal methods was not considered as service specifications were not described in formal way and due to time constraints.

Each TP was specified using a test management tool developed by CSELT and called MITS (Multi Integrated Test System). Test specification layout includes a short test purpose description, some information about network nodes under test, execution environment configuration, test details and the expected final status. Test details include a detailed view of the expected information flow (application operations, messages and parameters) and events at the examined network and user interfaces.

The developed test suite was structured into two main test groups (Service Handling and User Procedures), and comprised about 250 test cases. Each test group was structured in subgroups covering the items indicated in Table 2:

Table2 Test group included in the Test Suite

Test group	Items
Service Handling	Mobility Management Call Handling Restoration Handover Centralized Voice Mail
User Procedures	Public/Private switch proc. Cordless usage

- Mobility Management, Call Handling and Handover: tests for mobility management, call control and handover verification procedures.
- Restoration: tests for network nodes (databases) out-of-service restoration procedures.
- Centralized Voice Mail: tests for verification of the correct behaviour of the centralized voice mail, when involved.
- User Procedures: tests for handset use verification in private environment and change of network access method proofs (form public to private and vice-versa) performed by the CM user.

The tests were mainly executed with manual procedures, using standard terminals and then comparing the results of monitoring tools at the various interfaces. For some tests, a prototypal DECT GAP terminal emulator provided by Necsy was useful to reproduce user's actions and force events. The emulator supports the following functions:

- monitoring of the radio interface with one or more RFP (Radio Fixed Part), with complete decoding capabilities;
- manual forcing of attach, handover and anomalous situations;

- file storage of collected data;
- displaying of radio-electric parameters and RFP info channels.

This test session was probably one of the most fruitful ones, as it was essential not only to verify the interoperability of all systems, but also how service could be perceived by customers, providing a number of useful suggestions to improve user procedures, timing, and messages on terminal display.

4.6 UT100 evaluation under traffic condition

Load testing activity aimed at obtaining the following main goals:

- to give a preliminary system performance evaluation of Italtel UT100 local exchange, focusing the attention on the specific subsystems which support wireless access capabilities, DECT call control, mobility management procedures and local data base;
- to observe behaviour and stability of the network, by simulating traffic conditions in the laboratory environment.

Tests reproduced traffic scenarios involving both mobile terminals and POTS subscribers, and various services offered to the subscriber, like the CVM in case of called subscriber not reachable, or the possibility to switch to the home-cordless environment.

The test plant network configuration was identical to that used for end to end service integration testing and shown in Figure 2.

The activity required the utilization, in the laboratory environment, of both portable terminals (GAP profile) and a new DECT automatic traffic generator called DTLS © (DTLS is a trademark of Necsy). The generator can perform location registration procedures and simulate originating/terminating call scenarios from/to a relevant number of DECT terminals. DTLS has a modular architecture, with a variable number of radio modules which simulate mobile terminals, and a variable number of modules emulating POTS subscriber lines. A control unit allows to program a testing session with the desired traffic mix, collects data and produces session reports. Useful data are available, like call loss rate, failure causes, call tracing, and so on. The session was also useful to improve the prototypal version of the tool and identify new requirements for more extensive test campaigns.

The activity allowed to identify problems related to traffic conditions, which affected the quality of service offered to the CM subscriber and helped in their resolution. Nevertheless, the complexity of service architecture, and the fact that neither terminals, nor traffic tools are completely certified gave rise to troubles in correctly finding out which subsystem causes malfunctions to happen, in order to help a timely deployment of solutions.

5 CONCLUSIONS

This paper has described the testing approach chosen for the validation of a CM solution, based on DECT radio access technique, integrated into the Italtel UT100 local exchanges. The objective was not only to gain the confidence that the service is technically robust and acceptable to end users, but also the assurance that CM service would not adversely affect normal PSTN network operations. A side objective was also to verify the adequacy of testing tools and procedures for CM service.

The CM testing approach must cope with new challenges with respect to "traditional" switch based service testing needs, involving various technical areas as: integration of wireless and wireline access to switching system, interaction between different and complex mobile and network protocols in the network, use of newly released portable handsets and introduction of new network systems. This requires a combination of testing strategies already in use for IN and GSM testing, but with an extra level of complexity in network integration.

To cope with CM service testing requirements, a complex test plant configuration was designed and set-up, combining both radio (DECT) and network (PSTN, IN, GSM) skills in an unique team and paying special care in highlighting interoperability issues. Different test configurations were then used, sometimes in parallel testing sessions. Some simplifications and a significant saving of time were possible by a strict cooperation with the Italtel system integration department, which avoided unnecessary test duplications. This interdisciplinary approach and the establishment of such a complete test environment were proven essential to achieve the testing objectives in the scheduled time.

Protocol testing was performed extensively only on a subset of the network elements, leaving to the end-to-end service validation tests the important task of verifying the entire CM network as a whole. In the lack of standards for CM service, a new test suite needed to be developed, while for protocol aspects preliminary TTCN ATS were used.

During the activity evolution, some critical aspects had to be tackled, as the lack of adequate coverage in the available ATS for DECT GAP testing specifications. As for testing tools, several of them had to be prepared with in-lab developments. In general, there is still a lack of adequate testing tools for DECT radio interface as traditional DECT tools are only partially adequate for network integration testing and more appropriate for radio measurements. This means that identifying, tracing and debugging problems related to DECT upper layers had to rely mostly on the experience and intuition of the operators. Some prototypal DECT radio interface monitor and simulator instruments were then procured and have been extremely improved as a result of this activity. Nevertheless, a number of key issues have to be further studied for more intensive testing activities as those which could be required by a large deployment of the systems in the network, in particular for DECT traffic generation at very high traffic rates (e.g. thousands of Busy Hour Calls), and automatic execution of end to end service tests.

6 ACKNOWLEDGEMENTS

The authors wish to thank the colleagues who have contributed to the testing activities and the revision of the paper, in particular Silvio Grande and Stefano Gollinucci, from CSELT, and Renzo dal Molin and Giuseppe della Valle from Italtel.

7 REFERENCES

[1] Eynard et alii "Local Mobility and Personal communication services: market aspects and network solutions", ISS 95.

[2] Faraci, Pietroiusti, Santagostino "IN based approach for the provision of PCS services using PSTN infrastructure", ICIN'94.

[3] ETSI RES3 T, prTBR 22.

[4] ETSI prETS 300 444 "Radio Equipment and Systems (RES); Digital European Cordless Telecommunications (DECT) Generic Access Profile (GAP)", ETSI, Sophia Antipolis, Valbonne, F, February 1995.

[5] E. Guarene, P. Marchese "Advanced testing tools for the Italian IN", IWPTS 1991, Leidschendam, NL.

[6] G.P. Fici, P. Marchese, S. Orlando "SCP performance testing and evaluation", ICIN'94, Bordeaux, F, October 1994.

[7] L. Dragoni, L. Gabrielli, P. Marchese "Intelligent Network Service Testing", ICCC Conference on IN "The path to global Networking", Tampa, USA, May 1992.

[8] R0912-REP-DECT (Italtel confidential documentation), Italtel, Castelletto di Settimo Milanese (MI), I.

[9] ETSI Recommendation GSM 09.02 v. 3.8.0 "Mobile Application Part Specification", ETSI, Sophia Antipolis, Valbonne, F, January 1991.

8 ABBREVIATIONS

ADPCM: Adaptive Differential Pulse Code Modulation
AuC: Authentication Centre
ATS: Abstract Test Suite
BHC: Busy Hour Calls
BRA: Base Rate Access
CC: Call Control
CCFP: Common Control Fixed Part
CCS7: Common Channel Signalling number 7
CM: Cordless Mobility
CS-1: Capability Set 1
CSELT: Centro Studi e Laboratori Telecomunicazioni s.p.a.
CTM: Cordless Terminal Mobility
CVM: Centralized Voice Mail
DECT: Digital European Cordless Telecommunications
DLC: Data Link Control layer
DTMF: Dual Tone Multiple Frequency
ETS: Executable Test Suite
ETSI: European Telecommunication Standard Institute
FP: Fixed Part
GAP: Generic Access Profile
GSM: Global System for Mobile telecommunications
HDB: Home Data Base
HLR: Home Location Register
IN: Intelligent Network
INAP: Intelligent Network Application Part
ISDN: Integrated Services Digital Network
ISO: International Standard Organisation
ISUP: ISDN User Part
IUT: Interface Under Test
LE: Local Exchange
MAC: Media Access Control
MAP: Mobile Application Part
MITS: Multi Integrated Test System on personal computer
MM: Mobility Management
MSRN: Mobile Station Roaming Number
MTP: Message Transfer Part
NWKL: NetWorK Layer
OA&M: OperAtion & Maintenance
PHL: PHisical Layer
POTS: Plain Ordinary Telephone Services
PP: Portable Part
PSTN: Public Switched Telephone Network
RFP: Radio Fixed Part
SCCP: Signalling Connection Control Part
SCP: Service Control Point
SSP: Service Switching Point
TBR: Technical Basis for Regulation
TCAP: Transaction Capability Application Part
TP: Test Purpose
TS: Test Suite
TTCN: Tree and Tabular Combined Notation
VDB: Visitor Data Base

23
Generation of conformance test suites for B-ISDN signalling relevant to Multi-party testing architecture

Jin Young Choi, Beom kee Hong
Electronics and Telecommunications Research Institute(ETRI)
161 Kajong-Dong, Yusong-Gu, TAEJON, 305-350, KOREA
E-mail : jychoi@testopia.etri.re.kr, bkhong@testopia.etri.re.kr
TEL : +82-42-860-5858, FAX : +82-42-860-5440

Abstract

In this paper, we introduce the development experiences of conformance test suites for B-ISDN signalling protocol Q.2931. We developed Abstract Test Suites for the user and network side respectively, and made them executable by designing and realizing a test system. In our procedures, we applied not only the Single-party testing but the Multi-party testing for each test suite. Particularly in the Multi-party testing for the network side, we made the test suite by applying the method to control and observe two ports of the IUT concurrently. We also verified the test suites by internal testing with reference systems. These are able to increase the possibility of the interoperability among the systems implemented on different environments.

Keywords

Conformance Test, ATS, ETS, Multi-party testing, B-ISDN

1 INTRODUCTION

A communication protocol is a set of rules that defines all possible interactions among communicating systems. As communication protocols are more complicated and require higher reliability, it is necessary for an implementation to be checked for conformity to its standard specification to ensure that such implementations communicate reliably. This procedure is known as conformance testing(Sarikaya, 1989).

Up to the present there have been some Abstract Test Suites(ATSs) for conformance testing of communication protocols such as X.25 used in packet network and Q.931 in N-ISDN. However, standardization activities for ATS of the B-ISDN protocol with which various multimedia services will be introduced are at an early stage. The ATS for B-ISDN protocol is not standardized yet except the ones for ATM layer and AAL protocol. This paper is concentrated on the generation of test suites for B-ISDN signalling protocol which defines call control procedures for end-to-end connection. Since this protocol is not symmetric for the user side and the network side, the ATS is needed for each side. The test methods considered for the development of the ATS should be also different for both cases. Particularly for the network side protocol, consideration of a different method rather than the Single-party testing which is used conventionally is needed, that is the Multi-party testing.

ISO/IEC 9646 deals with the conformance testing methodology, the development of ATS using Tree and Tabular Combined Notation(TTCN), the execution of test, the testing result analysis, and so on(ISO/IEC 9646-1 Information technology - Open System Interconnection - Conformance testing methodology and framework - Part 1,2,3, 1994). There are two possible conformance testing architectures according to the test configuration: the Single-party testing architecture and the Multi-party testing architecture. The Single-party testing can be applied when an Implementation Under Test(IUT) is required, by the test purpose, to communicate with only one real other system. The Multi-party testing can be needed when an IUT is required, to communicate with multiple other real systems concurrently by means of the same or different service providers. In practice in order to perform conformance testing, it is necessary to select a proper testing architecture depending on testing environmental restrictions.

In this paper we describe the methodology and architectures that we have considered for the development of Q.2931 protocol's ATS and some procedures and techniques that we have taken during the development of ATS and the implementation of ETS. In particular we realized the ETS over the platform developed by ourselves and verified some conformance testing functions. This paper consists of as follows. We describe characteristics of B-ISDN signalling protocol in section 2. In section 3, we address testing architectures that we have adopted for conformance testing, and the ATSs for user side and network side of Q.2931 protocol are described in section 4. Section 5 describes the implementation of ETS from the ATS. Finally, section 6 concludes the paper.

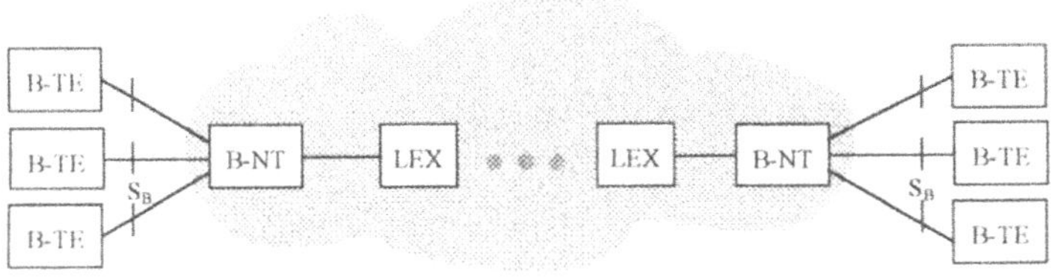

B-NT Network termination for B-ISDN
B-TE Terminal equipment for B-ISDN
LEX Local Exchange for B-ISDN

Figure 1 A reference configuration of B-ISDN.

2 CHARACTERISTICS OF B-ISDN SIGNALING PROTOCOL

B-ISDN is a digital communication network that utilizes broadband transmission and switching technologies to interconnect the concentrated or distributed subscribers and the service providers, and to support integrated services with wide-bandwidth distribution. Figure 1 shows a reference configuration of B-ISDN. It contains the following:

- Functional groups : B-NT, B-TE, and LEX
- Reference points : S_B.

B-ISDN is composed of several protocols such as physical layer, ATM layer, AAL, and high layer for signalling, and each layer has its own functions. Among these protocols, Q.2931 protocol is a B-ISDN Digital Subscriber Signalling System No. 2(DSS2) User-Network Interface(UNI) layer 3 specification for basic call/connection control(ITU-T Draft Recommendation Q.2931, 1994). This protocol specifies the procedures for the establishing, maintaining and clearing of network connections at the B-ISDN user network interface. The procedures are defined in terms of messages exchanged and applied at the interface between a B-ISDN terminal and a B-ISDN public network(reference point S_B and T_B coincident) as well as at the interface between a B-ISDN customer network and a B-ISDN public network(i.e. at the T_B reference point). The flow of messages for Q.2931 call/connection control is shown in Figure 2.

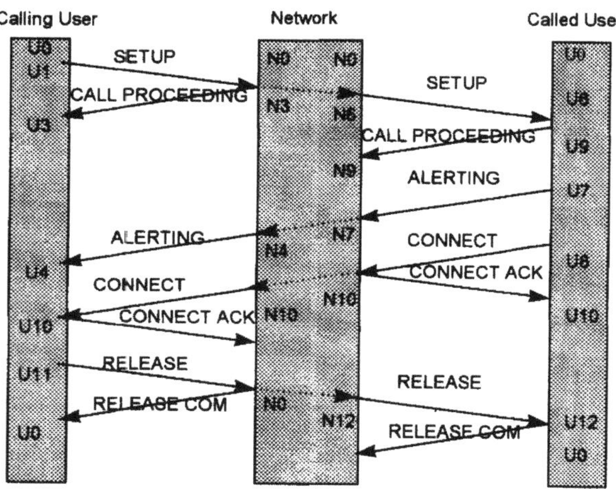

Figure 2 Flow of messages for Q.2931 call/connection control.

The main feature of this protocol is that the procedures for user side and network side are asymmetric. First of all, the behaviors of the calling user entity and the network entity of the destination interface are defined differently. Also the behaviors of the called user entity do not correspond with the behaviors of the network entity of the originating interface. For example, when the calling user sends a SETUP message to the network, the calling user enters the U1(Call Initiated) state. On the other hand, when the network sends a SETUP message to the called user, the network enters the N6(Call Present) state. In each case, the transition state and the reaction of the called user and the network are also different. A CALL PROCEEDING and ALERTING messages transmitted from the called user to the network as a response to the SETUP message are optional. Also some information elements included in messages are mandatory in the user-to-network or the network-to-user direction, but optional in the opposite direction. Therefore if we consider the conformance testing for Q.2931 protocol, it is obvious that a common test suite can not be applied for the user and network side. In particular, in case of the conformance testing of the network side, it is more meaningful to control and observe both interfaces of the network side simultaneously because in case of testing only one interface of the network side we should implement the UT function in the SUT or the called user system in the tester to play a role of corresponding counter part. Furthermore, the testing on the other interface should be performed again. On the other hand, if we control two interfaces in the tester at the same time, we can reduce those overloads and perform the testing more efficiently. This leads to consider a different testing architecture from the one of the user side. We describe the testing architectures adopted to the user and network side conformance testing in the next section.

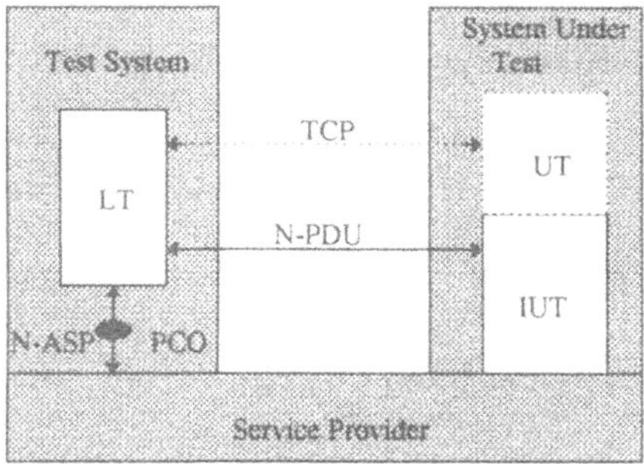

Figure 3 Applied Single-Party testing architecture.

3 TESTING ARCHITECTURES

In this section, we describe testing architectures considered for the development of Q.2931 user side and network side ATS(ISO/IEC 9646-1 Information technology - Open System Interconnection - Conformance testing methodology and frame work - Part 2, 1994).

3.1 Single-party testing architecture

We applied the Single-party testing architecture to the Q.2931 user side ATS configuration because it is sufficient to perform the test as exchanging messages between the test system and the IUT using only one connection. There are four methods in this architecture, and we decided to adopt Remote test method shown in Figure 3 as it could be applied to various systems flexibly manufactured by different vendors. In this configuration, provision is made for the case where it is not possible to observe and control the upper service boundary of the IUT. The test events at the LT PCO are specified only in terms of N-Abstract Service Primitive(N-ASP), and N-Protocol Data Unit(N-PDU). Some requirements for Test Coordination Protocol(TCP) and UT may be implied or informally expressed in the ATS, but no assumption shall be made regarding their feasibility or realization. The Remote test method can reduce the overloads to implement additional software or hardware for testing in the SUT. The IUT represents the Q.2931 user side entity and the System Under Test(SUT) can be B-TE or the user side of B-NT system.

3.2 Multi-party testing architecture

In case of intermediate systems, multiple protocols behave at multiple ports simultaneously. Therefore the Single-party testing method which was often used previously cannot be applied anymore. For the testing of these systems, it is required that the test method shall control and observe multiple ports at the same time, which can be achieved by the Multi-party testing.

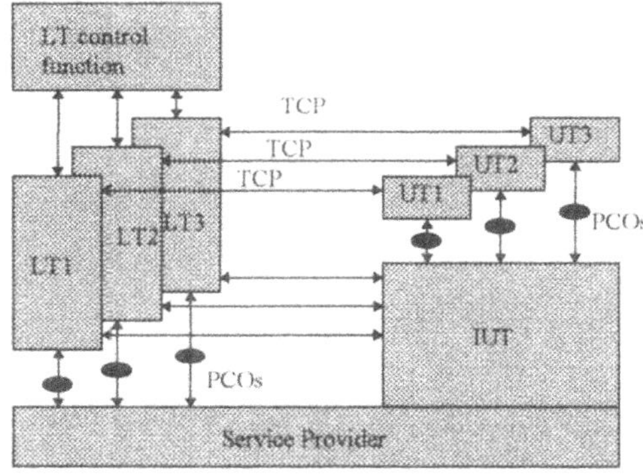

Figure 4 General Multi-Party testing architecture.

Q.2931 network side protocol running on an intermediate system behaves like those mentioned above by making a local call/connection. Therefore the testing must be performed as the procedure to stimulate an input to the IUT and to observe an output at the same port or other corresponding port. In case of Q.2931, since we consider only point-to-point configuration, it is sufficient to test only two ports in the IUT at the same time. Figure 4 shows the general architecture of the Multi-party testing, and Figure 5 is the applied testing architecture for the Q.2931 network side ATS configuration. This is also considering Remote test method. Two Lower Testers are located in the same test machine and the coordination function of those testers are more simplified. In this configuration, each LT represents one of the real systems with which the IUT needs to communicate. Each LT communicates with the appropriate part of the IUT, by observing and controlling ASPs and PDUs as in the Single-party testing. The UT functions are not specified nor are any assumptions made regarding their feasibility or realization. This can reduce the testing overload on the SUT. The IUT represents the Q.2931 network side entity having more than two ports and the SUT can be the network side of B-NT system or switching system. The logical call/connection from LT1 to LT2 through SUT forms a local call/connection between two users. That is, if the IUT is the network side of B-NT system, LT1 and LT2 perform the roles of terminal equipment B-TE.

If we apply the Single-party testing to the Q.2931 network side ATS configuration, the following overloads would be forced for the testing of the network side of the IUT. Also we think that the Multi-party testing architecture like this could be extended to the interoperability testing among systems implemented on different environments.

- Implementation of the UT functions in the SUT
- Implementation of the called user system in the tester to play a role of corresponding counter part
- Increasing of the number of test cases by performing the test on the other interface of the network side.

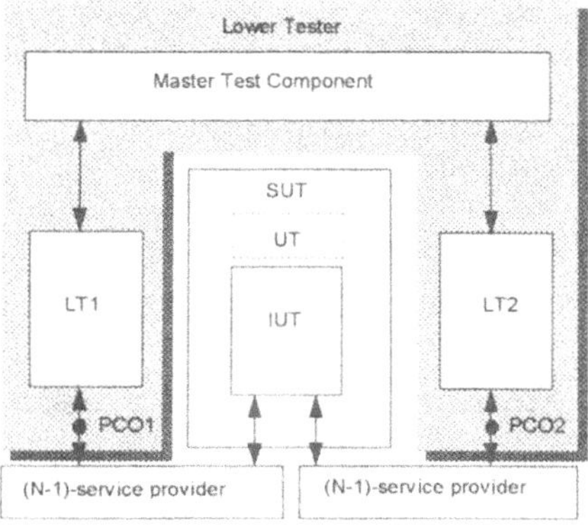

Figure 5 Applied Multi-party testing architecture.

4 DEVELOPMENT OF ABSTRACT TEST SUITE(ATS)

4.1 Overview

We developed the ATS for conformance testing using the following procedures. First of all, we made the Protocol Implementation Conformance Statement(PICS) proforma(ITU-T Draft Recommendation Q.931bis PICS and Abstract Test Suite for DSS1 Layer 3; Circuit Mode, Basic Call Control Conformance Testing, 1994). This is used for static conformance testing and composed of the following items.

- Major Capabilities
- Subsidiary Capabilities

- Call States
- Supported Messages Received/Transmitted by the user(by the network)
- Message Structure
- Information Elements Received/Transmitted by the user(by the network)
- Supported Timers
- Additional Information for Interoperability Information Elements Structure.

Then we extracted the test purposes by analyzing Q.2931 specification. In this process, we generated the Extended Finite State Machine(EFSM) describing the procedures of the protocol. Conformance Kit developed by Netherlands PTT was used as a development tool to get the raw level of the dynamic behavior part of the only user side ATS(Erik Kwast, Harma Wilts, Hans Kloosterman and Jan Kroon, 1991). But the results from the Conformance Kit did not resolve non-deterministic problems, timer operation and data flow parts of the protocol. So we made out test behaviors concerning these problems manually and the EFSM was only used as an input to the Conformance Kit. The test purpose document was used as the high-level test design document during the creation of tests in TTCN. We also made the state transition table based on the EFSM for the user side and network side ATS. This table represents an initial state, input/output event and a transition state. It also helps us extract test purposes from specification. Table 1 represents a part of the state transition table for the user side ATS. Each test purpose was expanded into detailed TTCN test behavior in the test suite. The writing of the test suite was a skilled manual activity.

Table 1 A part of the state transition table for the user side ATS

State / *Event*	*U0(Null)*	*U1(Call initiated)*	*U3(Outgoing Call Proceeding)*
SETUP	CALL_PROC, U9, ALERTING, U7 or CONNECT, Set T313=Now+4,U8	Ignore U1	Ignore U3
SETUP without MIE	RELEASE_COM (Cause = #96),U0	-	-
SETUP with MIECE, (IE flag = 0)	RELEASE_COM (Cause = #100),U0	-	-
SETUP with UnrIE (IE flag = 0)	STATUS(CS = U6,Cause = #99) CALL_PROC, U9, ALERTING, U7 or CONNECT, Set T313=Now+4, U8	-	-
SETUP with OIECE (IE flag =0)	STATUS (CS = U6,Cause = #100) CALL_PROC, U9, ALERTING, U7 or CONNECT, Set T313=Now+4, U8	-	-

MIE : Mandatory Information Element
MIECE : Mandatory Information Element Content Error
UnrIE : Unrecognized Information element
OIECE : Optional Information Element Content Error

4.2 User side ATS

The ATS for conformance testing of Q.2931 user side protocol consists of Test suite overview part, Declarations part, Constraints part, and Dynamic part(ISO/IEC 9646-1 Information technology - Open System Interconnection - Conformance testing methodology and frame work - Part 3, 1994). In this test suite, PCO was declared only one, named L, because Remote test method considering the Single-party testing was applied.

In Test suite overview part, the structure of the test suite and test purposes were addressed. Declarations part defined some parameters, constants, variables, PDUs, and timers. A message, especially, has some Information Elements(IEs) and an IE also has some fields. Therefore the declaration of one message should have a structured type. For example, in Table 2 Connection Identifier IE in the ALERTING message was declared as a type CIdinfo substructure because it splits into some fields again. This scheme is also applied in Constraints part. This point is the main feature of Q.2931 messages compared to other protocol's PDUs below layer 2 and it forces more efforts to describe the ATS. Dynamic part of the test suite consists of behavior descriptions of test cases, preambles, postambles, and some test steps. The behavior of a test case was described as the procedure sending an input, receiving an output and verifying the transition state. We generated 397 test cases and these test cases can be divided into valid test cases, invalid test cases, and inopportune test cases.

Table 2 An example of PDU declaration

PDU Type Declaration		
PDU Name : ALT		
PCO Type : LSAP		
Comment : alerting, u $\leftrightarrow$ n		
Field Name	**Type**	**Comments**
PD	OCTETSTRING[1]	protocol discriminator, M
CR	CRinfo	call reference, M, OCTETSTRING[4]
MT	MTinfo	message type, M, OCTETSTRING[2]
ML	MLinfo	message length, M, OCTETSTRING[2]
CId	CIdinfo	connection identifier, O, OCTETSTRING[4..9]
NId	NIdinfo	notification indicator, O, OCTETSTRING[5]

The test suite structure of the Q.2931 user side ATS is as follows.

- Active IUT testing(Call/Connection control procedures at the originating interface) - for each state such as Null state(U0), Call Initiated state(U1), Outgoing Call Proceeding state(U3), Call Delivered state(U4), and Release Request state(U11)
  - valid : General, Call reference, Timer
  - invalid : Mandatory IE, Optional IE

- inopportune : Message sequences

- Passive IUT testing(Call/Connection control procedures at the destination interface)- for each state such as Null state(U0), Call Received state(U7), Connect Request state(U8), Incoming Call Proceeding state(U9), and Active state(U10)
  - valid : General, Call reference, Timer
  - invalid : Mandatory IE, Optional IE
  - inopportune : Message sequences.

4.3 Network side ATS

The overall structure of the network side ATS is the same as the user side ATS. The description scheme of each part in TTCN is also similar. The differences between them are the state classification and the number of PCO. In the user side ATS, the state in the test suite corresponds to the state in the specification and the number of PCO was one. However in case of the network side ATS, there are some points to be considered more carefully.

As described in section 3.2, it is proper that conformance testing for Q.2931 network side requires a different testing method rather than the Single-party testing which is usually used. So we adopted the Multi-party testing composed of two LTs. In this configuration, the tester controls and observes two PCOs at the same time which correspond two ports of the IUT.

The most important activity is the creation of a set of test purposes which explains the proposed test method, architecture and gives a precise definition of the purpose of each test. Test purposes of the user side ATS were generated according to the call state defined in the specification. But in the network side, we considered two port states simultaneously and made 6 combined states by the reasons mentioned in section 3.2. Because two ports have the dependent relation and we should control two ports simultaneously, some combinations of two port states does not exist together at some time points. There are 11 call states in the network side protocol and theoretically there can be 121 combination states for two ports. But we reduced combination states to 6 combined states as follows by the above reasons. Each denotes the port 1 and port 2 state respectively and the port 1 is the originating side of a call/connection for convenience. Some states such as Call Initiated, Connect Request, and Release Request are not considered explicitly by the limitations of Remote test method and assumed an implicit state transition.

- Null(N0)/Null(N0)
- Outgoing Call Proceeding(N3)/Call Present(N6)
- Outgoing Call Proceeding(N3)/Incoming Call Proceeding(N9)
- Call Delivered(N4)/Call Received(N7)
- Active(N10)/Active(N10)
- Null(N0)/Release Indication(N12).

Table 3 represents an example of Dynamic part described under these considerations. We described events occurred at two PCOs sequentially considering the order of them in the ATS not using concurrent TTCN. This is due to the testing architecture used for and additionally this makes the implementation of the ETS easier. We generated 136 test cases for the network side ATS. The overall test suite structure is similar to the one of the user side ATS. It is divided into 6 state groups and each group has several subgroups such as valid, invalid, and inopportune test groups.

Table 3 An example of Dynamic part

Test Case Dynamic Behavior

Test Case Name : TC102
Test Group : q2931_Net/pt/N_3_6/
Purpose : OUTGOING CALL PROCEEDING-CALL PRESENT STATE TESTS. VALID TEST EVENTS. Ensure that on receipt of a valid ALERTING message from LT2 the IUT responds with an ALERTING message to LT1 and no responds to LT2 and moves to the relevant state 4 and 7 respectively.
Defaults Reference : general_default

Nr	Behavior Description	Constraints Ref.	Verdict
1	+ pr_N_3_6		
2	LT2 ! ALT **START** T310MAX, **START** no_output_timer_2	AL1('1'B,'0'B,'00'B)	
3	LT1 ? ALT **CANCEL** T310MAX	ALr1('1'B)	(PASS)
4	? **TIMEOUT** no_output_timer_2		(PASS)
5	+ cs('000100'B, '000111'B)		
6	+ po_N_0_0		
7	+ um		
8	-> Nr4		
9	? **TIMEOUT** no_output_timer_2		(PASS)
10	LT1 ? ALT **CANCEL** T310MAX	ALr1('1'B)	(PASS)
11	+ cs('000100'B, '000111'B)		
12	+ po_N_0_0		
13	+ um		
14	-> Nr10		
15	? **TIMEOUT** T310MAX		(FAIL)
16	+ po_N_0_0		
17	+ um		
18	-> Nr3		
19	? **TIMEOUT** T310MAX **CANCEL** no_output_timer_2		(FAIL)
20	+ po_N_0_0		

Detailed Comments :
pr_N_3_6 is the preamble test step to Outgoing Call Proceeding/Call Present state.
ALT is the ALERTING message.
cs() is the test step for checking state.
po_N_0_0 is the postamble to Null/Null state to make the IUT stable.
um() is the test step for an event to receive an unexpected message.

5 IMPLEMENTATION OF EXECUTABLE TEST SUITE(ETS)

5.1 Test system design

In order to realize the described test suites, we designed and developed the test system of which the block diagram is as Figure 6(K.Y.Kim, W.S.Kim, B.K.Hong, 1995). The system is composed of test management module, protocol testing module for two ports, low layer protocol processing module for each port, timer processing module, message distribution module, and IUT access module for each port. The functions of each module are as follows.

- test management module
  It performs, controls, and manages the conformance testing.
- protocol testing module for two ports
  It generates testing events for IUT and analyzes events received from IUT.
- low layer protocol processing module for each port
  It performs emulating functions for low layer protocol and conveys testing events above and below.
- timer processing module
  It processes information about timers.
- message distribution module
  It sends testing events to the related port of IUT and delivers the received event to low layer protocol processing module of the corresponding port.
- IUT access module for each port
  It provides the physical access point to the IUT. It consists of two ATM processing boards and one physical interface with two STM-1 ports.

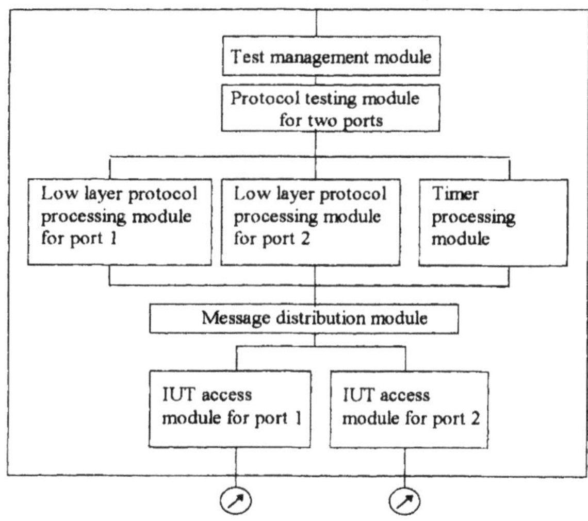

Figure 6 System block diagram.

Among these modules, the protocol testing module for two ports maintains information related each port. It also performs the role to exchange input/output with the related port through low layer protocol processing module and IUT access module for each port according to the testing purpose. Therefore in case of the Single-party testing of the user side ETS, only module of port1 shall be used. Furthermore, for the IUT having more than two ports, the system could be extended as having multiple modules for them. The timer processing module solves timing problems by controlling all timers used in the test suite. It has timer values related to the execution of a test case and identifies each timer when a timer expires. Also because the Multi-party testing approach is used in the network side ATS, the coordination problem of LTs can be occurred. But two LTs are located in the same tester, the synchronization problem could be solved more simply. At each port the timer values of arriving events are observed and recorded in each event, and the timer is managed in the test system. So the protocol testing module can distinguish each event according to the order they arrive. By using this system, we realized the implemented ETS into the platform developed by ourselves.

Table 4 The constitution of the user side ETS

Active IUT Test case		*Passive IUT Test case*	
State	*number of test cases*	*State*	*number of test cases*
Null(U0)	1	Null(U0)	25
Call Initiated(U1)	53	Call Received(U7)	43
Outgoing Call Proc.(U3)	52	Connect Request(U8)	48
Call Delivered(U4)	47	Incoming Call Proc.(U9)	43
Release Request(U11)	42	Active(U10)	43

5.2 User and Network side ETS

We implemented executable test suites from the ATS by using a TTCN to C translator and adding required API libraries such as encoding/decoding a message, sending/receiving a message, timer operation, managing test results, and so on. A public software called VINCE was modified as a platform performing Q.2931 simulation and low layer service providing functions.(Advanced Research Projects Agency(ARPA/CSTO) and the Naval Research Laboratory(NRL/CCS) VINCE version 1.2, 1994) Conformance testing functions of Q.2931 user side are divided into the passive IUT testing and the active IUT testing. The passive IUT testing is that the initiation of a test is caused by the tester. By triggering an input to the IUT a test case is started. In the active IUT testing, the IUT starts a test by sending an message(e.g. the SETUP message). This can be achieved by some user operation. Table 4 shows the constitution of the user side ETS implemented. On the other hand the network side ETS has two ports and the port corresponding to PCO1 is the originating side of a call/connection. But occasionally some test cases are started from PCO2. The distinction of the PCO for received messages is done by using the flag assigned by Message distribution module in Figure 6. Table 5 shows the constitution of the network side ETS implemented.

Functions implemented in the ETSs are as follows.

- declaration functions

  They declare all information needed for initiation of testing.
- test case execution functions

They execute the selected test case.
- test parameter input functions
  They determine test parameters required to perform testing.
- sending message creation functions
  They create message to be sent to the IUT.
- received message analysis functions
  They analyze received messages and decide verdicts.

Table 5 The constitution of the network side ETS

state	*N0/N0*	*N3/N6*	*N3/N9*	*N4/N7*	*N10/N10*	*N0/N12*
number of test cases	25	27	30	28	12	14

5.3 Experiments with Test suites

With these ETSs we needed to verify our test suites by removing errors, but there were no any formal methods. As an alternative, we decided to use any commercial or validated reference implementation. As a reference implementation, we utilized a public software called VINCE which was conformed to ATM Forum 3.1. But its functions are only for valid control flow parts and error handling procedures are not supported completely. Therefore we modified and supplemented the functions a little to be fitted to our specification of ITU-T Q.2931. Some features of error handling procedures were taken into account through these works.

In case of the verification of the user side ETS, we used the modified program directly. The program was connected to the test system with the ETS. All test cases were executed and the verdicts of pass or fail were generated. However we could not execute the network side ETS with the program as it was since the program did not have intermediate system functions. So we developed a simulating function working as an intermediate system and inserted it into the program. Similarly the program was connected to the system for two ports and all test cases were executed. Through these processes, we detected some faults in ATSs or ETSs and reflected them to corresponding suites.

We also applied the network side ETS to B-NT system that was developed by our institute. The testing scenario was as follows. The test system constituted a local call with B-NT system. Two LTs behaved like terminal equipment and were connected by B-NT through two ports. According to test procedures described in the ETS, two LTs stimulated inputs to B-NT and analyzed received outputs from B-NT. For each test case, the testing results were obtained. As a whole, the results of testing were on the good side.

6 CONCLUSION

Up to now we introduced our development procedures of conformance test suites for B-ISDN signalling. We developed the ATS for both the user side and the network side which was written by TTCN. In each case, we applied the Single-party testing and the Multi-party testing architecture respectively. Particularly in the Multi-party testing for the network side, we made out the test suite by applying the method to control and observe two ports of the IUT concurrently. In order to make the ATSs useful in the real world, we also realized ETSs in our test system. Through our verification procedures we detected some faults and reflected them correctly. The overall procedures described in this paper can be represented as Figure 7.

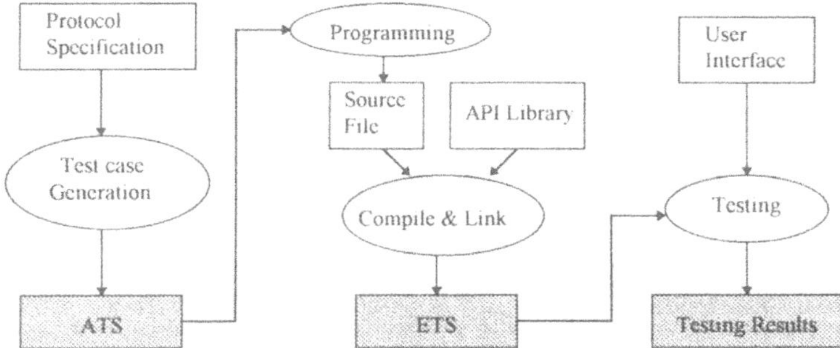

Figure 7 Overall testing procedures.

In the processes of developing the ATS and ETS, we had much skilled manual activities. In case of the ATS, we had to describe all definitions and constraints of messages relating to each test case. Though some definitions of messages were used commonly in several test cases, constraints were different for all test cases which had different test purposes. Particularly in the network side ATS, we had to consider all possible interaction between LTs through two ports for the description of the dynamic behavior part. We used a tool kit to implement the ETS, but it was necessary to provide API libraries supporting the ETS program at least.

With the testing environment mentioned in this paper, we have a plan to perform the conformance testing for the protocol implementation of B-TE, B-NT, ATM MSS, and switching system consisting of our B-ISDN testbed in the near future. We believe that these experiences can be applied to the development of test suites for more complicated protocols and they can increase the possibility of the compatibility and interoperability among the systems implemented on different environments.

7 REFERENCES

Advanced Research Projects Agency(ARPA/CSTO) and the Naval Research Laboratory(NRL/CCS). VINCE version 1.2. (1994)

Erik Kwast, Harma Wilts, Hans Kloosterman and Jan Kroon. (1991) User Manual of the Conformance Kit.

ISO/IEC 9646-1 Information technology - Open System Interconnection - Conformance testing methodology and framework - Part 1. (1994) : General concepts.

ISO/IEC 9646-1 Information technology - Open System Interconnection - Conformance testing methodology and framework - Part 2. (1994) : Abstract Test Suite specification.

ISO/IEC 9646-1 Information technology - Open System Interconnection - Conformance testing methodology and framework - Part 3. (1994) : The Tree and Tabular Combined

Notation(TTCN).
ITU-T Draft Recommendation Q.2931. (1994) B-ISDN Digital Subscriber Signalling No. 2(DSS2) User Network Interface Layer 3 Specification for Basic Call/Connection Control.
ITU-T Draft Recommendation Q.931bis PICS and Abstract Test Suite for DSS1 Layer 3; Circuit Mode, Basic Call Control Conformance Testing. (1994)
K.Y.Kim, W.S.Kim, B.K.Hong. (1995) Experience with the design of B-ISDN Integrated Test System. *Proc. of IWPTS '95.*
Sarikaya. (1989) Conformance Testing : Architectures and Test Sequences. *Computer Networks and ISDN Systems,* 17, 111-126.

About the Authors:

Jin Young Choi is a member of engineering staff at the ETRI since 1993. He studied Industrial Engineering at the Hanyang University, Korea, and received a MS from the Korea Advanced Institute of Science and Technology in 1993. Current research interests include protocol conformance test, traffic of ATM network.

Beom Kee Hong is a senior member of engineering at the ETRI since 1982 and a project leader of B-ISDN protocol test. He studied computer science at the Hongik University, Korea, and received a MS from the Hongik University in 1984. Current research interests include protocol conformance test, ATM network testbed.

24

Experiences with concurrent TTCN for GSM/SS7 testing

Elyane Prevot[1], *Michel D'Hooge*[2], *Patrice Labbe*[2]

[1]*France Télécom - CNET*
38-40 avenue du Général Leclerc, 92131 Issy-Les-Moulineaux
E-mail : bruno.chatras@issy.cnet.fr
Tel : +33 1 45-29-58-53 Fax : +33 1 46-29-31-42

[2]*Trialog*
25 rue du Général Foy, 75008 Paris
E-mail : 100411.2233@compuserve.com
Tel : +33 1 44-70-61-00 Fax : +33 1 42-94-80-64

Abstract

Global System for Mobile communication (GSM) is a European standard which defines telecommunication services for mobile subscribers. At the end of 1995, the European Telecommunication Standard Institute (ETSI) released the Mobile Application Part (MAP) version 2 protocol which supports signalling between GSM phase 2 network entities. It also provided the corresponding conformance test purpose specification.

The testing of this protocol involved numerous interfaces and thus required multi-party testing.

The introduction of concurrent TTCN reduced the testing specification complexity.

This paper describes a practical experience with concurrent TTCN in the GSM environment. It demonstrates the importance of completeness in test purpose specification and the advantage of allocating substantial resources to both ATS development and ATS validation to ensure ATS quality. It also presents a TTCN style guide in order to improve maintenance tasks. It describes some TTCN deficiencies and proposes enhancements.

Keywords

Conformance Testing, GSM, TTCN

1. INTRODUCTION

The Global System for Mobile communication (GSM) is a European standard for digital cellular systems which defines telecommunication services for mobile subscribers. Numerous operators have implemented this standard in many countries and provide the so-called GSM pan-European service.

The first version of this standard (known as GSM phase 1) was officially released in 1991 by ETSI, the European Standardisation Institute for Telecommunication. Currently available implementations are still based on this initial version. However, since then ETSI has defined GSM phase 2 by enhancing the GSM standard to provide additional services.

Since it is unlikely that all GSM phase 1 operators will upgrade their network entities at the same time, interoperability between GSM network entities of different phases is a strong requirement to ensure successful network operation from a subscriber's point of view. Conformance testing of the GSM protocols is the best way to increase the likelihood that mixed networks can interoperate to provide the best quality of service to GSM subscribers.

This paper presents experiences in the conformance test specification of the GSM phase 2 Mobile Application Part protocol (MAP). The introduction describes briefly the GSM network architecture and protocol stack. It also presents the base standard, MAP, to give an idea of the characteristics of the tested protocol. Section 2 discusses different aspects of test specification. Sections 3 and 4 describe the ATS and ETS production as well as our practical experience with concurrent TTCN. A comparison with GSM phase 1 testing is also given. The conclusion gives an overview of the man-power involved. A glossary of acronyms is provided at the end of the article.

1.1. GSM Network

GSM provides mobility and roaming facilities that enable subscribers to receive or send calls regardless of their location [Mouly]. The GSM network architecture and associated protocols are shown in Figure 1.

The ability to call a mobile subscriber at any time and anywhere in Europe implies the knowledge of its location in real-time. The Home Location Register (HLR), is an "intelligent" data base which stores the user's subscription and current location. This register is then interrogated by a Mobile service Switching Centre (MSC) each time a call is destined to a given subscriber. In other words, the HLR is able to provide routing information to transfer the incoming call to the called subscriber.

In order to locate a subscriber, the GSM service area is divided into elementary areas (cells) which in turn are logically grouped in location areas. MSCs associated with data bases, known as Visited Location Registers (VLR), manage those location areas. Thus, when a mobile subscriber moves from one area to another, it informs the MSC/VLR in charge of this area. The MSC/VLR will then inform the HLR of the current location of the subscriber.

The MSC/VLR also provides to roaming subscribers call related GSM services such as outgoing call handling, incoming call delivery, short message delivery, and supplementary services. Subscribers give their location to the network or ask for GSM telecommunication services via a Base Station Sub-system (BSS) by using radio access.

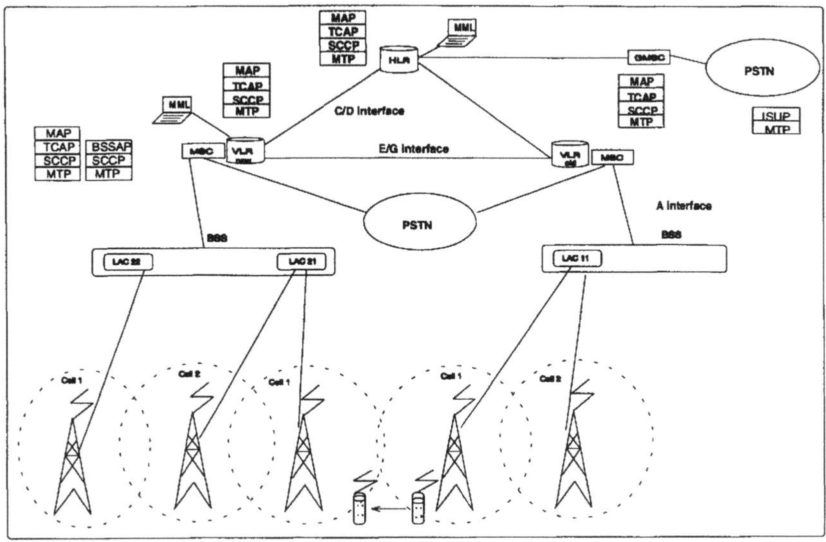

Figure 1: GSM network.

1.2. Base Specification: MAP

The Mobile Application Part (MAP) protocol supports the signalling between the GSM network entities, with the exception of the MSC/VLR which is supported by the Base Station Application Part protocol (BSSAP). MAP makes use of the Transaction Capabilities of the Signalling System N°7 protocol stack [Webster].

MAP is a complex protocol which makes use of 46 messages, described in ASN.1, each of which is used by MAP procedures to realise a given functionality. For instance, the procedure "Location Updating", which takes place between the VLR and the HLR in order to register the current subscriber location, involves the exchange of 6 messages: the request and acknowledgement of UpdateLocation, InsertSubscriberData, and ActivateTraceMode.

MAP version 1 provides the GSM phase 1 services while MAP version 2 supports the GSM phase 2 services. MAP version 2 also corrects some deficiencies in MAP version 1. It was frozen and published at the end of 1995.

One of the main GSM phase 2 characteristics is to allow a GSM phase 1 entity to interoperate with a GSM phase 2 entity. The objective of this compatibility is to allow soft migration from GSM phase 1 to GSM phase 2 by providing GSM operators with the option of implementing only part of GSM phase 2 functionalities. Thus the compatibility between versions has to be checked for GSM phase 2 entities. This imposes strong requirements on the MAP version 2 test specification.

2. TEST SPECIFICATION

GSM equipment suppliers are in the process of implementing GSM phase 2. Thus, there is a need for conformance testing of the GSM phase 2 network entities to ensure successful network operation. As we can see from Figure 1, testing the MAP protocol is a complex task due to the number of interfaces involved and the diversity of protocol used in those interfaces. This means that the MAP conformance testing of every GSM network entity is a big task, and it is often said that suppliers and operators are reluctant to pay for it. Thus in order to ensure conformance testing, consortia were created, such as the MoU/CONIG for the MAP version 1 conformance testing, and ETSI work items were planned.

ETSI published the first stable draft of the MAP version 2 test specification at the beginning of 1995. Due to lack of resources and TTCN expertise, ETSI defined only the test purposes. It was planned that the ATS specification in TTCN be done by an external body, such as an ETSI Project Team or a CTS project. The idea was to provide the test purposes as input to this external body with the support of ETSI to solve open issues on MAP protocol. Unfortunately, no funds were allocated for the derivation of the TTCN ATS from the ETSI test purposes. Thus France Télécom-CNET decided to develop the TTCN ATS on its own by using the services of Trialog, an external software house.

This project started with the definition of the ATS specification methodology and the analysis of all possible problems by taking a sample of representative test purposes. It then continued with the writing of all test cases. Section 3 gives an overview of the ATS development methodology. GSM phase 2 testing is based on the phase 1 experiences [Berkhout, Koster]

2.1. Test Purposes

The MAP version 1 conformance test specification was a set of 1000 tests. ETSI produced around 1600 test purposes for the MAP version 2 conformance test specification, the increase in number coming mainly from additional functionalities introduced in GSM phase 2. Table 1 gives an overview of number of test purposes allocated per procedure.

Table 1: Test purposes allocation

	Valid Behaviour Tests		*Invalid Behaviour Tests*	
	Number		*Number*	
Inter-Working between versions	57	(90%)	6	(10%)
Operation and Maintenance	166	(73%)	62	(27%)
Location Management	159	(77 %)	44	(23%)
Call Handling	301	(96 %)	11	(4 %)
Handover	294	(71 %)	120	(29 %)
Supplementary Services	93	(71%)	38	(29%)
Short Messages	150	(72 %)	57	(28 %)

ETSI based the MAP version 2 conformance test specification on the MAP version 1 test specification output. Enhancements was made by increasing the number of values tested in message fields, and syntactical error checking was reduced to the most significant cases.

The test suite structure was modified to be entity oriented instead of procedure oriented. This calls for a test suite structure with one group per GSM entity, i.e. HLR, MSC/VLR, and SMS-GMSC/IWMSC (short message service for the gateway and interworking MSCs). Then every entity group is sub-divided into several groups, each of which is related to one procedure (e.g. location management, call handling). Every procedure group is divided as well into two sub-groups: valid behaviour testing and invalid behaviour testing. Each of these is sub-divided in turn to take care of several testing aspects such as data variation, structure variation, and syntax errors.

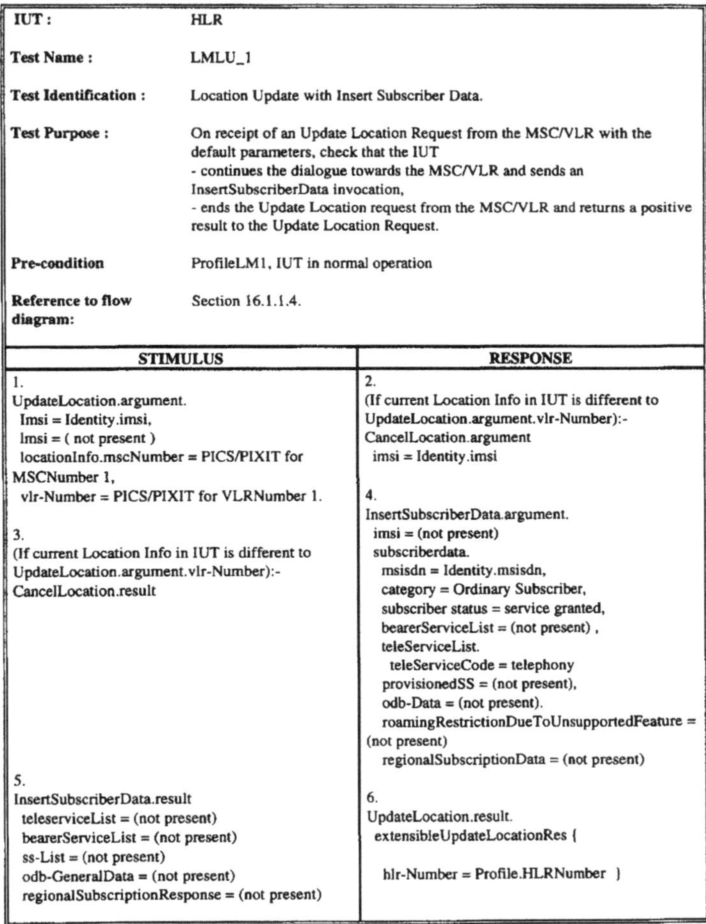

IUT :	HLR
Test Name :	LMLU_1
Test Identification :	Location Update with Insert Subscriber Data.
Test Purpose :	On receipt of an Update Location Request from the MSC/VLR with the default parameters, check that the IUT - continues the dialogue towards the MSC/VLR and sends an InsertSubscriberData invocation, - ends the Update Location request from the MSC/VLR and returns a positive result to the Update Location Request.
Pre-condition	ProfileLM1, IUT in normal operation
Reference to flow diagram:	Section 16.1.1.4.

STIMULUS	**RESPONSE**
1. UpdateLocation.argument. Imsi = Identity.imsi, Imsi = (not present) locationInfo.mscNumber = PICS/PIXIT for MSCNumber 1, vlr-Number = PICS/PIXIT for VLRNumber 1.	2. (If current Location Info in IUT is different to UpdateLocation.argument.vlr-Number):- CancelLocation.argument imsi = Identity.imsi
3. (If current Location Info in IUT is different to UpdateLocation.argument.vlr-Number):- CancelLocation.result	4. InsertSubscriberData.argument. imsi = (not present) subscriberdata. msisdn = Identity.msisdn, category = Ordinary Subscriber, subscriber status = service granted, bearerServiceList = (not present) , teleServiceList. teleServiceCode = telephony provisionedSS = (not present), odb-Data = (not present). roamingRestrictionDueToUnsupportedFeature = (not present) regionalSubscriptionData = (not present)
5. InsertSubscriberData.result teleserviceList = (not present) bearerServiceList = (not present) ss-List = (not present) odb-GeneralData = (not present) regionalSubscriptionResponse = (not present)	6. UpdateLocation.result. extensibleUpdateLocationRes { hlr-Number = Profile.HLRNumber }

Figure 2: Test purpose specification example.

The MAP version 1 test purposes were defined using a prose description. Although the writers were TTCN experts rather than MAP experts, the final product was of medium quality from both a TTCN and MAP point of view. The TTCN errors were essentially due to the fact that no TTCN editor tools or compilers were available at that time, so the test cases were written directly in MP format.

For phase 2, ETSI decided to define the test objectives in a way that gives enough information to make the TTCN specification easier since it is likely that the ATS specifiers will not be GSM protocol experts. The test purposes have thus been formalised in such a way that every Stimulus (tester behaviour) and every Response (IUT behaviour) are explicitly given. The Stimulus and Response messages are described to look like ASN.1 value notation. Figure 2 gives an example of this type of test purpose specification.

The Preamble and Postamble are implicitly specified by giving the required IUT status in terms of subscriber profile (e.g. roaming restrictions, supplementary services, location information). A table which defines the default Stimulus and Response message contents is at the beginning of each section in order to improve the readability of the specification.

2.2. Test Architecture

Previous work on GSM Mobile Network System testing was limited by the impossibility of using a multi-party abstract test method and concurrent TTCN [Koster], [Ahtiainen], [Berkhout]. This project had the good fortune to start when the concurrent standard and tools became available. The test architecture, i.e. the test method, of this project was defined following ISO 9646 recommendation [ISO9646]. The chosen method is "remote" and "multi-party".

The "remote" aspect comes from the fact that the SUT (System Under Test) is in its operational configuration, so it is not possible to access directly the upper and lower bounds of the IUT. As a consequence, the UT (Upper Tester) is in the SUT while the LT (Lower Tester) is in the test system. The later interacts with the IUT "remotely", i.e. through indirect ways that should have the same effect on the IUT as direct actions. For example, in order to call a procedure belonging to the MAP upper interface, the operator is asked to enter a Man Machine Language (MML) command which calls the procedures to be tested.

The "multi-party" aspect is due to the fact that some of the GSM entities in which MAP protocol is implemented have numerous interfaces. As shown in Figure 3, the MSC/VLR has interfaces with the HLR (MAP), the base station (BSSMAP or DTAP), the fixed telephone network (ISUP), and the operator (MML), among others. A PCO (point of control and observation) must be defined for each interface. The LT must deal with all these interfaces. This requires the use of the "multi-party" test method in conjunction with Concurrent TTCN, which are defined in the first Amendment to ISO 9646 and explained below.

3. ATS PRODUCTION/DERIVATION

In order to be as close as possible to the test suite structure presented in section 2.1, one ATS per entity was produced. Each ATS structure follows the test suite structure defined by ETSI.

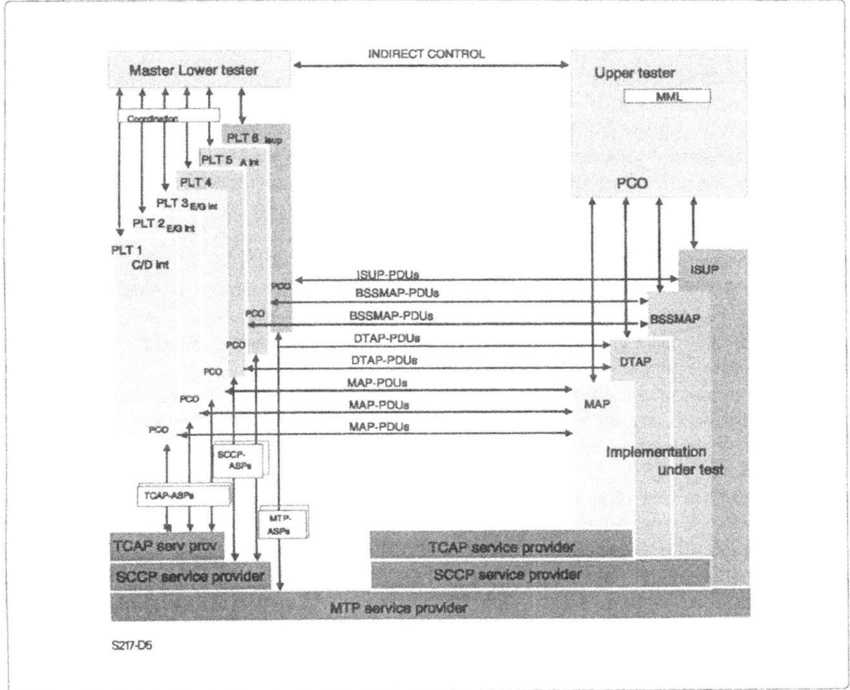

Figure 3: Multi-party testing of MSC/VLR.

3.1. Concurrent TTCN

As shown in the previous section, in order to validate a MAP implementation, it is necessary to examine simultaneously the PCOs located at all the interfaces. If we use the IS version of TTCN, which does not support concurrency, each test case must take into account all the messages on all the interfaces at the same time [Walter 92]. This leads to very complex test cases.

With concurrent TTCN, it is possible to define the messages sent or received separately on each interface and to describe the synchronisation between them. Since a new version of the TTCN compiler is available with concurrent TTCN facilities, it was possible to take advantage of these new facilities in this project.

3.2. TTCN Style Guide

Since a test suite is written by one group of people and often debugged or maintained by another, it is very important to have clear and self-explanatory code. To meet this goal, a style guide was developed for this project to simplify the debugging and maintenance tasks. This system proved to be very successful. If the same had been done for MAP version 1 test specification, maintenance work would had been facilitated and a number of resources would had been saved.

Test Suite Fragmentation

Although the tabular form of TTCN divides a test suite into convenient tables, the machine processable form is a giant single file for an entire test suite. While developing the MP form of the test suite, a hierarchy of directories and groups of small files were used to divide the suite into more manageable fragments. A file containing the test suite skeleton uses include directives to incorporate the MP code stored in other files.

Naming Convention

The EWOS/TA/93/006 [EWOS] explains the importance of naming conventions and gives guidelines on this point. However, the examples given use underscores, which lengthen names, and the names are mostly in uppercase, which lessens readability. Therefore we developed naming conventions which better suit our needs.

A strict and exhaustive naming convention was developed to ease the naming and the reading of identifiers for each test element, ranging from simple constants to test steps. Each identifier is divided into 2 or 3 semantically different parts. The following rules describe the basis of these conventions.

A prefix identifies the type of the identifier and its domain of application. For example, *v* stands for *variable*, *c* for *constant*, *px* for *pixit* and so on. This is useful for recognising the function of an identifier where a number of different types could be used. This is followed by a short, usually abbreviated, but easily understandable description of the function. For example, *VlrAddr* stands for *Visitor Location Register Address*. A postfix can be added to distinguish between identifiers which would be identical when using only the first two fields of the naming convention. Capital letters are used at the start of each part as separators. However, an underscore can be used to separate parts to improve readability.

pxIutAddrMsc	a pixit for the IUT address for the MSC mode.
pxIutAddrGmsc	a pixit for the IUT address for the GMSC mode.

Special conventions apply to constraint identifiers since they represent a large part of the test suite. The first character of the prefix defines whether it is used in an indication or request. This is useful since the syntax of the two types of constraint is different. The protocol used is also part of the prefix, where M stands for MAP, D for DTAP, etc.

qM_ATMa_OmcId	request, MAP, Activate Trace Mode Argument, OMC Id field included.
iM_ISDr_ss	indication, MAP, Insert Subscriber Data Result, Supplementary Service field included.
iD_LocUpdAcc_noTmsi	indication, DTAP, Location Updating Accept, no TMSI field.

For test steps, the prefix contains the name of the entity simulated, and if applicable the role played.

Msc1AnchPerformHO	Entity = Msc #1; Role = Anchor MSC; Service = Perform Handover
Hlr1DSD_UError	Entity = Hlr #1; Role = N/A; Service = Delete Subscriber Data; Specialisation = Waiting for a negative result.

Local Tree Hierarchy

In order to see the basic structure of a test case at a glance, each PTC (Parallel Test Component) is represented as a local tree in the case and not as a test step. Each of these local trees is in turn divided into three local trees which represent the preamble, the body, and the postamble. Within these trees, test steps can be called. In this way, the creation of many test steps which cannot be reused is avoided. Figure 4 shows an example of a concurrent TTCN test case.

3.3. Suggestions for TTCN Improvements

During our experience in writing test suites in TTCN, we encountered certain difficulties with the language. The concurrent aspects of TTCN are useful and pose no problems, but often the basic TTCN syntax is awkward. The following sections describe our suggestions for improvements in basic TTCN that would make programming easier.

Test Case Dynamic Behaviour

Test Case Name : LMLC1
Group : MOB/LM/LC/BT/VB/NT/SO/BS/
Purpose : New location cancellation request.
On receipt of an update location request from the new VLR with the default parameters, check that the IUT – begins a new dialogue towards the Previour VLR and sends a cancelLocation invocation
Configuration : cfgVlr1Vlr2
Default :
Comments :

Nr	Label	Behaviour Description	Constraints Ref	Verdict	Comments
1		CREATE(PTC_Vlr1:Vlr1_LMLC1,PTC_Vlr2: Vlr2_LMLC1)			
2		+MtcRunAa(cpVlr1,preamble)			
3		+MtcRunABba(cpVlr1,cpVlr2,testBody)			
		Vlr1_LMLC1			
4		ACTIVATE(Vlr1Default)			
5		cpVlr1 ? CmSTART	STRT_b		preamble
6		+Vlr1UpdateLocation(pxImsiLMLC1,iM_I SDa_lu(pxMsisdnLMLC1))			
7		cpVlr1 ! CmVERDICT	VRD_b(R)		
8		cpVlr1 ? CmSTART	STRT_b		
9		+Vlr1CancelLocation(pxImsiLMLC1, man)			
10		cpVlr1 ! CmVERDICT	VRD_b(R)		
		Vlr2_LMLC1			
11		ACTIVATE(Vlr2Default)			
12		cpVlr2 ? CmSTART	STRT_b		
13		+Vlr2UpdateLocation(pxImsiLMLC1,iM_I SDa_lu(pxMsisdnLMLC1))			
14		cpVlr2 ! CmVERDICT	VRD_b(R)		

Detailed Comments :

Figure 4: Example of Concurrent TTCN test case.

Lack of Modularity

One of the more important limitations of the TTCN language comes from the fact that an Abstract Test Suite is defined, following ISO 9646-3, as a single, monolithic program. An ATS is defined as a sequence of declarations, test cases, test steps, and so on. Although each concept is defined separately, for a big test suite this kind of structure is neither sufficient or efficient. Tools have been developed to get around this aspect, with the possibility of navigating easily in a large test suite, but it would seem much better to improve the language in order to introduce new modular capabilities.

Most programming languages allow a program to be divided into modules, such as files and procedures. When developing a test suite, it would be much better to be able to deal with "modules" or "sub-suites", each one dealing with one aspect of the IUT to test. Declarations and tests steps defined in other modules could be declared as "external". Before test execution, all these modules would be linked together in order to produce the complete ETS.

Furthermore, is it impossible at this stage to think of an "Object TTCN", in order to climb two evolutionary steps of software technology at a time? TTCN could benefit a great deal from object oriented aspects such as static independence between test suite components, declarations with classes and object instances, inheritance mechanisms, and dynamic communication between PTCs and MTCs (Master Test Components).

Timer as a Parameter

Currently in TTCN it is not possible to pass a timer as a parameter to a step, a default, or a local tree. As a result, if a step is to be used with two different timers, one solution is to duplicate the step with the timer as the only difference. Another is to pass a variable as a parameter which identifies the timer to use, and the choice of which behaviour line to use is based on the value of the variable. It would simplify the code if a timer could be passed as a parameter.

Select Expression to Parameterize Constraints

Because constraints cannot be parameterized with a select-expression, we made the selection in the dynamic part by qualifying the two alternate receive events accordingly. Since it is known before run-time which one is expected, it would be conceptually cleaner and practically more concise to have the following type of construct:

```
<contraint&attributes> ::= IF <selectexpression id>
                           THEN <contraint&attributes>
                           ELSE <contraint&attributes>
```

Qualified Attachment

Qualifying a step would be convenient and concise. The format could be as follows:

+T [Q]

The qualifier Q would be ANDed to the qualifiers of first column events when the step is expanded. Non qualified events would be considered to have an implicit constant qualifier [TRUE]. Qualifying steps with non-qualifiable events (e.g. GOTO) in the first column would lead either to a syntax error or to a different policy. For instance:

+T [Q]

would be equivalent to :

[Q]
    +T

when T has non-qualifiable events in the first column.

Optional Step Attachment

Some test cases would be easier to write if test steps could be made optional. In the following example, brackets indicate an optional step.

+T1
    +T2
+T2

would be equivalent to :

+[T1]
    +T2

The example given is rather simple, but complexity increases exponentially when the sequence of optional steps is longer. An alternative would be to encapsulate all that follows an optional step into an intermediate tree, but this leads to deep encapsulation and poor readability.

3.4. Practical Experience

ATS derivation was carried out by a team of 3 people over approximately 11 months. The time necessary to produce the ATS is shown in the following table. This does not include time for tasks such as management, "peripheral" activities like methodology definition, or quality control.

The first test case family is a special case, because it covers all MAP protocol aspects for GSM v1 and v2, and the productivity is not representative for standard test case derivation. The two other families, with a production rate of 2.4 to 3 test cases per day, are a more reliable basis to use to determine the time needed for ATS derivation.

It is important to note that the times below include all operations related to ATS derivation: test purpose and protocol analysis, TTCN writing, syntactical adjustment after compilation, technical review, and debugging after ETS execution. Thus it not the time needed simply to write a suite in TTCN.

Table 2: Time necessary to produce the ATS

Test case families	*Number of test cases*	*Man/days*	*Test per day*
Inter-Working	63	75	0.84
Operation and Maintenance	228	78	2.9
Location Management	203	81	2.5
Call Handling Procedure	312	103	3
Handover	414	173	2.4
Supplementary Services	131	50	2.8
Short Messages	207	72	2.9
TOTAL	1558	630	2.5

4. ETS GENERATION AND VALIDATION

The Executable Test Suite (ETS) generation and validation consisted of running TTCN test cases first with a simulated IUT and then a real one. The chosen tester was the Tekelec Octopus tool which provides both ETS development and execution environments.

4.1. Test Tool

Octopus is a test execution software environment which provides the means to produce ETS from TTCN ATS in a quasi-automatic way [Ahtiainen]. The production is automated when the protocol is defined with the ASN.1 notation. This is the case for the MAP protocol. Octopus also implements concurrent TTCN defined in TTCN-IS 9646-3, AM1. This environment was mainly developed for SS7 testing, but other signalling protocols (Internet, OSI, etc.) can be tested with a minimum software development. Octopus provides three main tools.

- A tool to derive ETS from ATS which is called a "TTCN to C compiler ".
- A tool to configure the underlying protocol stack used by the test suite, which is the protocol stack beyond the different test suite PCOs.
- A tool to execute an ETS through a graphical user interface. This tool provides tracing, pics/pixit definition, and test case selection facilities.

4.2. Generation

One ATS per GSM network entity was defined, so three ETSs were produced. ETS generation comprised several steps on the Octopus platform.

1. The first step was to generate the executable C code files.
2. The second step was to interface the test suite and the underlying protocol stack by mapping the ATS Abstract Service Primitives to the underlying protocol stack primitives.

3. The third test step was to write the C code functions for the test suite operation. A number of test suite operations were defined, which involved a good deal programming effort. This was necessary mainly for the MSC/VLR test suite because the structure of information elements of the BSSAP protocol is different from MAP's even though they both carry the same information. Thus mapping functions were needed. This issue shows that a harmonisation of interworking protocols would facilitate the testing process.
4. The last step consisted of setting PIXIT and PICS parameters and the test configuration. A great number of PICS/PIXITs were defined due to the protocol's complexity and thus this step required substantial effort. For instance the HLR test suite requires the creation of different types of subscriber profiles in order to make data variation of the MAP InsertSubscriberData message. If the purpose of a test case is to modify some HLR data then a different subscriber type in the HLR data base is needed for each test case. Indeed, since a test case execution verdict is unsure, the HLR data base status at the end of the test case is also unsure. A postamble can be defined to put the HLR back in the initial state, but it is uncertain what goes wrong within the test case if it fails. Thus, we avoid this uncertain solution. This is one big problem of protocol conformance testing which involves data bases.

4.3. ETS Validation

The ATS validation is often skipped when developed within an official project such as an ETSI or CTS project, because there is no volunteer or no allocated funding for this task. However, in order to ensure a high quality ATS, this project dedicated some resources to this task. Since it is not easy to review an ATS only by making reading review, this task is often postponed to the ETS validation. We followed three approaches to validate an ETS.

- The first one was to make a reading review of ATS TTCN test cases while waiting for the availability of a real IUT or a reference implementation.
- The second way was to simulate the IUT within the tester by using the Octopus simulation facility. This simulation requires development of another test suite to play the role of the IUT. The test cases of this test suite provide a kind of simplified reference implementation but are more than a symmetric realisation of the MAP test suites.
- The third approach was to run some test cases against a real MAP entity on a platform. The IUT was in fact a partial implementation of the MAP version 2 protocol. Thus, only a few capabilities were available and thus testable.

It was not realistic to evaluate 1600 test cases, so a different strategy was adopted. A sample of representative test cases was selected to have a good coverage of the test suite. The ETS validation was performed in parallel with the ATS development. The ATS development was made procedure by procedure for each entity or test suite. This approach allowed the evaluation of test cases related to one procedure while test cases of another procedure were being specified. In this way, the evaluation process was optimised. It started when the first set of test cases related to one procedure were available, which was about two months after the start of the project. It ended one month after the delivery of the last set of test cases.

This evaluation and debug tasks costs around 5 man-months for the three test suites of 1600 tests. Table 3 gives an overview of resources dedicated to every ETS evaluation method mentioned above.

Table3: ETS evaluation man power

	HLR	*MSC/VLR*
Reading review	3 man-weeks	7 man-weeks
IUT simulation	3 man-weeks	5 man-weeks
Real IUT	1 man-week	1 man-week

Test suite evaluation leads to frequent modifications of the TTCN ATS, which requires several compilations. Since an ATS is seen as a whole unit, no partial compilation can be made. This highlights a practical deficiency in TTCN. It would be useful if the tester or the TTCN editor could provide a tool which extracts either a test case or a group of test cases and the TTCN components used to generate the equivalent MP. The support of this functionality in the Octopus environment is under study.

5. CONCLUSION

The GSM MAP phase 2 protocol conformance testing specification was an extended project and involved numerous resources. Around 35 man-months were allocated to the production and validation of 1600 TTCN test cases.The project proved that the detailed ETSI test purpose specification substantially facilitated test case specifier's task.

This experience shows that the introduction of the concurrent TTCN of the 9646-3 standard simplified and improved the readibility of TTCN ATS specification. The dedication of resources to the TTCN programing style at the very early stage of the project ensured the simplification of debugging and later maintenance tasks.

Finally, even though the application of concurrent TTCN to a huge ATS highlights some TTCN deficiencies, this amendment fulfilled most of our requirements. Therefore, we encourage the use of TTCN as a conformance test specification language and we hope ISO will continue to enhance this standard to answer the needs of test specifiers.

6. ACKNOWLEDGMENTS

The authors would like to thank their colleagues Barbara Raither, Vincent Belaïche (Trialog), and Bruno Chatras (France Telecom - CNET) for their useful comments during the elaboration of this paper.

7. REFERENCES

A.Ahtiainen, B. Chatras, M. Hornbeek, S. Kesti (1993). Experiences with Octopus automated TTCN translation tools applied to GSM/SS7, in *Proceedings of IWPTS 6th* (ed.O.Rafiq).

S.R. Berkhout, M.F. Witteman (1991). Application of multi-party conformance test methods to the GSM mobile network system, in *Proceedings of IWPTS 3rd* (ed. I. Davidson and D.M.Litwack), Elsevier Science Publishers, IFIP 91.

EWOS (1993). EWOS/TA/93/006 : The TTCN Style Guide and Quality Criteria,

ITU (1994). ITU-T Recommendation X.290 to X.296: OSI Conformance testing methodology and framework for protocol recommandations for ITU-T applications (Also published as ISO/IEC 9646).

ISO Test Methods and their Applicability to the GSM Mobile Network System, Rüdiger Köster, Markus Dicks, DETECOM

ETSI SMG. ETS-300.599 : European digital cellular telecommunications system (Phase 2). Mobile Application Part (MAP) specification (GSM 09.02).

M.Mouly, M.B.Pautet (1992). *The GSM System for Mobile Communications.* Michel Mouly and Marie-Bernadette Pautet Editors.

T.Walter and B.Plattner (1992). An operational semantics for concurrent TTCN in *Proceeding of IWPTS 5th* (ed. G.v Bochman, R.Dssouli and A.Das).

T.Walter and F.Kristoffersen (1994). TTCN test case correctness validation, in *Proceedings of IWPTS 7th.*

N.Webster (1995). The Testing of BT's Intelligent Peripheral using abstract test suite from ETSI, in *Proceedings of IWPTS 8th* (ed. A.Cavalli and S.Budkowski).

BIOGRAPHY

Elyane Prévot graduated from ENSSAT in 1988. She has been involved in the standardization and implementation of test specifications for GSM networks since 1990 when she joined France Telecom - CNET. From 1994 to 1995 she chaired the ETSI ad-hoc group on Mobile Application Part test specification for GSM phase 2.

Michel d'Hooge graduated from Supelec in 1993. Since then he has been a software development specialist at Trialog. He was involved in this project since its conception.

Patrice Labbé is a real-time software and network expert who has more than 15 years of professional experience. He has been involved in numerous avionics, automotive, and telecom projects, often for methodological or test/validation purposes.

GLOSSARY

ASN.1	Abstract Syntax Notation one
ATS	Abstract Test Suite
BSS	Base Station Sub-system
BSSAP	Base Station Sub-system Application Part
BSSMAP	BSS Management Part
DTAP	Direct Transfer Part
ETS	Executable Test Suite
ETSI	European Telecommunication Standard Institute
GSM	Global System for Mobile communication
HLR	Home Location Register
ISUP	ISDN User Part
IUT	Implementation Under Tester

ITU	International Telecommunication Union
LT	Lower Tester
MAP	Mobile Application Protocol
MoU/CONIG	Memorandum of Understanding/Conformance Network Interface Group
MTC	Master Tester Component
MSC	Mobile Switching Centre
MTP	Message Transfer Part
OMC	Operation and Maintenance Center
PCO	Point of Control and Observation
PSTN	Public Service Telecommunication Network
PTC	Parallel Test Component
SCCP	Signalling Connection Control Part (SS7 protocol)
SMS-GMSC	Short Message Gateway MSC : Gateway MSC dedicated to the routing of incoming short message
SMS-IWMSC	Short Message Interworking MSC : dedicated to the routing of outgoing short message
SS7	ITU Signalling System number 7
SUT	System Under Test
TCAP	Transaction Capability Part (SS7 protocol)
TTCN	Tree and Tabular Combined Notation
UT	Upper Tester
VLR	Visited Location Register

25

Assessment of automatic generation methods of conformance test suites in an industrial context

Laurent Doldi, Vincent Encontre

VERILOG, 150 rue Nicolas Vauquelin, BP 1310, F-31106 Toulouse Cedex, France

(doldi, encontre)@verilog.fr

Jean-Claude Fernandez

INRIA, Vérimag, Miniparc Zirst, rue Lavoisier, F-38330 Montbonnot Saint-Martin, France

Jean-Claude.Fernandez@imag.fr

Thierry Jéron

INRIA, IRISA/Pampa, Campus de Beaulieu, F-35042 Rennes, France

Thierry.Jeron@irisa.fr

Sylvie Le Bricquir, Nathalie Texier

CAP SESA REGIONS, rue de la Rigourdière, F-35517 Cesson-Sévigné Cedex, France

(slebricq, ntexier)@ecom.cgs.fr

Marc Phalippou

FRANCE TELECOM / CNET, Technopole Anticipa, 2 avenue Pierre Marzin, F-22307 Lannion Cedex, France, phalippo@LANNION.cnet.fr

Abstract

In this paper we describe three experiments in the automatic generation of conformance abstract test suites for protocol testing. The methods and tools have been applied to a same industrial protocol, the DREX protocol. This protocol has been specified in SDL language. Abstract conformance test suites have been generated in TTCN language. The study proved that the technology transfer to obtain a commercial tool is now possible.

Keywords

SDL (Specification and Description Language), ISO 9646 standard, IUT (Implementation Under Test), Test purposes, Abstract test suites, TTCN (Tree and Tabular Combined Notation), MSC (Message Sequence Charts).

1 INTRODUCTION

Conformance testing is a major issue of strategic importance for the acceptance of OSI and other products based on standards for open systems. The central point in a conformance testing process lies in the availability of a test suite which must be closely related to the protocol (or system) specification. In spite of an important research effort in this area, this problem has not yet received a satisfactory solution. Many methods have been published, but there is no available commercial tool today able to process industrial SDL specifications in order to compute test suites in a really automated way.

This paper presents the conclusions of an experimental study whose objective was to evaluate and compare several available test generation techniques. Three prototype tools have been applied to the SDL specification of a real industrial protocol, and their outputs have been analysed and compared with manually written test suites. This experiment shows that by taking the best part of state-of-the-art research on test generation, it is possible to define now the architecture of an operational test generation tool. Our contribution includes in particular the following key points:

1. produce useful tests: many test generation methods are criticized by practitionners because they define non usable tests (too many, not checking the right parts,...). To avoid this trouble, we started from an existing industrial protocol, and we compared the results of our experiment with previously manually generated test suites. This guarantees that our results are really close to the users' needs.
2. use efficient algorithms: many test generation methods are unable to process industrial specifications for performance reasons (state explosion, computation time,...). We investigated how to reuse powerful algorithms and techniques (in particular on-the-fly computations) from protocol verification domain. In this way we beneficiate from many years of experience in this area, and we avoid reinventing the wheel.
3. combine complementary methods: we discovered advantages and drawbacks in each one of the three test generation methods which we have studied. The way we propose to combine them for building an "ideal" tool (see section 4) takes a best profit from all these combined ideas.
4. produce qualitative and quantitative gains: the use of formal methods increases the confidence in the automatically generated test cases. The automatic methods require time to specify the protocol and to formalize test purposes. But this is amply balanced by the benefit in time of automatic methods.

This paper is organized as follows: section 2 summarizes the context and the objectives of this study. Section 3 gives some details about the three tools which have been used in our experiment. Section 4 presents the results of the experiment. Finally section 5 evaluates what could be expected from a combination of the best part of each tool and analyses the perspectives for an industrial transfer of the results in order to obtain a commercial tool.

2 CONTEXT AND OBJECTIVES

2.1 Context and objectives of the study

The experiments related in this paper have been conducted during an industrial contract for the Délégation Générale pour l'Armement (DGA) of the French Ministry of Defence. Partners of this contract are VERILOG, CNET (Centre National d'Etudes des Télécommunications), CAP SESA, and two research groups of INRIA (Institut National de Recherche en Informatique et Automatique). This study started in November 1994 and ended in November 1995.

The goal of the study is to prove that the automatic generation of conformance test sequences from a formal specification of a protocol and test purposes is feasible and profit-earning in industrial context.

2.2 Methods and tools

Three methods and tools have been studied and/or developed : TGV, TOPIC V2 and TVEDA V3. TOPIC V2 is a prototype of VERILOG based on a previous prototype TOPIC (Montiel, 1994). TGV (for Test Generation from Verification techniques) from INRIA is based on verification techniques such as synchronous product and on-the-fly verification. TGV did not exist at the beginning of the study. TVEDA V3 from CNET (Phalippou, 1994) incorporates two distinct generation techniques; the technique used during this study is based on syntactic transformation of the specification.

From the methodological point of view, the experiment is in keeping with the general pattern of abstract tests suites specification as described in ISO 9646 standard and take also into account the processes used by practitioners. We use a formal description of the protocol, given in the SDL language (CCITT, 1988). We also use test purposes to select test cases. A test case is a set of sequences of interactions related to a particular test purpose.

In order to compare the methods and tools, these three tools had to generate abstract test suites, starting with the same SDL specification and test purposes in natural language for the DREX protocol described below.

2.3 The DREX protocol and its specification

The protocol used for our expriments is called the DREX protocol. This protocol has been chosen for three principal reasons:

- the protocol had to be an industrial one in order to prove feasibility of automatic test generation methods on realistic specifications,
- an SDL specification of a similar protocol was already available, written by CAP SESA,
- hand written test suites had already been produced, also by CAP SESA and this would serve as a basis for comparison with automatically generated test suites.

The DREX protocol is a military protocol which alllows the access to the transit network of the French Army defined in the framework of Integrated Services Military Networks. It is the analogous of the D protocol in the civil framework, with military features.

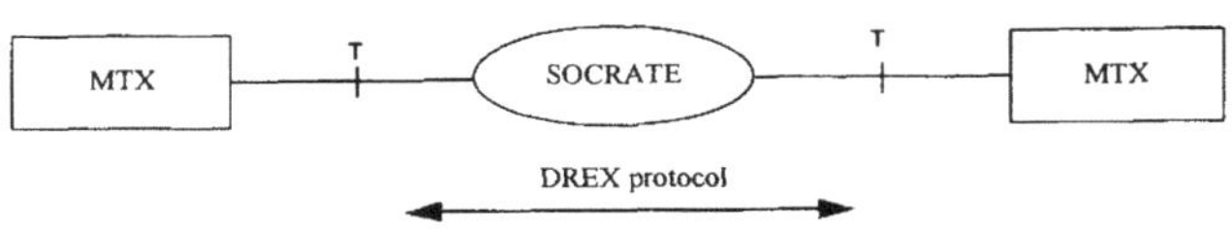

Figure 1 Reference configuration of DREX

The DREX protocol runs on a network called SOCRATE and connects several MTX (telecommunication means of military bases) at the T interface (see Figure 1). The DREX protocol contains a functional description of the offered services and the specification of the user interface for the call command. The SDL specification models the behaviour of the DREX protocol on the entire network SOCRATE.

Only a subset of the military supplementary services offered by the DREX protocol have been specified in SDL. These services are Priority, Roving User, Call Forwarding, Implicit Partioning of Users, Safety Path and User to User Signalling. A generic SDL specification has been written and instantiaced for each service. So, there is as many SDL specifications as services. This means that tests have been produced for each service separately and the interactions between services have not been studied. The SDL specification models only one connection between two MTX. They are called DR for the calling MTX and DE for the called MTX. The size of the SDL specification corresponding to each service is about 2000 lines.

2.4 Test architecture and test purposes

The test method chosen is called transversal method. The implementation of the protocol inside the network is seen as a black box. The tester models the behaviour of two MTX (seen as users from the network point of view), DR and DE connected to the SOCRATE network by two PCOs (Points of Control and Observation) at interface T. The communication between the MTX and the network is supposed to be asynchronous. The test configuration is shown in Figure 2.

The test purposes used as an input of our experiments are written in natural language. These test purposes described the context of the test and an abstraction of the intended behaviour. The context of the test is concerned with properties of the subscription record of the users or properties of the network. An example of context is « The called MTX does not have access to this service » or « The network is saturated ». The remainder of the test purpose describes sequencing properties between sendings and receptions of the tester and network, as well as properties on parameter of messages. For example, « The called MTX receives a SETUP message containing the parameter User to User Information (IUU). It sends an ALERT message with parameter IUU. Verify that the network sends to the calling MTX an ALERT message without parameter IUU ».

These test purposes have been used by the automatic tools for the test cases selection.

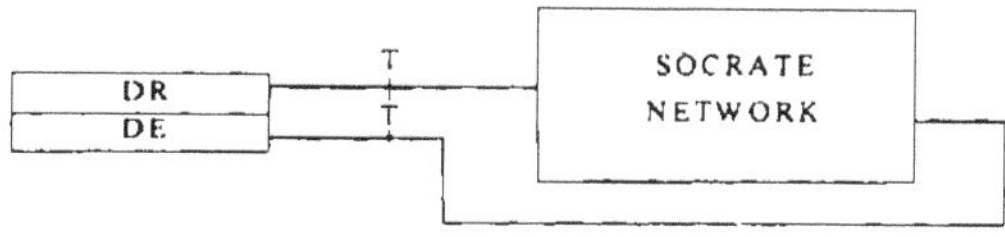

Figure 2 Test configuration for DREX

3 PROTOTYPES DESCRIPTION

3.1 TGV

Overview

TGV is a prototype tool for the generation of test suites developed by Inria. It is based on techniques that have proved useful in verification such as on-the-fly verification, product of transition systems and behavioural equivalence (see (Fernandez, 1996 and 1996b) for more details). It takes as input the state graph produced by GEODE of an SDL specification constrained by the test context and an automaton formalizing the behavioural part of a test purpose. The transitions of the test purpose automaton are labelled with some actions of the tester corresponding to visible actions of the IUT.

The TGV package outputs test suites given in TTCN standard language, distinguishing the control (behaviour) and data (constraints) parts. As usual, the intermediate form TTCN-MP can be graphically displayed using TTCN-GR.

TGV has been developed in the toolbox CADP (Caesar Aldebaran Distributed Package). This was particularly useful because CADP is an open environment which does not depend on a particular language. Another advantage of CADP is the availability of libraries for the management of graphs and memory.

The external view of the TGV package is illustrated in Figure 3.

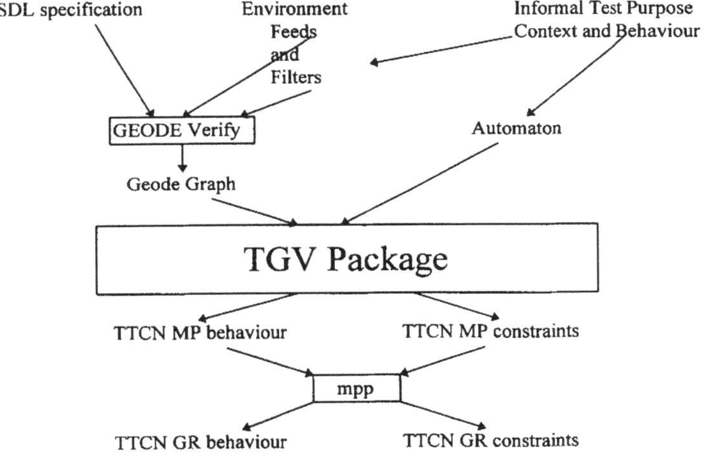

Figure 3 External view of TGV

SDL model

The SDL model is the same as the one used by the other tools. Its state graph is first generated by GEODE using exhaustive simulation. The exhaustive simulation also uses as inputs feeds and filters. Feeds are used to close the SDL specification with a restricted set of messages sent by the environment in order to limit the size of the state graph generated by GEODE. Filters specify some transitions that are considered as not firable. These filters are used to select the behaviours of the protocol within a context specified in the test purpose (see below).

The simulation of the model also assumes the *reasonable environment hypothesis* which says that, each time the environment sends a message to the network, it waits until stabilization. This means that no new message can be sent by the environment until it receives all specified sendings of the protocol.

Test purposes

A test purpose defines a property on some observable interactions of the IUT. The constraint part gives some property on the state of the implementation. It is modelled by a set of filters.

The behavioural part which describes an ordering between interactions is modelled by an automaton using the Aldebaran format. This automaton is a direct acyclic graph with a set of distinguished accepting states with no successor.

Test generation process

The generation of a test case can be decomposed in several functional parts which are performed by different tools, as shown in Figure 4.

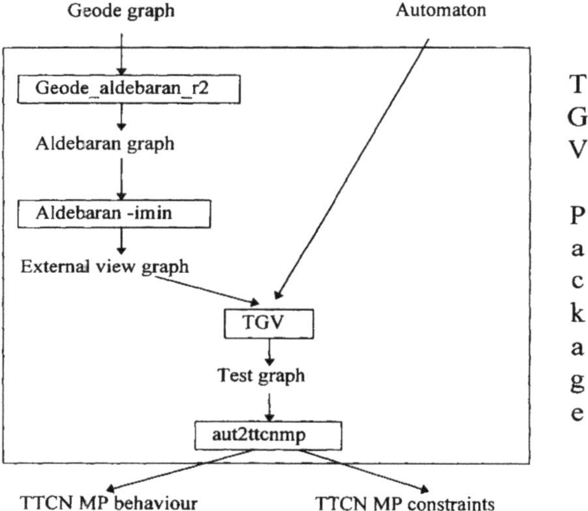

Figure 4 Test suite generation with TGV

Step 1: State graph generation with GEODE. The inputs to the GEODE exhaustive simulator are the SDL specification, feeds and filters. The output is a graph in GEODE format representing the whole set of behaviours of the specification in the context given by feeds and filters.

Step 2: Translation and mirror image. In this step, the graph generated by GEODE is translated into a graph in the Aldebaran format which represents the external view of the behaviour of the specification in the test environment. This is done to take into account the fact that the tester cannot interact direcly with the implementation but with a test environment in which the implementation is placed. But the specification models the *internal view* of the system, i.e. the behaviour of the system with its internal actions, without considering the way it interacts with the environment. Thus we have to put the specification in the same environment as it should be taken into account in the test generation. For our experiment, the communication between the tester and the implementation is asynchronous. This may introduce concurrency in the behaviour as we consider two interaction points. This has been modelled in the translation.In the following, we will consider the environment point of view: outputs are *controlable* actions initiated by the environment (which may be the tester) and sent to the IUT whereas inputs are *observable* actions, initiated by the IUT and received by the environment. Internal actions of the specification are replaced by invisible transitions denoted by τ. Only interactions with the environment are to be captured. As we now consider the tester view, sending are to be replaced by receptions and vice versa.

Step 3*: *Minimization with Aldebaran. In this step, the graph produced at the preceeding step is minimized, using the Aldebaran tool, with respect to the τ*a equivalence. The minimization roughly means that only observable actions are preserved and states with same observable behaviours are collapsed. Collapsing is in fact not necessary, the important thing here is only the hiding of τ actions. This is crucial as we intend to make on the fly generation of test cases and on the fly minimization is not efficient. After this step, the minimized graph produced represents the external view of the protocol behaviour. We call it the *external view graph*.

Step 4: Test graph generation. This is the kernel of the tool. The algorithm is based on a depth-first traversal of a kind of synchronous product between the external view graph and the automaton representing the test purpose. In this product, states are composed of a state of the graph and a state of the automaton. A transition t is firable in (s_g; s_a) and leads to (s_g'; s_a') in two cases. Either t is firable in the graph in state s_g and leads to s_g' and t is firable in the automaton in state s_a and leads to s_a', or t is firable in s_g and leads to s_g' but t is not firable in s_a, in which case $s_a' = s_a$.

During the traversal, two different things are done:

- the algorithm checks that at least one sequence allows to reach a state in which the automaton component is an accepting state. In case of success, this ensures that the test purpose is at least realizable. We could impose this for all sequences but this one was taken because in some cases the test context could not be modelled by filters. Thus some sequence did not satisfy the test purpose.
- meanwhile, a skeleton graph is synthesized. This graph contains sequences of the synchronized product that contain an accepting state of the automaton. In this graph, there is never the choice between two outputs of the tester because the tester controls outputs.

Some transitions are decorated with verdicts with the following informal meaning. (PASS): the test purpose is satisfied by the current sequence. But a sequence leading to the initial state (Postamble) must be applied in order to carry on another test case. It is a temporary verdict as the application of the postamble may produce FAIL verdicts. PASS: a definitive verdict meaning that the initial state has been reached after a (PASS) verdict. The sequence between (PASS) and PASS is a Postamble. FAIL: means non-conformance of the IUT on inputs of the tester. For the conformance relation chosen this means that after a sequence of interactions valid in the specification, this input does not corresponds to a valid output of the specification. INCONCLUSIVE: verdict produced when an input of the tester is allowed in the specification but cannot lead to a (PASS) or leads to a behaviour that is not considered in the test case because testing cannot be exhaustive in practice.

After that, the graph is decorated with timers. Timers are used in test cases when a reaction of the implementation is expected but one does not want to wait for an unbounded time because an error that is not observable may have occurred. The difficulty in the management of timers is that concurrency and non-determinism should not be treated in the same way. Timers are managed by TGV in the following way. A timer **tm** is associated to each possible input of a message **m**. The timer **tm** is started in the last transition which necessarily precedes the reception of **m**. In a state where the input of **m** is possible, a timeout of **tm** is also possible. In each transition sequence, a timer **tm** is cancelled in a transition following its start and as soon as the reception of **m** is no more possible in the future. This is were concurrency and choice must be distinguished.

Step 5: TTCN generation. In this last step, the tool aut2ttcnmp takes as input the test graph generated at the preceding step, and produces two files : one describing the behaviour part of the test case (obtained by unfolding the test graph in a tree), the other describing the constraints part. Both files are in TTCN MP format and are translated into files in TTCN GR format by **mpp**, a tool from VERILOG.

3.2 TOPIC V2

Overview

Developments of TOPIC V2 have started in the RACE II/TOPIC project, based on the results of the RACE II/PROVE project. A product version of this prototype called TTCgeN is now included in the VERILOG ObjectGEODE product line.

TTCgeN generates abstract tests in TTCN-IS MP from (i) the SDL model describing the protocol to be tested and (ii) test objectives (or purposes) formalized with MSCs.

The TTCgeN test generation algorithm is based on the exhaustive exploration of SDL model behaviour. This exploration is made by going through a subset of the overall state graph. This subset is obtained by constraining the complete protocol model behavior with the help of the test purposes. TTCgeN is thus based on automatic exhaustive simulation techniques.

SDL model

The SDL model describes the protocol to be tested, not its environment. This model is open, which means that it exchanges signals with the environment via dedicated channels.

In order to lower the number of possibilities regarding possible data conveyed by signals incoming from the environment, test purposes include data samples able to be received by the protocol model.

Like for TGV, the reasonable environment hypothesis has been put on the environment. Of course the delay between two emissions from the environment must be very short compared to timeouts involved within the protocol model.

No other restrictions are imposed on the SDL model, except that a finite state graph (constrained by test purposes) can be derived from this SDL model.

Test purposes

A test purpose must be described with the help of an MSC. Figure 5 shows an example of an MSC describing a test purpose. Signal sequence a and b is a preamble to set the IUT in the desired testable state. The next sequence shows that after having received signal x, the IUT should send back signal y carrying data value 15.

This test purpose MSC can be described by the user through two different ways by:

1. using the ObjectGEODE MSC Editor directly or,
2. generating this MSC from within the ObjectGEODE SDL Simulator after a simulation session corresponding to the desired signal sequence. This particular way guaranties that the test purposes are correct vs. the SDL model.

Test generation process

On a conceptual point of view, the generation of a test consists of the following operations : to find in the overall state graph all the event sequences corresponding to the test purpose (PASS verdict); and to complement these sequences with all valid but non expected events (INCONCLUSIVE verdict), and non valid events (FAIL verdict).

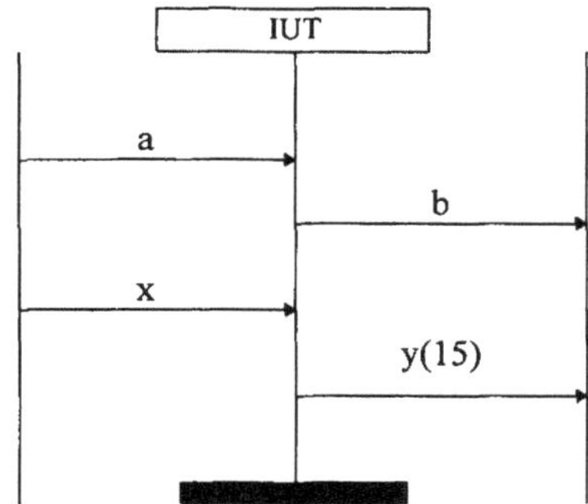

Figure 5 Example of an MSC describing a test purpose

The overall test generation process is described in Figure 6.

One of the major difficulties is to find these sequences as most efficiently as possible, that is by exploring as less states as possible. This can be achieved doing so-called co-simulation between the SDL protocol model and a test purpose previously translated as an observer. An observer is described in a language called Geode Observation Automata Language (GOAL) (Algayres, 1995), derived from studies conducted at CNET. In the test generation process, GOAL observers are automatically derived from MSCs.

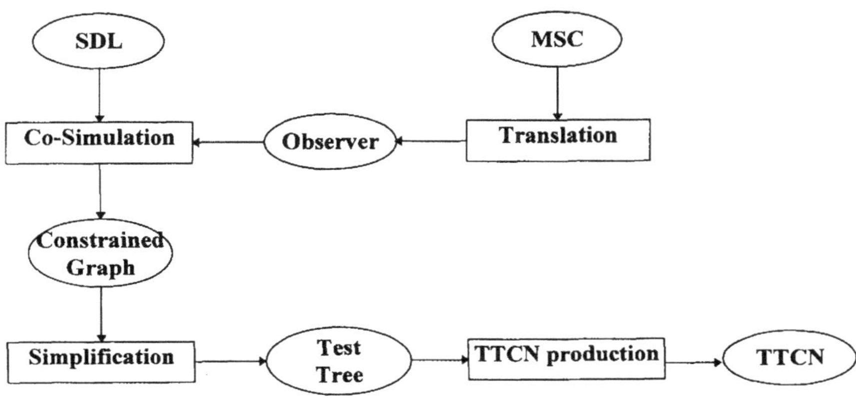

Figure 6 Overall test generation process with TOPIC V2 (TTCgeN)

Co-simulation allows the exploration of only a subset of the overall state graph, thus enabling the use of this technique on infinite graphs, i.e. on SDL models that cannot be exhaustively simulated. The subset is built based on exploration constrained by the GOAL observers.

The constrained graph resulting from co-simulation contains sequences (i) leading to the PASS verdict and the preamble, (ii) valid but not expected answers from the IUT (INCONCLUSIVE verdict), and (iii) valid but not expected stimuli from the Tester.

Simplification is then applied to the resulting constrained graph for removing the later sequence case.

Lastly, TTCN production is in charge of (i) translating constraints (signal parameters) in TTCN, (ii) generating PDU declaration, (iii) complementing the test sequence with the FAIL verdict in case of no response from IUT (use of timers) or wrong response from IUT (?OTHERWISE).

3.3 TVEDA V3

Overview

TVEDA V3 is a test generation tool which has been developped by CNET (France Telecom) since 1989. TVEDA produces TTCN test suites from Estelle or SDL formal specifications. TVEDA incorporates two distinct test generation techniques, which have been studied successively:

1. the "syntactic" technique produces test cases by a simple syntactic transformation of the specification. In this technique, emphasis is put on heuristics which allow an automatic computation of the test purposes (see section *Test purposes*). On the contrary, no semantic computation (e.g. simulation) is done on the specification. Therefore the test cases which are produced are incomplete, and must be manually completed. This technique has been described in (Phalippou, 1990).
2. the technique "by simulation" has been developped more recently. In addition to the automatic computation of the test purposes (same heuristics as in the syntactic technique), this new method uses advanced exhaustive simulation functions in order to compute complete compilable TTCN test suites. The simulation functions are not part of the TVEDA tool, but are provided by coupling TVEDA with available verification tools (such as GEODE, from VERILOG company, for SDL language (Algayres, 1993). This technique has been described in (Clatin, 1995).

In this project only the syntactic method has been used, because the technique by simulation was not available when the project started (end of 1994).

The syntactic technique of TVEDA is currently being industrialized by an european consortium, financed by the European Commission, named CATG, whose prime contractor is SEMA GROUP. The resulting industrial tool will be called TTCN Maker, and will be publicly available by the end of 1996.

SDL model

TVEDA takes as input a formal specification written in SDL. The syntactic technique puts some constraints on the SDL constructs which can be handled by the tool. These constraints are of two kinds:

1. restrictions on SDL constructs. The tool ignores several constructs, such as abstract data types, dynamic process management, etc.
2. restrictions on SDL style. The syntactic technique generates test cases for a single SDL process. Several processes can be accepted, but TVEDA will produce distinct test cases for each one of the processes, and then merge the test cases into a single test suite. This means that the tool cannot compute the combined behaviour of the various processes. In addition, to be relevant for the syntactic method, each transition of the specification must

start with an input from a given stable state, produce eventually outputs, and end into a stable state. This means that spontaneous transitions as allowed by SDL 92 are not allowed by TVEDA. Complex procedures (i.e. containing inputs or outputs) are not allowed either.

Test purposes

One of the main features of TVEDA, which is not provided by the other tools described in this paper, is the automatic computation of the test purposes. For doing this, the tool incorporates some heuristics, based on the syntactic structure of the SDL specification. Basically, one test case is produced in order to check each branch of the SDL (the resulting coverage is thus close to the "branch coverage" often used for the test of sequential programs). This basic test selection mechanism can be modified by the user of TVEDA in the following ways:

1. restrict more or less the part of the specification which is tested: choice of states, input signals, procedures, etc.
2. enumerate more or less data values which appear in the SDL branches: test several values of parameters on input signals, or on the contrary merge test cases corresponding to several branches if the observable behaviour is the same in these branches, etc.

These heuristics have been defined after observation of the stategy used by human experts in order to define the test cases. They have been validated by applying TVEDA to several real protocols and then by comparing TVEDA test purposes with the test purposes of standardized manually generated test suites for the same protocols. For instance, in the case of ISDN D protocol (level 2 user side, level 3 user and network side), more than 90% of the manual test purposes have been correctly generated by TVEDA.

The test purposes produced by TVEDA look like : "Check that the IUT sends a <message and significant parameter values> in response to a <message and significant parameter values> in state <state>. The IUT is supposed to enter state <state>".

Test generation process

Once the test purposes have been selected, TVEDA computes the behaviour of the test cases. This is done in several steps.

Step 1: Analysis of the SDL specification, and computation of an abstract extended finite state machine (EFSM). During this step, the test architecture is taken into account (which signals can be controlled, which signals can be observed: this is the definition of the PCOs). The level of abstraction at which the specification is tested can be influenced, through the decision to "control" or not variables or signals: this mechanism, which has been adjusted in an empirical way, is very important, and has a significant influence on the output of the tool. It is a characteristic of TVEDA, hardly studied in classical test generation methods. More information can be found in (Phalippou, 1990).

Step 2: Production of test cases in an internal abstract format. Incomplete test cases are produced, which have the following structure :

- call to a preamble which brings the IUT into the start state of the tested transition. This preamble is not completely computed with the "syntactic" technique, since this would require computation of the SDL behaviour (on the contary it is produced by the technique "by simulation").
- send and receive events corresponding to the tested transitions. At this level, non determinism is taken into account. Non determinism can result from uncontrolled variables (with the consequence of decisions in which it is not possible to know which branch will be executed) or from uncontrolled events (which have visible consequences at the PCOs). Constraints are also produced in an incomplete way, using heuristics

where possible and informal comments in the remaining cases.

- call to a check sequence which is supposed to check the final state of the tested transition and bring back the IUT in its initial state. Here again this check sequence is not completely produced by the "syntactic" technique (it is produced by the technique "by simulation").

Step 3: Formatting of TTCN suite. Elements which are purely linked to TTCN are added to the internal abstract test suite. This includes translation of the constraints into ASN.1 format, computation of the test suite structure, of the test case names, production of the different TTCN parts (overview, declarations, constraints, test cases, test steps), and finally production of hypertext links which relate each test case with the corresponding tested part of the formal specification.

The output of TVEDA is a test suite in standard TTCN-IS MP format.

4 EXPERIMENTS RESULTS

4.1 Contribution made by the different prototypes

The experimental work carried out in this study has shown that the three prototype test generators worked as expected on a real protocol. The test generation times are short and the memory requirements are low. The main advantages of each prototype with regard to the others are reviewed below:

Advantages of TOPIC V2

- graphical test purpose (MSCs),
- generation of declarations.

Advantages of TGV

- generation algorithms based on the technology of on-the-fly verification,
- generation of tests taking into account asynchronism between IUT and tester,
- high-level control of timers.

Advantages of TVEDA V3

- test purposes computed automatically (this has allowed 2 new test cases to be discovered in the context of the experimental work),
- generation of declarations,
- generation of a full suite (and not of isolated tests),
- takes into account the level of abstraction at which the specification has to be viewed from the point of view of the test.

4.2 Quantitative results

Figure 7 represents a comparison of the effort required for each phase and for each test generation method. At first sight, this diagram seems to show that automatic generation is of little value. The following comments demonstrate that this is not the case:

- more than half of the workload is given over to writing the SDL model : as well as enabling automatic generation of tests, this model allows the quality of the specification of the protocol to be improved by means of syntactic and semantic verification and by simulation carried out using a tool such as GEODE ; the effort needed to adapt the specification to the TVEDA format would not have been necessary if the specification had been written directly for TVEDA,

- about a third of the workload is given over to the finishing work on the test suite in order for it to be accepted by a TTCN compiler (addition of declarations, postambles, semantic verification, etc...) : with a 'full' tool rather than the prototypes used in this study, hardly any of this finishing work would be required,
- this comparison relates to 54 tests on the DREX protocol : if we extrapolate the results, the value of the automatic test generators becomes more obvious. This is described in the 'Perspectives' section.

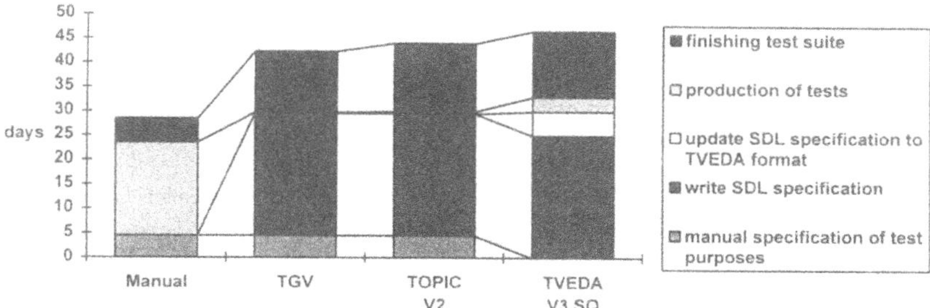

Figure 7 Effort required to obtain an abstract test suite

Also, the figures shown allow for the fact that the people carrying out the experiments were specialists in their tools, but they were not necessarily specialists in all the domains involved in this study, namely the SDL language, the DREX protocol, and testing. In spite of this, tests which are totally comparable with manual tests have been generated.

4.3 Qualitative results

The comparison of the test suites automatically generated with hand-written test suites has enabled some errors to be detected in the manual tests, which shows the value of the automatic methods for producing correct tests.

This comparison has also enabled the prototypes to be refined as the tests were conducted by incorporating rules into the algorithms which had not originally been thought of.

The use of automatic test generation methods offers significant advantages:

- an SDL model of the protocol is obtained which can be simulated and verified by powerful tools available on the market such as ObjectGEODE; this allows us to ensure that the specification of the protocol is correct, which is difficult without a formal model,
- when refining the protocol, it is less dangerous and less costly to update the SDL model (which can be simulated to ensure that it is still working) and then to rerun the generator, than it is to modify a TTCN test suite by hand,
- an automatic generator makes the tests safer by avoiding the human error inevitable in a manual approach and this saves a great deal of time, even if the figures cannot show this ; it also enables more tests to be generated, which gives better coverage of the protocol (the coverage seems to be correct for the DREX CS, but for other protocol layers it may be envisaged that achieving adequate 'manual' coverage would be difficult).

5 PERSPECTIVES

Not everything has yet been resolved in the prototypes, but this experiment and the comparison between the prototypes have brought to light their advantages and limitations and shown how to design a tool incorporating the best ideas from TOPIC V2, TGV and TVEDA V3.

Two steps towards the industrialization of such a tool have been identified.

First step: A 'full' tool

The 'full' tool based on TOPIC V2 integrates:

- generation algorithms based on the technology of on-the-fly verification of TGV and TOPIC V2,
- the automatic computation of declarations and postambles,
- the control of the abstraction level and the full suite building proposed by TVEDA V3,
- the taking into account of asynchronism between IUT and tester, and the high-level control of timers proposed by TGV.

This 'full' tool requires the specification of test purposes with the help of MSCs.

Second step: An 'ideal' tool

The 'ideal' tool integrates the features of the 'full' tool and:

- the automatic computation of test purposes with an optimization of the method proposed by TVEDA V3,
- the editing of generation parameters via a user interface,
- the interactive piloting of the test generation by the user to allow the incremental building of the test suite.

Figure 8 shows an extrapolation of the production effort for 895 tests on the DREX protocol, for each phase and for test generation with the manual method, with the 'full' tool and with the 'ideal' tool.

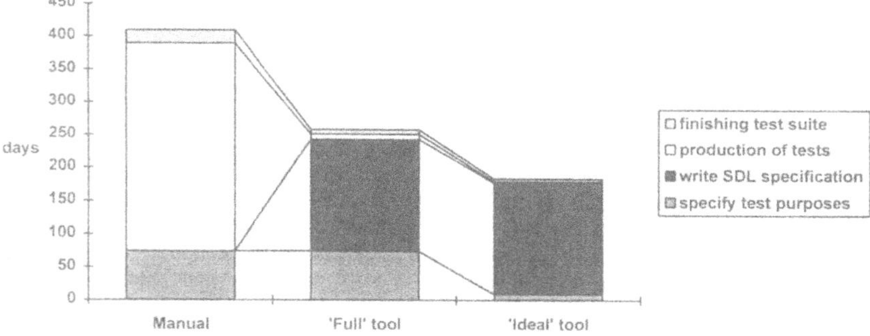

Figure 8 Effort for 895 tests, with a 'full' and an 'ideal' tool

For manual test generation a rule of three has been used except for the effort of the finishing work on the test suite, which has been estimated at 20 days.

With the 'full' tool the workload of writing and debugging the SDL model is estimated at 168 days. A rule of three has been used to calculate the specification time for test purposes and the implementation time for the generation of tests (TOPIC V2 figures). The effort of the finishing work on the test suite is estimated at 7 days.

With the 'ideal' tool the workload for writing and debugging the SDL model is the same as for the 'full' tool. The effort of configuring the tool to generate the test purposes is estimated at 9 days. The implementation time for the generation of tests and the finishing work on the suite is estimated at 6 days ; in fact this enables the tests to be regenerated if a problem has appeared.

Regarding productivity, the current generation tools (described in this paper) already provides a productivity improvement of +25% over conventional (manual) test writing techniques, while the 'full' tool should provide +38% productivity gain and the 'ideal' tool +55%.

6 ACKNOWLEGMENTS

We wish to thank the DGA/STSIE for financing this project, particularly Cpt. Plessix, Mrs. Courtier and Mr. Caporini who launched the project and were responsible for it, as well as the other DGA participants : Mr. Le Pape and Mr. Ropars from DGA/CELAR.

We wish also to thank the other partners in the experiment related here. In particular, C. Jard and G. Viho from IRISA/Pampa, J. Sifakis from Vérimag and P. Barbier from CAP SESA.

7 REFERENCES

CCITT/SGx/WP3-1 (1988) Specification and Description Language, SDL, in *CCITT Recommendation Z.100.*

Clatin, M., Groz, R., Phalippou, M. and Thummel, R. (1995) Two approaches linking a test generation tool with verification techniques, in *Proceedings of IWPTS'95,* Evry.

Fernandez, J.C., Jard, C., Jéron, T. and Viho, G. (1996) Using on-the-fly verification techniques for the generation of conformance test suites, *to appear in Conference on Computer-Aided Verification (*CAV '96*),* New Brunswick, New Jersey, USA.

Fernandez, J.C., Jard, C., Jéron, T. and Viho, G. (1996b) An experiment in automatic generation of test suites for protocols with verification technology, *to appear in Science of Computer Programming, Special Issue on Industrial Relevant Applications of Formal Analysis Techniques* (ed. J.F. Groote and M. Rem), Elsevier Science publisher.

Algayres, B., Lejeune, Y., Hugonnet, F. and Hantz, F. (1993), The AVALON Project: A VALidatiON environment for SDL/MSC descriptions, in *Proceedings of 6th SDL Forum,* Darmstadt.

Algayres, B., Lejeune, Y., Hugonnet, F. and Hantz, F. (1995), GOAL: Observing SDL behaviour with GEODE, in *Proceedings of 7th SDL Forum,* Oslo.

Montiel, J., Roth, R. and Donaldson, A.J.M (1994). Methods for QoS Verification and Protocol Conformance Testing in IBC - Results and Further Recommendations. Race Project R2088 TOPIC, Deliverable 15, DocR2088/DAT/TMS/DS/P/015/b1.

Phalippou, M. and Groz, R. (1990) Evaluation of an empirical approach for computer-aided test cases generation, in *Proceedings of IWPTS'90,* Washington.

Phalippou, M. (1994) Test sequence generation using Estelle or SDL structure information, in *Proceedings of FORTE'94,* Berne.

INDEX OF CONTRIBUTORS

KEYWORD INDEX